AF615713

A DICTIONARY OF ASTRONOMY

A DICTIONARY OF
ASTRONOMY

Dr Robert E. W. Maddison F.S.A.

EDITOR

VALERIE ILLINGWORTH

Hamlyn

LONDON · NEW YORK · SYDNEY · TORONTO

First published in 1980 by
The Hamlyn Publishing Group Limited
London · New York · Sydney · Toronto
Astronaut House, Feltham, Middlesex, England

ISBN 0 600 32996 8

Filmset in 10 on 10·5 Linotron 202 Bembo by
Tradespools, Frome, Somerset
Printed in Italy

The picture on the front and back of the jacket is of the Orion Nebula

CONTENTS

aberration 1. The apparent displacement of a star produced by the motion of the Earth in its orbit and by the finite velocity of light. It was discovered by J. Bradley who concluded (1728) that the effect of aberration depends solely on the ratio of the Earth's velocity in its orbit to the velocity of light. Distinction is made between ANNUAL ABERRATION, which results from the revolution of the Earth in its orbit, and DIURNAL ABERRATION, which results from the rotation of the Earth on its axis. Annual aberration manifests itself to an observer by the apparent motion of a star in an ellipse over the course of a year. The maximum displacement of the star, given by the semi-axis major of the ellipse, is equal to 20·47 seconds of arc. This angle (α) is the CONSTANT OF ANNUAL ABERRATION and is determined from the relationship

$$\tan \alpha = v/c,$$

where v is the Earth's orbital velocity and c is the velocity of light. For a star at the pole of the ecliptic the ellipse becomes a circle; for a star on the ecliptic the ellipse becomes a straight line.

Diurnal aberration involves a very much smaller displacement of a star, to the east, which at transit amounts to a maximum of 0·32 seconds of arc – for an observer at the equator – and is zero at the poles. It is caused by the Earth's easterly rotation, which carries an observer in a circle with a velocity that is greatest at the equator and zero at the poles.

2. A defect, such as blurring, distortion, or false coloration, that can occur in the image produced by a lens or curved mirror. The major aberrations are CHROMATIC ABERRATION, SPHERICAL ABERRATION, **coma**, and **astigmatism**. CHROMATIC ABERRATION is produced (only by lenses) as a result of unequal refraction by the lens of the individual elementary colours in white light. Each colour is brought to a focus at a different distance from the lens, causing the image of an object to be fringed with prismatic colours. SPHERICAL ABERRATION results from the unequal refraction of the light rays by different zones of a spherical lens or mirror. The rays are not brought to the same focus, causing a lack of definition about the edges of the image. Both chromatic and spherical aberration can be minimized by using a suitable combination of lenses.

ablation Wearing away or erosion of an object, e.g. of a glacier by melting, of a rock by water, or of a meteorite by friction on passing through the Earth's atmosphere.

absolute magnitude The apparent magnitude that a star would have if it were placed at a distance of 10 parsecs from the Earth: it is the distance at which an object would have a parallax of 0·1 seconds of arc. The absolute magnitude (M) may be derived from the **apparent magnitude** (m) and the **parallax** (π) by the following formula:

$$M = m + 5 - 5 \log \pi$$

where π is in seconds of arc. The absolute magnitude of stars is usually between -5 and $+15$. The Sun has an absolute magnitude of $+4{\cdot}8$. See also **magnitude**.

absolute temperature Temperature which

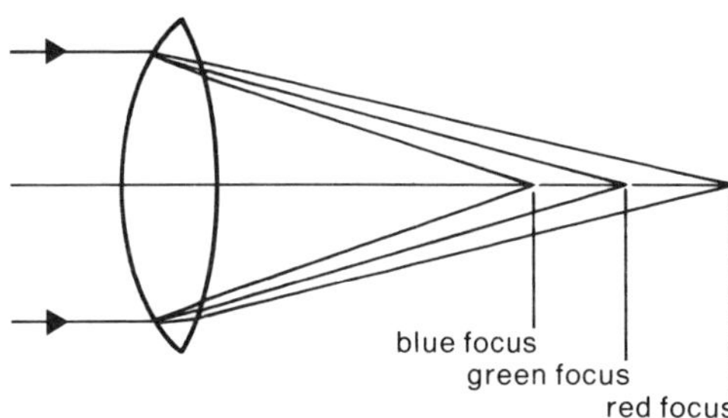

Two examples of aberration. Above: chromatic aberration of a convex lens. Right: spherical aberration of a spherical mirror.

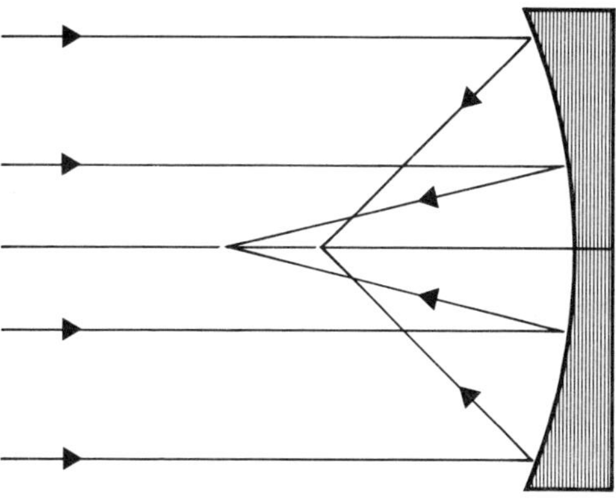

is measured from the zero on a scale of temperature formulated by thermodynamic reasoning. The international Kelvin Scale, in which a temperature interval of one kelvin (K) equals one degree Celsius (°C), is in general use. The relationship between a Kelvin temperature (T) and the temperature (t) on the Celsius scale is

$$T = t + 273{\cdot}15$$

absolute zero The natural but unattainable lower limit of temperature. It is the starting point of the Kelvin Scale, a thermodynamic temperature scale, the unit of which is the **kelvin**.

absorption spectrum Lines or bands produced when parts of a continuous spectrum are cut off by an intervening medium. The dark lines in the spectrum correspond with the emission lines in the emission spectrum of a substance. See also **spectrum**.

accretion The accumulation of matter by a star or other celestial object. The space between celestial objects is not completely void, but contains gas (mainly hydrogen) and dust grains at a very low concentration. A star on moving through a cloud of this matter will gather some of it to itself and so increase in mass. Matter can also be transferred from one component of a close binary system to another by an accretion process. Accretion is considered to be an important factor in the evolution of stars, planets, and comets.

accretion disc see **black hole**

Achernar (α Eri) A hot blue star that is the brightest star in the constellation Eridanus.

achromatic lens A lens which is designed to minimize chromatic **aberration**. The result is usually achieved by using a doublet of two lenses made from different types of glass such that two specially chosen wavelengths of light come to the same focus, and at the same time the residual aberration is reduced to a minimum.

Adams, John Couch (Laneast, Cornwall, June 5, 1819 – Cambridge, January 21, 1892) An English astronomer who, while still an undergraduate at Cambridge, decided to investigate 'the irregularities in the motion of Uranus . . . in order to find whether they may be attributed to the action of an undiscovered planet beyond it . . .'. By September 1845 the problem was virtually solved: an unknown planet was perturbing the motion of Uranus and therefore preventing the accurate prediction of its future position. Independently, and without knowledge of Adams's work, U. J. J. Le Verrier was engaged on the same problem and came to the same conclusion. He was fortunate in enlisting the aid of J. G. Galle of the Berlin Observatory to search for the unknown planet, which the latter with his assistant, H. L. D'Arrest, located on September 23, 1846, within one degree of the predicted position. The honours of predicting the existence of the Trans-Uranian planet Neptune, as it came to be called, were shared by Adams and Le Verrier. Adams also made important contributions to celestial mechanics, of which his treatment of the Moon's secular acceleration may be mentioned.

aerolite A stony **meteorite**. More than 90% of all meteorites are of this kind.

aether A hypothetical medium postulated as permeating all matter and space. This ubiquitous material had been abandoned in the 18th century, but was revived by Thomas Young (1773–1829) early in the 19th century in connection with the wave theory of light. This all-pervading luminiferous aether was assumed to be the medium in which the vibrations of light were propagated. The experiments of A. A. Michelson (see **Michelson, Albert**) and E. W. Morley to discover the existence of aether-drift relative to the Earth gave negative results. It was Einstein (see **Einstein, Albert**), who in one of his memoirs of 1905 made the luminiferous aether 'superfluous' and the aether-drift problem meaningless. Later, in 1920, in relation to his General Theory of Relativity, he said that space is endowed with physical qualities, and in this sense there exists an aether, but that does not make the aether equivalent to a medium.

airglow A faint glow arising in the Earth's atmosphere. Molecules of gases in the upper atmosphere, where their density is low and solar radiation is strong, split up and are excited or ionized by ultraviolet radiation or by charged particle collision. On returning to their normal state a faint light, normally visible only on clear moonless nights as a grey luminous background to the stars, is emitted. The appearance is greatly enhanced during twilight when the Sun's radiation traverses long stretches of emitting layers in the atmosphere.

Airy disc see **diffraction**

Airy, George Biddell (Alnwick, July 27, 1801 – Greenwich, January 2, 1892) English astonomer who was Astronomer Royal and Director of the Greenwich Observatory. His most significant accomplishment was the reorganization of that Observatory, which he equipped with instruments of his own design and where he created magnetic and meteorological departments. He determined the density of the Earth by the pendulum experiment and also instituted regular photographic observations of the Sun.

al-Battani (*Latin:* Albategnius, Albatenius; before 858 – 929) One of the greatest astronomers of Islam, he made accurate astronomical determinations, e.g. of precession and of the inclination of the ecliptic.

albedo The fraction of solar light falling on an element of a diffusedly reflecting surface, i.e., one that is not mirror-like, which is reflected from it. It varies from unity to zero. The SPHERICAL ALBEDO now used in astronomy is defined as the ratio of the amount of light reflected in all directions by the surface of a sphere illuminated by parallel rays of light to the total amount of light incident on the sphere. This condition is virtually satisfied for bodies in the solar system receiving light from the Sun. Albedo is important for astrophysics as it enables conclusions to be drawn about the nature and constitution of the surface of celestial bodies, especially planets and satellites.

Alfvén, Hannes (born 1908) A Swedish physicist who, in 1954, put forward a theory of the origin of the solar system which depended on the effects of electromagnetic forces.

Alfvén waves Hydrodynamic waves discovered by H. Alfvén. They move perpendicularly through a magnetic field with a velocity known as ALFVÉN VELOCITY.

Algol (β Per) The second brightest star in Perseus, the variable brightness of which has long been known. J. Goodricke (see **Goodricke, John**) in 1783 was the first to suggest that the variability resulted from two stars in binary motion, the fainter and larger one from time to time passing in front of the brighter one. This star is now the prototype of a class of eclipsing variables or eclipsing binary stars known as ALGOL-TYPE VARIABLES.

A line J. von Fraunhofer's designation for one of the absorption lines in the solar spectrum; it is in the infrared at 7100 Å. See also **Fraunhofer lines**.

Almagest An encyclopedia of astronomy compiled by Ptolemy in about A.D. 140. It is thought to be based mainly on the work of Hipparchos, whose star catalogue it incorporates. It is the most ancient accurate description of the heavens, and remained authoritative until the middle of the 16th century, as nothing else was available.

almanac An annual publication containing information on the calendar. Larger compilations include up-to-date astronomical, ecclesiastical, historical, political and other information. For astronomical and navigational purposes the best-known item is *The Nautical Almanac*, first published in 1765 (for the year 1767), which was amalgamated with *The American Ephemeris* from the issue for the year 1960. It is published in the UK under the title of *The Astronomical Ephemeris*, and in the USA as *The American Ephemeris and Nautical Almanac*. There are the subsidiary items *The Abridged Nautical Almanac*, *The Air Almanac*, and *The Star Almanac for Land Surveyors*.

alpha particle or **α-particle** The nucleus of a helium atom. It consists of two protons and two neutrons. Its mass is 4·00260 amu.

Alphonsine Tables Astronomical tables produced at the order of Alphonso X of Castile about 1272 by Judah ben Moses and Isaac ibn Sid. They were known in Paris by 1292, became popular in the 14th century, and were frequently printed in the 15th and 16th centuries. They were eventually displaced by the **Rudolphine Tables** computed by J. Kepler (see **Kepler, Johannes**).

altazimuth An instrument, or mounting for an instrument, that can move in both azimuth and altitude, i.e. about both horizontal and vertical axes. See also **telescope, mounting of**.

altitude The angular distance of a celestial body above the observer's horizon. It is measured in degrees from 0 to 90 along the vertical circle: positive towards the zenith, negative towards the nadir. See also **coordinates**.

Amalthea Satellite V of **Jupiter**. See also **satellite**.

Ananke Satellite XII of **Jupiter**. See also **satellite**.

anastigmat see **astigmatism**

Anaxagoras (5th cent. B.C.) A Greek philosopher whose cosmogony was based on the theory that the universe was formed from the original chaos by a vortex established by Mind. He explained solar and lunar eclipses by interposition of the Moon and the Earth.

Andromeda Galaxy (M31) A spiral galaxy that lies in the northern constellation Andromeda and is the most distant celestial object visible to the naked eye. Its distance is about two million light years. Cepheid variable stars and novae have been observed in this galaxy, which is accompanied by two elliptical galaxies, M 32 and NGC 205.

Andromedids A meteor shower which appears in November and has its radiant in the constellation Andromeda near the star Gamma. The shower is also known as the BIELIDS because it is associated with **Biela's Comet**, which began to break up in the 1840s.

The Andromeda Galaxy.

ångström or **ångström unit** (*symbol*: Å) A unit of length used by Ångström in his atlas of the solar spectrum. It has been generally used for expressing wavelength, although it is gradually being superseded by the nanometre (10^{-9} m): 1 Å = 10^{-10} m. It is still used with the SI in the specialized field of spectroscopy.

Ångström, Anders Jonas (Lögdö, Sweden, August 13, 1814 – Uppsala, June 21, 1874) A physicist who was a pioneer in applying spectroscopy to astronomy and who made a spectral analysis of the Sun. He recognized the relationship between the Fraunhofer lines in the Sun's absorption spectrum and the discontinuous emission spectrum of incandescent gases. His greatest work, *Recherches sur le Spectre Solaire* (Uppsala, 1868), was accompanied by an atlas, *Spectre normal du soleil*, containing information on the wavelengths of about one thousand spectral lines. It remained a standard of reference for many years. He was the first to examine the spectrum of the aurora borealis.

angular diameter The apparent diameter of a celestial body expressed in angular measure. It is the angle subtended at the observer by the true diameter of the body under observation. If the distance of the celestial body from the observer is known, then its true diameter can be calculated.

angular velocity The rate at which a body changes its angular separation from a fixed point or direction. The rate at which the angle is swept out is measured in radians per second: π radians equal 180°.

annual aberration see **aberration**

annual parallax The angle subtended at a celestial object by the semi-axis major of the Earth's orbit, which is 1 A.U. It is measured by determining the semi-axis major of the **parallactic ellipse** traced on the celestial sphere. The reciprocal of the annual parallax in seconds of arc is the distance of the object in parsecs.

annular eclipse see **solar eclipse**

anomalistic month The time between two successive passages of the Moon through one of the apsides (see **apsis**) of its orbit. It is equal to $27^{d}.55455$ mean solar time.

anomalistic year The time between two successive passages of the Earth through one of the apsides of its orbit. It is equal to $365^{d}.25964$ mean solar time.

anomaly A term used in celestial mechanics in connection with the mathematical description of the orbital motion of one celestial body about another. There are three anomalies. TRUE ANOMALY, in the case of a planet, is the angle which the radius vector of the planet makes with the line of apsides of its orbit. Its value varies as the planet moves round its orbit. In the diagram it is the angle *PSX*. ECCENTRIC ANOMALY, in the case of a planet, relates to the motion of a point round the circle circumscribing the planet's elliptical orbit. This point is found by dropping a perpendicular from the planet to the semi-axis major and producing it to cut the circumscribing circle. In the diagram it is the angle *PCX′*.

MEAN ANOMALY is measured in the same way as the true anomaly, but its value is given by an imaginary planet traversing the orbit with constant velocity in the same periodic time as the real planet, in accordance with Kepler's Second Law (see **Kepler's Laws**), traverses the orbit with nonuniform velocity.

Anomaly is reckoned from 0° to 360° from **perihelion** in the direction of motion.

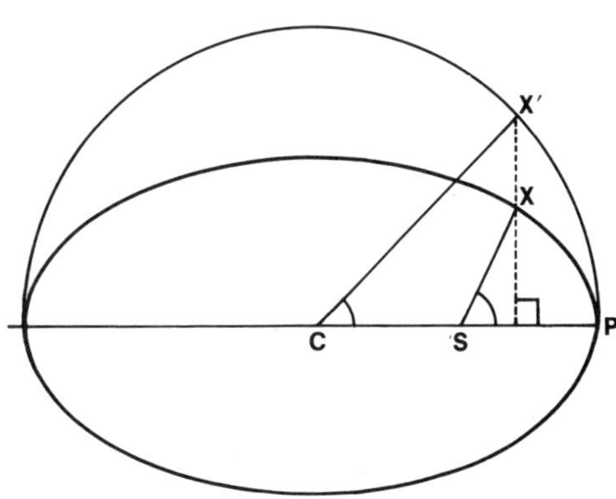

True anomaly and eccentric anomaly.

ansa (plural: ansae) Either of the apparent extremities of Saturn's rings, which appear like handles projecting from the planetary disc under certain conditions of observation. The term is also applied to the extremities of a lenticular galaxy.

antapex That point on the celestial sphere in the constellation Columba away from which the Sun and the entire solar system appear to be moving. It is diametrically opposite to the **apex**.

antimatter Hypothetical matter composed of antiparticles. Antiatoms composed of a nucleus of negatively charged antiprotons and neutral antineutrons with surrounding shells of positively charged antielectrons, or positrons, are conceivable. Matter consisting of such antiatoms would be stable by itself, but in contact with ordinary matter all would be dispersed as lighter particles and energy. Antimatter has not been detected in the universe, though it plays an important part in some modern cosmologies.

Antoniadi, Eugène Michael (Constantinople, 1870 – Paris, February 10, 1944) An astronomer of Greek parentage who became a naturalized French subject. He was noted as a most skilful observer of the inner planets, especially of Mars and Mercury. He published *La Planète Mars* (Paris, 1930) and *La Planète Mercure et la Rotation des Satellites* (Paris, 1934). He devised a system known as the **Antoniadi scale**, by which observers can record the seeing conditions under which their observations are made.

Antoniadi scale The scale of seeing conditions devised by Eugène Antoniadi. It distinguishes five conditions of seeing: **1.** Perfect; complete steadiness. **2.** Slight tremulous motion, with calm periods lasting several seconds. **3.** Moderate, with greater tremulous motion. **4.** Poor, with constant troublesome disturbance. **5.** Very bad; not possible to make rough sketches.

apastron The point in the relative orbit of one component of a binary system at which it is farthest from the other star.

aperture The clear diameter of the objective lens or primary mirror of a telescope.

aperture synthesis A technique used in radio astronomy for increasing the resolving power of a system of small radio telescopes, thus making them equivalent to a much larger instrument. The technique involves distributing and combining electronically a number of small fixed radio dishes, or an array of antennae, at suitable points in a line and utilizing the Earth's rotation to make the assembly sweep out an area equivalent to a much larger radio telescope.

apex That point on the celestial sphere in the constellation Hercules towards which the Sun and the entire solar system appear to be moving at a velocity of 19–20 km/second relative to the nearby stars.

aphelion The point in the orbit of a planet or of a comet at which it is at its farthest distance from the Sun.

aplanatic lens A lens in which spherical and chromatic aberration have been eliminated as far as possible.

aplanatic telescope A telescope whose lens system is so designed that the image is corrected for spherical aberration, chromatic aberration, and coma.

apochromat A lens especially corrected to a high degree to eliminate chromatic aberration.

apogee The point in the orbit of a body around the Earth at which it is at its farthest distance from Earth.

apojove The point in the orbit of any one of the satellites of Jupiter at which it is farthest from that planet.

Apollo A series of US manned spacecraft proposed by NASA in 1960 for orbital flights round the Earth and Moon. In May 1964, shortly after the successful flight of Yuri Gagarin in space, the US president stated that it should be the national aim to achieve a manned landing on the Moon by the end of the decade. The Apollo spacecraft consisted of a command module for the three men, a service module containing the propulsion engine, and a lunar module for conveying two of the crew between the Moon's surface and the orbiting command module. The first manned flight was by Apollo 7, launched on October 11, 1968, and lasted nearly 11 days. Apollos 8, 9 and 10 made flights around the Moon. Apollo 11, launched on July 10, 1969, culminated on July 20th in the successful Moon landing when Neil Armstrong and Edwin Aldrin became the first men to step on the Moon. They returned safely to Earth on July 24th, bringing back 20 kg of lunar samples. Subsequent Apollos, 12 and 14–17, have brought back further samples and left behind five ALSEPs (Apollo Lunar Surface Experimental Packages) which have transmitted a stream of information about lunar physical conditions.

apparent magnitude The brightness of an object as seen from Earth. J. F. W. Herschel (see **Herschel, John**) found that the apparent luminosity of a star of first magnitude was about 100 times that of one of the sixth magnitude. It was decided to adopt a scale of magnitudes in which this ratio is exactly 100:1. This means that 100 stars of magnitude 6 taken all together are equal in brightness to a single star of magnitude 1; that is to say, the ratio 100:1 applies to a *difference* of five magnitudes, and this is a logarithmic scale. So, on passing from one magnitude to the next, the ratio of brightness is the fifth root of 100, which is 2·512. The difference in magnitude between two stars of luminous intensity I_1 and I_2 is given by the formula put forward by N. R. Pogson (see **Pogson, Norman**) who clearly defined magnitude:

$$m_1 - m_2 = 2{\cdot}5 \log (I_2/I_1)$$

A difference of 1, 2·5, 5, 10 magnitudes thus corresponds to a ratio of brightness of 2·512, 10, 10^2, 10^4 respectively. Zero on this scale of magnitudes is fixed with reference to stars of the **North Polar Sequence**. The scale is extended to very bright bodies, so that very bright ones have negative magnitudes. The apparent visual magnitude of Vega is 0; that of Sirius is −1·5; that of the Sun −26·8. As the atmosphere absorbs one-fifth of the light that it receives from space, all magnitudes are in fact 0·22 lower than observed.

The difference between the apparent and absolute magnitudes is known as the **distance modulus** and enables the distance of a star to be calculated:

$$m - M = 5 \log r - 5 = -(5 \log \pi + 5)$$

where r is the star's distance in parsecs and π is its parallax. See also **magnitude**.

apparent retrogression An effect caused by an inferior planet having a faster orbital speed relative to the Earth, or a superior planet having a relatively slower speed. As a result, the planet appears, if observed over a sufficiently long period of time, to reverse the direction of its apparent motion against the background of stars. After some time the planet passes through a stationary point and reverts to its former direction of motion. This motion of a planet in its apparent orbit produces what is called the LOOP OF RETROGRESSION.

apparent solar time Local time, based upon the apparent diurnal movement of the Sun. Noon occurs when the Sun reaches its maximum altitude, i.e. when it crosses the observer's meridian. Time reckoned in this way is subject to considerable variations, due mainly to the non-uniform motion of the Sun. The accuracy was increased by using the concept of **mean time**, which is based on the motion of a hypothetical 'mean Sun'.

apparition The appearance in the sky of a comet or the period during which a planet is observable. The CIRCLE OF APPARITION is that part of the sky in any given latitude within which the stars are always visible.

Appleton layers see **F layers**

appulse The apparent close approach of two celestial bodies whose directions of motion converge (as observed on the celestial sphere) though the bodies in question are in reality remote from each other.

apse Alternative form of **apsis**.

apsis (*plural:* apsides) Either of the two

points in the elliptical orbit of a planet or satellite at which it is at its greatest or its least distance from the primary body about which it revolves. The apsides of the Earth are its aphelion and perihelion; the apsides of the Moon are its apogee and perigee. The LINE OF APSIDES is the line connecting the two apsides of a planet or a satellite.

Aquarids Two meteor showers with their radiants in the constellation Aquarius. See **Delta Aquarids**; **Eta Aquarids**.

Arend object One of the **Trojans**, as yet unnamed. It was discovered by S. Arend of the Royal Observatory of Belgium at Uccle in September 1950. It is minor planet 1950 SA.

areocentric Having Mars as the centre.

areography The description and mapping of the surface features of Mars.

areology The study of the substance of Mars in the same way that geology is the study of the Earth's crust and strata.

Argelander, Friedrich Wilhelm August (Memel, March 22, 1799 – Bonn, February 17, 1875) A distinguished German astronomer who in 1837 became Professor at Bonn, where a new observatory was built under his direction. He acquired fame through his vast undertaking to prepare an atlas and catalogue of all stars down to magnitude 9·5 in the northern hemisphere. This immense work is known as the **Bonner Durchmusterung**. Many stars are still referred to by their identification number in this catalogue.

Ariel **1.** The name given to a series of UK scientific satellites launched by NASA in conjunction with a joint UK-US research programme. The first was launched in 1962 and together with the next three was devoted to studying the Earth's atmosphere. Ariel 5 launched in 1974 studied cosmic X-ray sources, and Ariel 6 launched in 1979 is observing cosmic rays and X-ray sources. **2.** Satellite II of Uranus. See also **satellite**.

Aries, First Point of The vernal equinox (or equinoctial point), i.e. the point at which the Sun crosses the celestial equator. The term originated in antiquity when the vernal equinox was in fact in the constellation Aries, but the term is still retained although the point is now in the constellation Pisces. The position of the point is not fixed: it moves slowly backwards (westwards) along the ecliptic, on account of the precession of the Earth's axis, making a complete circuit in about 25,800 years. This point serves as the origin for the co-ordinates right ascension and celestial longitude. As the position of this point with respect to the stars slowly changes with time, the term 'equinox' is used to define the reference system when giving the co-ordinates of a star. Catalogues of stars are always prepared for a definite equinox, for example 1850, 1900, 1950, or for the equinox of date when no correction has been made to bring the data to a recognized epoch.

Aristarchos of Samos (3rd cent. B.C.) A Greek astronomer who was the first to propose the heliocentric world system. He calculated the distances of the Sun and Moon from Earth, as well as their sizes. His method was sound but the results were inaccurate. He explained the immobility of the fixed stars by their great distance from the Earth's orbit.

Aristotle (Stagira, 384 B.C. – Euboea, 322 B.C.) Greek philosopher and encyclopedist. He extended the system of homocentric spheres of **Eudoxos of Cnidos** to 55 spheres so as to account for all the celestial motions. He attempted to estimate the size of the Earth, which he demonstrated to be spherical. He postulated that celestial bodies are made of aether, and are perfect and incorruptible. He wrote *De Caelo*, a work on astronomy.

Arizona meteor crater An enormous circular depression, almost 1300 metres in diameter and about 180 metres deep, near Canyon Diablo in Arizona. It was formed by the fall in the remote past of an enormous meteorite of nickel-iron, estimated to weigh about one million tonnes. Although large quantities of meteoritic nickel/iron have been recovered from around the crater, the main body of the meteorite has not been found: it is probably buried deeply below, and to one side of, the present floor of the crater. The crater was discovered in 1891.

arm population stars see **Population I stars**

array An arrangement of antennae used in radio astronomy. See **radio telescope**.

artificial satellite A man-made body moving in orbit around the Earth or some other body without propulsion. To achieve this result the body must be launched from the Earth's surface to such an altitude that it is in dynamic equilibrium with the Earth's gravitational attraction and the centrifugal force of its orbital motion. The first artificial satellite, Sputnik I, was successfully put into orbit by the USSR on October 4, 1957. Since then, very many have been placed in orbit for various purposes. Particulars of them are to be found in the periodical *Spaceflight* and in the annual volumes of *Whitaker's Almanac*. Artificial satellites have been put to a variety of uses – scientific, meteorological, navigational, communications, and military. Astronomical satellites have been launched to detect and analyse radiation from space that is unable to penetrate the Earth's atmosphere. The findings have revolutionized the fields of X-ray astronomy, gamma-ray astronomy, and ultraviolet astronomy. Valuable information has also been obtained about the Earth's magnetic field and atmosphere, the ionosphere, cosmic rays, meteorites, etc.

ascending node see **node**

ashen light A faint luminosity sometimes exhibited by the darkened area of Venus when it appears as a crescent, close to inferior conjunction. Its cause is not known for certain, but is probably the result of electrical disturbance in the planet's ionosphere. Its appearance is very similar to earthshine on the Moon, which is produced in a different way.

aspect The position of a planet relative to the Sun as seen from Earth. The following positions are distinguished, depending on the **elongation** of the planet. **Conjunction** occurs when the elongation is 0°, SEXTILE when it is 60°, QUADRATURE when it is 90° or 270°, TRIGONAL when it is 120°, and **opposition** when it is 180°.

aspherical lens A lens in which one of the surfaces is not spherical. The figure of one surface is frequently a parabola.

A stars see **stars, spectral classification of**

asterism A small group of easily recognizable stars, not necessarily forming a complete constellation.

asteroid The name coined by Sir William Herschel in 1802 for the celestial objects now known as **minor planets**.

astigmatism An aberration in lenses and mirrors that occurs when light falls obliquely on the system. Rays of light from different parts of the object are not brought to focus at one point but in two separate focal lines, one in the plane of the optical axis and the other in a plane perpendicular to it. A lens system designed so as to minimize astigmatism is called an ANASTIGMAT.

astrobiology or **exobiology** The study of environments elsewhere in the universe from the point of view of their potential for supporting life in some form. Also, the study of the effect of space travel on terrestrial organisms.

astrograph In general, a telescope used for obtaining photographs of star fields, mainly for astrometry. The term specifically means the series of special refractors designed for the **Carte du Ciel** project. The instruments were equatorially mounted refractors having an aperture of 33 cm and a focal length of 340 cm. The objectives were doublets corrected for colour and coma. The photographic plates had a field of side 2° on a scale of 1′ to 1 mm.

astrolabe An early astronomical instrument used to determine the altitude of celestial bodies. In its simplest form it consists of a graduated circular plate with an alidade or sighting vane pivoted at its centre. The instrument is suspended from a ring and in use is held vertically. The object to be observed is sighted through the alidade and the altitude is

A mariner's astrolabe of the 16th century.

read off on the graduated circle. More elaborate versions comprise three plates: one plate has a stereographic projection of the celestial sphere for a given latitude, the second plate is rotatable on the first and cut away so as to form a star map, and the third is a thick plate, with a graduated circular scale and an alidade for measuring altitude, in which the other two plates fit.

astrology The word originally embraced both NATURAL ASTROLOGY, which is the true science of astronomy, and JUDICIAL ASTROLOGY, which is the pseudo-science professing to predict events bearing on the destiny of human beings, individually or generally. Astrological conclusions, based on traditional interpretations, are summarized in horoscopes. The need of astrologers to have information on the motions of celestial bodies did much to assist the development of astronomical knowledge.

astrometric binary see **binary**

astrometry or **positional astronomy** The branch of astronomy concerned with the measurement of precise positions of celestial objects, determined either from photographic plates taken specially for the purpose or, more recently, by means of radio telescopes. Stellar parallaxes and proper motions are derived from differential positions obtained from measurements made at widely separated times.

astronautics In its widest sense, the science that deals with all the requirements for space flight. The successful development of space flight has enabled astronomers to obtain information about celestial objects, not only at closer range but also free from the disturbing effect of the Earth's atmosphere. Planetary probes, artificial satellites and space stations, and manned missions have yielded an immense amount of new information about the Sun, Moon, planets, and planetary satellites, and about the many different types of stars and other celestial objects. Much of this information concerns powerful sources of X-rays, gamma rays, ultraviolet and infrared radiation, which could only be fully investigated after the advent of rockets and spacecraft.

Astronomer at the Cape The Royal Observatory at the Cape of Good Hope was established by Order in Council in 1820. The following table lists the holders of the office of Astronomer at the Cape, in early records referred to as the Astronomer Royal at the Cape.

Name	*Held office*
Rev. Fearon Fallows (1789–1831)	1820–1831
Thomas Henderson (1798–1844)	1831–1833
Sir Thomas Maclear (1794–1879)	1833–1870
Edward James Stone (1831–1897)	1870–1879
Sir David Gill (1843–1914)	1879–1907
Sydney Samuel Hough (1870–1923)	1907–1923
Sir Harold Spencer Jones (1890–1960)	1923–1933
John Jackson (1887–1958)	1933–1950
Richard Hugh Stoy (1910–)	1950–1968

The title has now lapsed.

Astronomer Royal The Royal Observatory at Greenwich was founded in 1675 by King Charles II, who appointed J. Flamsteed (see

Flamsteed, John) as 'our astronomical observator' at a salary of £100 per annum. The following table lists those who have held the office of Astronomer Royal.

Name	*Held office*
Rev. John Flamsteed (1646–1719)	1675–1719
Edmond Halley (1656–1742)	1720–1742
Rev. James Bradley (1693–1762)	1742–1762
Rev. Nathaniel Bliss (1700–1764)	1762–1764
Rev. Nevil Maskelyne (1732–1811)	1765–1811
John Pond (1767–1836)	1811–1835
Sir George Biddell Airy (1801–1892)	1835–1881
Sir William Henry Mahoney Christie (1845–1923)	1881–1910
Sir Frank Watson Dyson (1868–1939)	1910–1933
Sir Harold Spencer Jones (1890–1960)	1933–1955
Sir Richard van der Riet Wooley (1906–)	1955–1971
Sir Martin Ryle (1918–)	1972–

Up to 1971 the Astronomer Royal was also Director of the Royal Observatory at Greenwich (later named the Royal Greenwich Observatory). Since that date appointments to the two offices have been made separately. The following have held the office of Director of the Royal Greenwich Observatory.

Name	*Held office*
Dr Eleanor Margaret Burbidge	1972–1973
Dr Alan Hunter	1973–1975
Dr Francis Graham Smith	1976–

Astronomer Royal for Scotland This office was created by Royal Warrant in 1834 to provide a Director for the Royal Observatory at Edinburgh. The post was formerly, and still is, held by the Regius Professor of Practical Astronomy in the University of Edinburgh. The following table lists those who have held the office.

Name	*Held office*
Thomas Henderson (1798–1844)	1834–1844
Charles Piazzi Smyth (1819–1900)	1845–1889
Ralph Copeland (1837–1905)	1889–1905
Frank Watson Dyson (1868–1939)	1905–1910
Ralph Allen Sampson (1866–1939)	1911–1937
William Michael Herbert Greaves (1897–1955)	1938–1955
Hermann Alexander Brück (1905–)	1957–1975
Vincent Cartledge Reddish (1926–)	1975–1980
Malcolm Sim Longair (1941–)	1980–

astronomical date In astronomy it is the practice to express a date in the form 'year, month, day', e.g. 1980 January 1, which can be followed by the time of an observation, e.g. 1980 January 1, 0957 G.M.T. Alternatively the time may be expressed as a decimal of a day, e.g. 1980 January 1·26584 U.T.

astronomical twilight see **twilight**

astronomical unit (*symbol:* A.U.) A fundamental unit of distance measurement in astronomy. It is the mean distance of the Earth from the Sun, and is equal to $149{\cdot}5979 \times 10^6$ kilometres (almost 93 million miles).

astronomy The scientific study of all aspects of the heavens and celestial bodies. It treats of the distances, magnitudes, masses, motions, compositions, and evolution of a very great variety of celestial bodies both in the solar system and beyond it – in our own Galaxy and in other galaxies. It is based on careful, accurate observations over long periods of time. The science now comprises various specialized branches including astrometry, astrophysics, celestial mechanics, cosmology, radio astronomy, and X-ray astronomy. There is an overlap with geophysics in respect of such subjects as aurorae, atmospheric physics, cosmic rays, and solar-terrestrial relationships.

astrophysics The study of the physical properties of celestial bodies. It is based mainly on the study of radiation from these bodies, and has developed through the application of photography, photometry, and spectroscopy as observational techniques.

atmosphere The envelope of gases that surrounds stars and certain planets and planetary satellites. The Earth's atmosphere is recognized as consisting of five distinct layers. The **troposphere** extends to a height of about 10–20 km, with the temperature falling to about 200 K. The composition by volume of the lower atmosphere is approximately 78% nitrogen, 21% oxygen, 0·9% argon, and 0·3% carbon dioxide. The remaining small proportion consists of the rare gases helium, neon, krypton and xenon, and pollutants. The **stratosphere** extends up to about 50 km, with a rise in temperature up to about 270 K. The **mesosphere** extends up to about 80 km, with the temperature again falling to about 170 K. This layer contains the D region of charged particles, which forms part of the **ionosphere**. The THERMOSPHERE extends to about 500 km, with the temperature again rising. This layer contains the E and F regions of charged particles, also part of the ionosphere. The EXOSPHERE merges into interplanetary space.

atmospheric dispersion The scattering of light in the Earth's atmosphere. The presence of suspended particles in the atmosphere causes much of the light they receive to be dispersed. The effect is not the same for all wavelengths of light. Blue is preferentially diffused, and as a result the normal appearance of the clear sky in daytime is blue.

atmospheric extinction The reduction in the brightness of light from celestial bodies as a result of absorption, dispersion, and scattering of the light as it passes through the atmosphere before reaching the observer. The extinction is greatest when the altitude of the body is low because light then has to travel through a longer path in the atmosphere. It increases with decreasing wavelength, causing the light to become reddened. The amount of extinction can vary greatly, so a correction for it has to be made in all measurements of brightness.

atmospheric refraction The alteration in the direction of a ray of light proceeding from a celestial body to an observer on Earth. The Earth's atmosphere decreases in density from the surface upwards. Consequently a ray of light from space will be deviated, or refracted, in accordance with the law of refraction. The greatest effect is produced when the celestial body is on the horizon, and makes the body appear to have a higher altitude than it has in reality; it will even make a body visible before it has in fact risen. The refraction is zero for light coming from the zenith. All astronomical observations of stars and other bodies must be corrected for atmospheric refraction in order to obtain the true position from the observed apparent position.

atom The smallest particle of matter that can take part in chemical reactions. Every kind of atom is considered to consist of a positive **nucleus** containing protons (positively charged particles) and neutrons (neutral particles of almost equal mass to protons). The net positive charge on the nucleus is called the ATOMIC NUMBER. The nucleus is surrounded by shells of **electrons** (negatively charged particles) in orbit about the nucleus. The number of orbiting electrons is equal to the number of protons, making the atom electrically neutral. An **element** consists entirely of atoms of the same kind. The combining of elements produces the material compounds or substances of our world.

atomic clock A type of clock which utilizes the emission or absorption of energy involved in the transition of an atom from one or other of its two stable states. The **second** is defined in the SI with reference to the caesium atom.

atomic mass unit (*abbrev.:* amu) One-twelfth of the mass of the nuclide ^{12}C. 1 amu = $1{\cdot}66 \times 10^{-24}$ g. The energy equivalent of 1 amu is 931 MeV.

atomic number see **atom**

aurora An illumination of the night sky that is common within the polar circles and hence often called **polar lights**. It is known as the AURORA BOREALIS in the northern hemisphere and as the AURORA AUSTRALIS in the southern hemisphere. The aurora appears in many forms, e.g. as a horizontal arc with rays radiating upwards, or as a hanging curtain exhibiting beautiful, flickering colours, usually red and green. Aurorae occur at altitudes between 90 and 200 km. The phenomenon is caused by electrical discharges in the upper atmosphere and is linked with terrestrial magnetic disturbances associated with solar activity. Charged particles from the Sun are attracted by the Earth's magnetic field and diverted towards the polar regions, where they interact with atmospheric components. Spectroscopic examination of auroral light reveals the existence of emission lines of atoms and molecules in the upper atmosphere. A characteristic prominent green line (5577 Å) comes from oxygen.

australite see **tektites**

autumnal equinox see **equinox**

axial inclination The angle between the axis of rotation of a planet and the perpendicular to the plane of its orbit; this is equivalent to the angle between the equatorial and the orbital planes.

axis The AXIS OF ROTATION of a celestial body or of a system of stars is the imaginary line about which the rotation takes place. The CELESTIAL AXIS is the prolongation of the Earth's axis to the celestial sphere. The **optical axis** of an optical system is the central axis of the system.

azimuth The angle measured in a horizontal plane between the vertical circle through a celestial object and the plane of the observer's meridian. It is usually measured eastwards (clockwise) from the north point from 0° to 360° but is sometimes measured from the south point. Azimuth and **altitude** are the co-ordinates used in the horizon system. See also **co-ordinates**.

Baade, Wilhelm Heinrich Walter (Schröttinghausen, March 24, 1893 – Göttingen, June 25, 1960) A German astronomer who worked at the Hamburg-Bergedorf Observatory and from 1931 to 1958 at the Mount Wilson and Palomar Observatories. He discovered the distant minor planet, Hidalgo, in 1920. He was distinguished for his work on the composition and structure of galaxies, which led to his theory for the existence of Population I and Population II stars. He showed that the cosmic distance scale between extragalactic bodies was incorrect and should be doubled. He resolved the nucleus of the Andromeda Galaxy (M 31) into its component stars and described the radio source in Cygnus (**Cygnus A**) as two galaxies in collision.

Babcock, Harold Delos (Edgerton, Wisconsin, January 24, 1882 – April 8, 1968) An American astrophysicist distinguished for his work in spectroscopy of the Sun and in solar magnetism. He discovered (with his son, H. W. Babcock) stellar magnetism; he measured the magnetic field of 78 Virginis, thereby revealing the connection between electromagnetism and relativity.

Babcock, Horace Welcome (Pasadena, California, September 13, 1912–) An American astronomer distinguished for his work on solar and stellar magnetic fields; he invented a solar magnetograph. He also investigated the rotation of the Andromeda Galaxy.

background noise All extraneous sound in a system which is producing, measuring, or recording signals. In radio astronomy background noise can arise from: (1) the receiving equipment (thermal noise); (2) synchrotron radiation (galactic noise); (3) fluctuations in the current of the detector (quantum noise); (4) star noise.

background radiation All the extraneous effects which cause interference when receiving or measuring a signal. The cosmic background radiation detected at gamma-ray, X-ray, and radio wavelengths is thought to arise from a large number of weak unresolved sources situated at great distances. The **microwave background radiation** is thought to be the remnant of the radiation of the early universe.

background stars The celestial bodies which are external to the solar system; they correspond to the 'fixed stars' of ancient times.

Baily, Francis (Newbury, April 28, 1774 – London, August 30, 1844) An English astronomer who repeated the experiment of Henry Cavendish (1731–1810) to determine the Earth's mean density, and who devised star catalogues and observed the solar eclipses of 1836 and 1842.

Baily's beads A very short-lived phenomenon seen during a solar eclipse and first described by Francis Baily, who observed it during the annular eclipse of May 15, 1836. Just before and just after totality the limb of the Moon can be seen within the boundary edge of the Sun. Light from the Sun shines through the valleys on the Moon's limb and thereby produces the appearance of a string of brightly shining beads.

Baker-Nunn camera A **Schmidt camera** specially designed for the photography of satellites.

Baker-Schmidt telescope Any of various modified forms of the **Schmidt camera** designed by J. G. Baker to eliminate defects such as astigmatism, coma, or distortion and give instruments of high performance.

Balmer lines One of the distinctive series of lines in the spectrum of hydrogen, occurring mainly in the visible region of the spectrum. This and the other series of hydrogen spectral lines are caused by the movement of electrons between various specific levels of energy in the hydrogen atom. The Balmer series of lines is a dominant characteristic of the spectra of A-type stars. The rule governing the wavelengths of lines in the spectrum of hydrogen was first formulated by J. J. Balmer (1825–98) in 1885. The first Balmer line, $H\alpha$, is at 6563 Å, $H\beta$ is at 4861 Å, $H\gamma$ at 4342 Å, and $H\delta$ at 4102 Å. The series tends to a limit at 3647 Å.

band A series of closely spaced emission or absorption lines in a spectrum produced by molecules.

bandpass filter A system for isolating a relatively narrow range of frequencies, used in radio communication and radio astronomy.

band spectrum see **spectrum**

bar A unit of pressure in the cgs system. 1 bar $= 10^6$ dyne $\text{cm}^{-2} = 10^5\ \text{N m}^{-2}$.

Barlow lens A planoconcave lens placed between the objective (or mirror) and the eyepiece of a telescope in order to increase the focal length of the objective, and hence increase the magnification. It can also improve definition.

Barnard, Edward Emerson (Nashville, Tennessee, December 16, 1857 – Williams Bay, Wisconsin, February 6, 1923) An American astronomer who was a pioneer of photographic astronomy. He discovered the fifth satellite of Jupiter (1892), several comets and dark nebulae, and the star which bears his name.

Barnard's star A star in the constellation Ophiucus. It has the largest known proper motion (10·27 arc seconds per year) and is the fourth nearest star. Careful study of the proper motion of this star enabled P. van de Kamp in 1963 to establish the presence of an invisible companion whose mass is about 1·5 times that of Jupiter. This particular star is considered to have a planetary system.

barred spiral galaxy see **galaxy**

barycentre The centre of mass of the Earth-Moon system. It lies within the Earth as the mass of that body is 81·3 times that of the Moon.

Bayer, Johann (Rain, Bavaria, 1572 – Augsburg, March 7, 1625) A lawyer who compiled the first complete star atlas, *Uranometria* (Augsburg, 1603). In this he introduced a system for naming stars according to their brightness, using a notation based on the letters of the Greek alphabet, e.g. Beta (β) Geminorum, Gamma (γ) Draconis. Stars of the same brightness were notated according to their position in the imagined figure of the constellation: the sequence of letters does not correspond to an exact gradation of brightness. This notation, known as BAYER LETTERS, was rapidly adopted. Bayer's attempt to name northern constellations from the Old Testament and southern constellations from the New Testament was unsuccessful.

Becklin-Neugebauer object An infrared object in the Orion nebula, discovered in

1966 by E. E. Becklin and G. Neugebauer. In the range 2–12 μm the distribution of spectral energy corresponds to a black body at 700K. It is presumed to be a collapsing protostar.

Beer, Wilhelm (Berlin, January 4, 1797–March 27, 1850) A banker and astronomer who is remembered for his collaboration with J. H. von Mädler (*see* **Mädler, Johann von**) in the production of the *Mappa Selenographica* (1834–36), at that time the most accurate lunar map. They were authors also of the descriptive work *Der Mond* . . . (1837) and of the first detailed chart of the surface of Mars.

Bellatrix The star γ Orionis in the constellation Orion. It marks the left shoulder of the quadrilateral figure of Orion. It is a blue giant star with an apparent visual magnitude of 1·6 and is about 140 pc distant.

bent pillar mounting see **telescope, mounting of**

Bessel, Friedrich Wilhelm (Minden, July 22, 1784 – Königsberg, March 17, 1846) A German merchant who turned astronomer and was director of the observatory at Königsberg from 1810 until his death. He was a pioneer in exact astronomical measurements as regards positional astronomy. He used Fraunhofer's recently devised heliometer to determine the parallax of a fixed star (61 Cygni in 1838). He suggested the existence of a planet (Neptune) beyond Uranus. From the perturbations in the proper motion of Sirius and of Procyon he deduced the existence of their companion stars, which were detected years after his death. He is best known as a mathematician. He introduced certain mathematical functions (now called BESSEL FUNCTIONS), which were used in the mathematical study of planetary orbits.

Betelgeuse (α Ori) The second brightest star in the constellation Orion. It is an orange-red supergiant star of variable magnitude. Its diameter as measured with an interferometer is over 400 solar diameters (about 560 million km).

Bethe, Hans Albrecht (Strasbourg, 2 July, 1906 – 1967) A German physicist, director of the Theoretical Physics Division, Los Alamos Science Laboratory. He developed a theory concerning the production of energy in stars, including the Sun (see **carbon-nitrogen cycle**). He also proposed a theory for the production of electron-positron pairs and (with W. H. Heitler) for the origin of showers of electrons and positrons in cosmic radiation.

biconcave lens A lens that is bounded by two opposing concave surfaces. The lens is diverging.

biconvex lens A lens that is bounded by two opposing convex surfaces. The lens is converging.

Biela's Comet A now-disintegrated comet of period 6·6 years that was discovered by W. Biela (1782–1856) in February 1826 and identified as one that had been seen in 1772 and 1805. Its return was predicted for 1832, and it was seen. It was apparently not seen in 1839, as it was too close to the Sun. At the return in 1846 the comet was observed to split into two portions, the smaller of which increased in brightness. The double comet continued in orbit but the two portions were slowly separating, and after two months were about 282,000 km apart. At the next return in 1852 the two portions were about 2·3 million km apart. They have not been seen since.

During the evening of November 27, 1872 there was a magnificent display of meteors seeming to fall from the direction of Gamma Andromedae. This swarm of meteors was seen precisely when the Earth crossed the orbital path of the vanished comet, and is the debris of the disintegrated comet.

Bielids see **Andromedids**

big bang theory One of the cosmological models (governed by Einstein's General Theory of Relativity) advanced to explain the origin of the universe. According to the theories of G. Gamow, which were developed from those of G. Lemaître, there was initially nothing but a primeval condensate of matter which exploded, at time zero, attaining a temperature of about 15,000 million K. This explosion, 10 to 20 thousand million years ago, began the expansion of the universe, which still continues. As the temperature dropped extremely rapidly particles of matter and radiation, i.e. photons, were produced by nuclear reactions. The matter somehow collected into immense clouds of gas and gradually developed into galaxies and member stars.

Calculations based on this theory and supported by observations indicate that the matter remaining from this explosion, and making up the bulk of most celestial objects, should consist mainly of about 90% hydrogen and 10% helium. The weak universal microwave background radiation detected, in 1965, in space is considered to be the residual radiation of the primeval explosion. Some theorists regard this explosion not as the instant of creation of a unique universe, but merely the beginning of a new era of expansion in the cyclical sequence of contraction and expansion.

billion In English usage, a million million, i.e. 10^{12}; in American usage, a thousand million, i.e. 10^{9}. To prevent confusion, the use of this word is best avoided.

binary A star which on close examination is found to consist of two stars in orbital motion about a common **centre of mass**, each component star describing an elliptical orbit of high eccentricity. In some cases the components are coloured differently from each other. The ratio of the sizes of the orbits is inversely proportional to the ratio of the masses of the component stars. This ratio can be determined by positional measurements or by spectrum analysis. The sum of the masses of the system can be calculated from Kepler's Third Law. The orbital periods of binaries vary enormously. Some are very short, as in the case of spectroscopic and eclipsing binaries (see below), and vary between hours and some few years. The periods of most visual binaries vary from about 20 to 100 years.

Binaries are usually classified as visual, spectroscopic, or eclipsing binaries, according to the means by which they are observed. In a VISUAL BINARY the components are sufficiently far apart, with an angular separation greater than about 0″·1, so that the system is observable with the telescope as a double star. Unless the period is short enough to be determined, it cannot be easily decided if one is dealing with a true binary system or with an **optical double**. Mizar (Zeta Ursae Majoris) was first observed by G. B. Riccioli in 1650 to be a double star.

A SPECTROSCOPIC BINARY is a system whose components are too close for their separation to be measured visually. The binary nature is revealed by a characteristic Doppler shift in the absorption lines of the stars' spectra, arising from the variation in their radial velocity. The brighter component of Mizar (ζ UMa) was the first star shown to be a spectroscopic double, by H. C. Pickering in 1889.

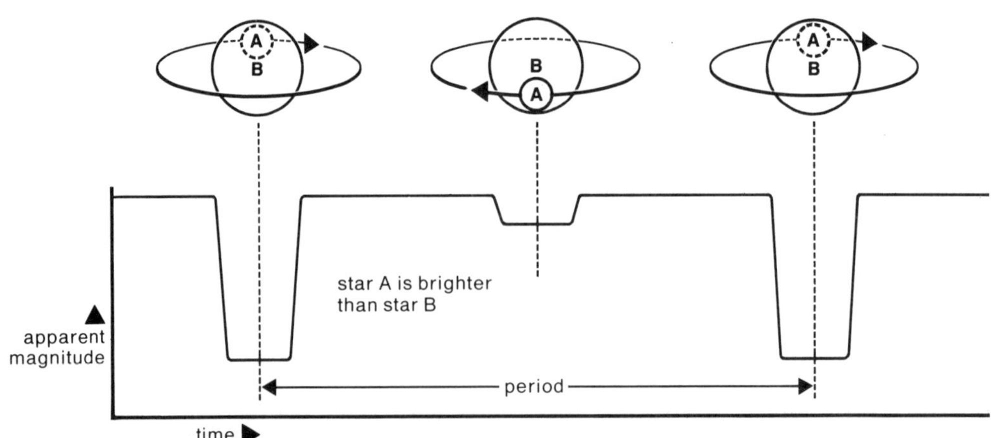

Light curve and relative orbit of an eclipsing binary.

An ECLIPSING BINARY is also known as a PHOTOMETRIC BINARY because the study of such a system depends entirely on the **light curve**. With an eclipsing binary, the relative orbit is viewed edgewise, or nearly so, so that one component can be observed to pass once in front of, and once behind, the more massive primary star, during each revolution. The light therefore appears to fluctuate. Most eclipsing binaries are also spectroscopic binaries, usually large stars in small orbits. Detailed analysis of the light curve enables much information to be deduced as to the nature of the system. Algol (Beta Persei) was the first star of this kind to be identified, by J. Goodricke in 1782.

Study of irregularities in the proper motions of stars has in some cases revealed the existence of invisible companions and enabled the orbits and masses of the components of such ASTROMETRIC BINARIES to be determined. Barnard's star is an example.

A star with three or more components is called a multiple star, Alpha Centauri being an example.

black body An ideal body which absorbs all radiation that falls upon it, and is a perfect emitter of radiation.

black-body radiation The radiation, of a continuous range of frequencies, which would be emitted by a black body and which is a function of temperature only. The spectral intensity distribution of the radiation is in accordance with **Planck's radiation law**.

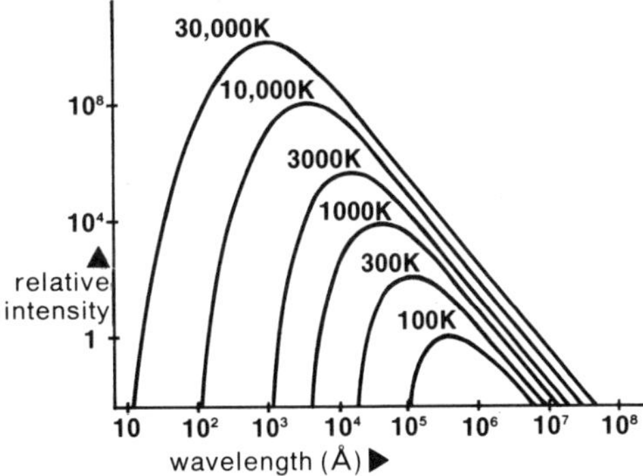

Variation of radiant intensity with wavelength for black bodies of different temperatures.

black drop A phenomenon caused by diffraction of light and produced when a small dark disc is almost touching the interior rim of a larger disc. It has the appearance of a dark ligament connecting the rims of the two discs, though in fact there is a small gap between the theoretical images of the discs. It is observed at the moment during transit that the disc of Mercury, or of Venus, has completely entered the Sun's disc and at the moment immediately before the planet starts to emerge.

black hole A localized region of space from which neither matter nor radiation can escape, i.e. from which the escape velocity exceeds the velocity of light. The boundary of this region is called the EVENT HORIZON. Its radius – the **Schwarzschild radius** – depends on the amount of matter that has fallen into the region: it increases linearly as the mass increases. Theory predicts that a black hole of stellar mass forms when a massive star undergoes total gravitational collapse. The gravitational collapse of a star of up to about 1·4 solar masses can be physically halted to produce a white dwarf. A slightly more massive object will collapse to form a neutron star. If, however, the mass exceeds about 3 solar masses, even after a supernova explosion has blasted away the outer layers of the star, the collapse can only continue. As the radius of the star contracts below its Schwarzschild radius, the object becomes a black hole and effectively disappears. A star of 3 solar masses has a Schwarzschild radius of about 9 km. Inside the event horizon of the black hole, space and time are highly distorted and the stellar matter is increasingly compressed until it forms an infinitely dense **singularity**. Black holes are thought to have an immense range of mass. SUPERMASSIVE BLACK HOLES of maybe a thousand million solar masses could be the source of energy in quasars and other types of active galaxy.

Since no light or other radiation can escape from black holes, their detection is extremely difficult. Any matter encountered by the black hole will most likely go into orbit first rather than being drawn directly into it. A rapidly spinning disc of matter, known as an ACCRETION DISC, therefore forms around the object. It can be shown that matter in an accretion disc will emit considerable energy, mainly in the form of X-rays. In a binary system in which the two component stars are very close gaseous matter can be transferred from the larger to the more compact component. If this compact object is a black hole an accretion disc will form and will emit X-rays. The search for black holes as components of X-ray binary stars is therefore in progress. The most likely candidate is the massive 'unseen' companion to a huge luminous supergiant star, which together form the intense X-ray binary **Cygnus X-1**.

blaze star The name given to T Coronae Borealis, the nova of May 12, 1866, whose magnitude rose very rapidly from 9·5 to 2, the star remaining visible to the naked eye for some days. It then declined in magnitude, rose again to magnitude 7 and finally dropped to magnitude 9·5, where it remains as a slightly variable star.

B line J. von Fraunhofer's designation for one of the absorption lines in the solar spectrum; it is in the red part of the spectrum at 6875 Å and is caused by oxygen in the Earth's atmosphere. See also **Fraunhofer lines**.

blink microscope see **comparator**

blooming A coating, usually magnesium fluoride, applied by electro-evaporation *in vacuo* to the surfaces of glass components (lenses, prisms) used in optical apparatus in order to minimize reflections and so increase the amount of transmitted light.

blue clearing The sudden, unusual and infrequent transparency of the Martian atmosphere. Normally, the atmosphere on Mars is impenetrable to blue and violet light, but when it becomes transparent the surface markings are revealed. The phenomenon was first observed by E. C. Slipher (see **Slipher, Earl C.**). It has been proposed that the 'blue clearing' should be called SLIPHER'S PHENOMENON.

BN object see **Becklin-Neugebauer object**

Bode, Johann Elert (Hamburg, January 19, 1747 – Berlin, November 23, 1826) A German astronomer who became director of the observatory at Berlin. He is remembered as the crusader for the 'marvellous relationship' followed in the distances between planets, which was pointed out by J. D. Titius (1729–96) in 1766 in his translation into German of C. Bonnet's *Contemplation de la Nature*. On the basis of this relationship (see **Bode's Law**) he suggested the existence of an unknown planet between Mars and Jupiter. Confirmation of this suggestion was obtained in 1800 onwards by the discovery of Ceres and other small bodies. It was suggested (wrongly) that these minor planets were fragments of a disintegrated major planet. Bode proposed the name 'Uranus' for the planet discovered by W. Herschel.

Bode's Law More properly the TITIUS-BODE Law, this is an empirical law which has received some modifications since its first enunciation. To each number in the progression 0, 3, 6, 12, 24, 48, 96 and 192 add 4, and then divide each by 10. The resulting numbers 0·4, 0·7, 1·0, 1·6, 2·8, 5·2, 10·0 and 19·6 are in close agreement with the solar distance in astronomical units of each planet as far as Uranus. Because of this agreement many attempts have been made to find a theoretical basis for Bode's Law, but they have so far failed.

bolide see **meteor**

bolometer A sensitive instrument for measuring radiant energy and used in astronomy to measure the total radiation from a star. There are now many types of bolometer, including semiconductor and supercooled devices, but most measure the change in electrical resistance that results from the absorption of radiation.

bolometric magnitude A theoretical figure which measures the total amount of energy radiated by a body at all wavelengths. It can be calculated from the visual magnitude.

Bond, George Phillips (Dorchester, Massachusetts, May 20, 1825 – Cambridge, Massachusetts, February 17, 1865) An

American astronomer who became director of the Harvard College Observatory. With his father, W. C. Bond, he discovered Hyperion – the eighth satellite of Saturn (1848) – and Saturn's crêpe ring (1850).

Bond, William Cranch (Portland, Oregon, September 9, 1789 – Cambridge, Massachusetts, January 29, 1859) An American astronomer who became director of the Harvard College Observatory. He was a pioneer of astronomical photography. With his son, G. P. Bond, he made intensive observations of the planets, Sun, and nebulae.

Bondi, Hermann (Vienna, November 1, 1919–) An Austrian mathematician, now resident in England, who is distinguished for his studies on the constitution of stars, on general relativity, and on the steady-state theory of the universe.

Bonner Durchmusterung A catalogue comprising F. W. A. Argelander's *Astronomische Beobachtungen*, Bonn, 1846–69, and the continuation by E. Schönfeld, *Astronomische Beobachtungen*, Bonn, 1886. In this catalogue, which contains data on 457,857 stars, the stars are numbered in declination zones from +90° to −22°, and are cited in the form 'BD +52° 1638'.

Brackett series The series of spectral lines, far in the infrared, associated with the fourth energy level of the hydrogen atom. It was investigated by F. S. Brackett.

Bradley, James (Sherborne, Gloucestershire, March, 1693 – Chalford, Gloucestershire, July 13, 1762) An English astronomer who in 1721 became Savilian Professor of Astronomy at Oxford and in 1742 succeeded E. Hailey as Astronomer Royal. He discovered the **aberration** of light and the **nutation** of the Earth's axis (1747), predicted by Newton. His catalogue of stars was published posthumously and formed the basis of F. W. Bessel's catalogue (1818).

Brahe, Tycho (Ottensen, Denmark, December 14, 1546 – Prague, October 24, 1601) A lawyer who turned astronomer. Under the patronage of King Frederick II of Denmark he built and equipped two observatories (Uraniborg and Sjärneborg) on the Island of Hven, which he left in 1597 following disputes with the new king Christian IV. He then settled in Prague, where J. Kepler became his assistant. Working in the pre-telescope era, he was expert in making accurate naked-eye astronomical observations with instruments such as mural quadrants, whose precision he greatly improved. He observed inequalities in lunar motion and developed lunar theory. In November 1572 he observed the great supernova in Cassiopeia. This observation, together with his study of the comet of 1577, demonstrated that the Aristotelian notion of the immutability of the heavens was untenable. He could not, however, accept the world system put forward by Copernicus, and so developed his own planetary theory. In this TYCHONIAN PLANETARY SYSTEM the planets move round the Sun, and the Sun itself, like the Moon, moves round the stationary Earth. The accurate series of observations made by

Tycho Brahe in his observatory at Hven.

Tycho Brahe were of inestimable value to Kepler in deriving his Laws of Planetary Motion.

Bremsstrahlung (German, meaning 'brake radiation') Electromagnetic radiation produced when energetic electrons are decelerated, for example when approaching the nucleus of an atom. It occurs in ionized gas clouds.

Brown, Ernest William (Hull, November 29, 1866 – New Haven, Connecticut, July 22, 1938) An astronomer noted for his investigations into planetary and lunar motions, the latter leading to the suggestion that the Earth's rate of rotation is variable.

B stars see **stars, spectral classification of**

bubble sextant An instrument designed mainly for use in aircraft, and hence sometimes called an AIR SEXTANT. Its principle of operation is exactly the same as for the marine sextant, except that it cannot be levelled by reference to the horizon. Alignment of the instrument is achieved vertically by means of a small bubble in a liquid contained in a glass dome placed at the top of the instrument.

butterfly diagram A diagram devised by E. W. Maunder (see **Maunder, Edward**). It illustrates the distribution of sunspots in solar latitude during the 11-year periodicity of their occurrence. Their progression towards the solar equator gives the diagram the characteristic pattern by which it is known.

calcium K-line The violet Fraunhofer line, of wavelength 3934 Å, that occurs in the spectrum of the ubiquitous element calcium. It is of particular use in the monochromatic examination of the Sun's disc for solar activity. It occurs prominently in the spectra of certain types of stars (G and F stars).

calendar A list of the days of the year, divided into days, weeks, and months, adapted to the requirements of daily life or to particular purposes. From remote times the divisions of the calendar have depended on the periods of the Sun and the Moon. The great civilizations of the past – Babylonian, Egyptian, Greek, Jewish, and Roman – devised calendars based on astronomical observation. In 46 B.C. a revised calendar was introduced by Julius Caesar, because failure to introduce every two years the intercalary month into the Roman calendar of 355 days had resulted in its becoming out of phase with the seasons. In this revised calendar, known as the JULIAN CALENDAR, the year was divided into 12 months. Every fourth year an extra day (LEAP DAY) was added to the usual year of 365 days. The average length of the year, i.e. 365·25 days is, however, longer than a tropical year (the interval of time between two successive passages of the Sun through the vernal equinox) by over eleven minutes. The Julian Calendar year thus slowly became out of phase with the seasons by over 18 hours a century. By the 16th century there was an obvious discrepancy between the time dates for the vernal equinox and for Easter, and those given by the calendar.

In 1582 Pope Gregory XIII introduced a revised calendar, called the GREGORIAN CALENDAR, in which the vernal equinox was restored to the date assigned to it in the Easter tables, namely March 21. This revision meant dropping ten days from the calendar, so October 5, 1582 became October 15, 1582. The average length of the calendar year was reduced by inserting the intercalary day every four years, except for the centurial years *not* divisible by 400. Thus the years 1600 and 2000 are leap years, but 1700, 1800 and 1900 are not. The Gregorian Calendar displaced the Julian Calendar and is now used by most countries. It was adopted in Great Britain and its territories in 1752, when September 2 was followed by September 14. The difference between the Gregorian calendar year of 365·2425 days and the tropical year is slight.

In about A.D. 525 Dionysius Exiguus, a chronologist, introduced the method of reckoning the years by reference to the Christian era. He fixed the beginning of the present calendar so that Christ was born in December A.D. 1.

Other proposals have been made for reforming the calendar. In the FRENCH REVOLUTIONARY CALENDAR the months were given fanciful names based on the seasons. Year 1 (An 1) began on September

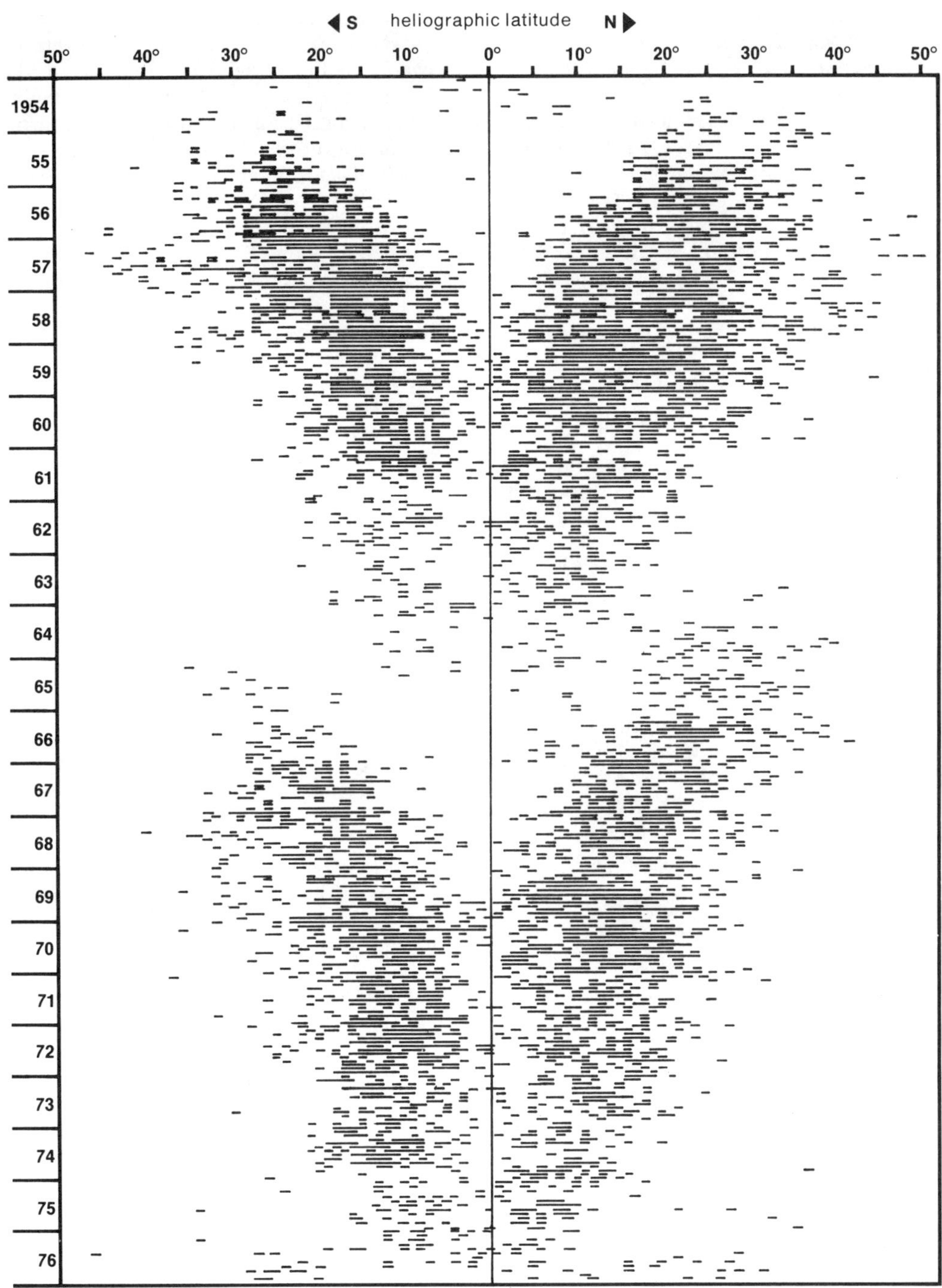

Maunder's butterfly diagram showing clearly the characteristic pattern.

22, 1792, Year XIV began on September 23, 1805, and the Revolutionary Calendar was discontinued on December 31 of that year. The Gregorian calendar was once more brought into use in France. There has been propaganda for a WORLD CALENDAR in which each year is the same. Each quarter of 91 days begins on a Sunday; each year begins on Sunday, January 1. The calendar is stabilized and made perpetual by ending the year with day 365 (equivalent to December 31), which is designated 'W' and called 'Worldsday'. Leap year day is added, when necessary, at the end of the second quarter (equivalent to June 31).

Callisto Satellite IV of **Jupiter**. See also **satellite**.

camera Photography has become an indispensable aid to astronomy. Telescopes may be designed for photographic use, in which case the plate-holder is at the focus of the instrument. Where long exposures are necessary there must be appropriate arrangements to allow for the Earth's rotation and to follow the apparent movement of the object in the sky. The design of instruments for astronomical photography is dictated by the amount of light that can be collected from the object(s) of the photograph. In the case of faint stars it is the aperture of the objective that is important. With extended objects, such as nebulae, not only has the light-gathering ability to be high, i.e. a large aperture is needed, but the focal length must be as short as possible so that a small bright image on high-resolution photographic plates can be obtained.

In recent years photoelectric methods have been developed for enhancing light intensity. In the IMAGE TUBE the image produced by an optical telescope is enhanced by electronic means so that a very much brighter image appears on a phosphor screen. This intensified image can then be photographed or fed into a computer. See also **Baker-Nunn camera**; **Schmidt camera**.

canali Italian name given to the network of dark markings apparently seen on the surface of the planet Mars and first described by G. V. Schiaparelli in 1877. They have since been observed and charted by many astronomers, particularly by P. Lowell at Flagstaff, Arizona. He believed them to be artificial waterways because of their regular nature and was convinced that they were the work of intelligent beings carried out for the purpose of irrigation. This notion undoubtedly gained credence because of the unfortunate significance of the word 'canals' used to translate *canali*. There has been much controversy about the nature of this network pattern, which the largest telescopes resolve into separate spots. Photographs taken by the Mariner and Viking space probes proved conclusively that the *canali* are neither artificial nor natural formations on Mars.

candela (*abbrev.*: cd) The unit of luminous intensity in the **SI**. It is the intensity in the perpendicular direction of a surface of 1/600,000 square metre of a black body at the temperature of freezing platinum under a pressure of 101,325 newtons per square metre.

Cannon, Annie Jump (Dover, Delaware, December 11, 1863 – April 13, 1941) An American astronomer distinguished for her photographic work at the Harvard College Observatory and for the *Draper Catalogue of Stellar Spectra*.

Canopus The star α Carinae, a supergiant and the second brightest star in the southern sky. It is not visible from Europe.

Cape Canaveral The site in Florida, USA, (28·29 N 80·30 W) of the John F. Kennedy Space Center from which US (and other countries') space satellites and probes are launched. The first successful US satellite, Explorer I, was launched there on February 1, 1958. The first international satellite, Ariel 1, which made an experimental study of the ionosphere, was launched there on April 26, 1962. The site was renamed Cape Kennedy in memory of President J. F. Kennedy, but the original name was subsequently restored.

Capella The star α Aurigae, a spectroscopic triple of giant stars, and a possible source of X-ray emission.

captured rotation see **synchronous rotation**

carbonaceous chondrite see **chondrite**

carbon-nitrogen cycle One of the sequences of nuclear reactions thought to be the source of stellar energy. This cycle was postulated in 1938 by H. A. Bethe and C. F. von Weizsäcker. An alternative process involves the **proton-proton reaction**. In the carbon-nitrogen cycle one helium nucleus is formed and four hydrogen nuclei are lost with the release of an enormous amount of energy. Nitrogen and oxygen are formed as intermediate products. The carbon behaves as a catalyst. The process takes place at a temperature of about $1{\cdot}5 \times 10^7$ K and is the major energy source in hot massive stars.

carbon stars Designated as C stars, these are a class of stars which replace the R and N classes in the Harvard Draper spectral system. They give spectra showing strong bands of CO, CN, or other compounds of carbon and also show the presence of much lithium.

A view of the Kennedy Space Center showing the Skylab Orbital Workshop.

Carme Satellite XI of **Jupiter**. See also **satellite**.

Carrington, Richard Christopher (Chelsea, London, May 26, 1826 – Churt, Surrey, November 27, 1875) An English astronomer who produced a catalogue of circumpolar stars. He made systematic observation of sunspots to determine the Sun's period of rotation and discovered the systematic drift of the Sun's photosphere. He was the first (independently of Hodgson) to observe in conjunction with the giant sunspot of September 1, 1859 the brilliant phenomenon of a **solar flare**, which was later followed by a violent terrestrial magnetic storm.

Carte du Ciel A photographic atlas accompanied by a catalogue of the stars. This enormous project to chart the entire sky was proposed at an International Conference held in Paris in 1887. Eighteen observatories throughout the world agreed to co-operate and a particular zone of the sky was allotted to each. A special photographic telescope, developed by Paul and Prosper Henry and having an objective aperture of 33 cm and a special mounting, was mass-produced for the work. The photographic plate has a field of side 2°, and the centre of each plate coincides with one corner of an adjacent plate. Each star is photographed three times on each plate by slightly shifting the plate-holder so as to produce the stellar images as a triangle. The work was started in 1896 and is still unfinished. It is most unlikely that the survey will ever be completed because modern astrographic photography is much superior.

Cartesian co-ordinates see **co-ordinates**

Cassegrain-Newtonian telescope A reflecting telescope which has the facility of being used either at the Cassegrain focus or at the Newtonian focus.

Cassegrain telescope A reflecting telescope system devised in 1672 by Cassegrain. The light received by the parabolic primary mirror is reflected back from a convex secondary mirror through a hole in the primary mirror. It is then brought to a focus – the CASSEGRAIN FOCUS – on the axis of the telescope where the image can be observed by the eye or by instruments.

Cassini A family of Italian origin that produced several notable astronomers.

CASSINI, GIOVANNI DOMENICO (Perinaldo, Italy, June 8, 1625 – Paris, September 14, 1712) was the first to make a correct evaluation of the dimensions of the solar system. He discovered the division in Saturn's rings, which bears his name, and also four of that planet's satellites. He improved the tables of Jupiter's satellites, which helped O. C. Rømer in his determination of the velocity of light. He prepared a complete map of the Moon, which was engraved.

CASSINI, JACQUES (Paris, February 18, 1677–Thury, France, April 15, 1756) engaged in geodesic measurements on the meridian of France. The results indicated that the Earth is an oblate spheroid. He determined the proper motion of the star Arcturus.

CASSINI, CÉSAR FRANÇOIS (Thury, June 17, 1714 – Paris, September 4, 1784) was director of the Observatoire de Paris. He was

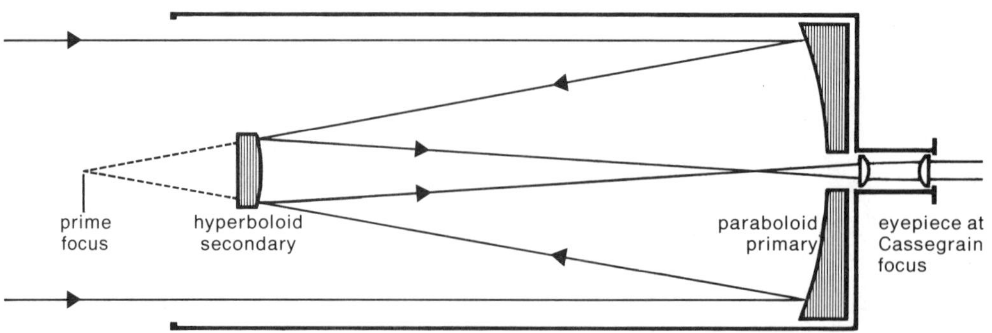

Light path in a Cassegrain telescope.

concerned in establishing a topographic map of France.

CASSINI, JEAN DOMINIQUE (Paris, June 30, 1748 – Thury, October 18, 1845) was director of the Observatoire de Paris.

Cassini division The gap between Rings A and B in the ring system of **Saturn**, discovered by G. D. Cassini in 1675.

Cassiopeia A (3C461) One of the strongest radio sources, believed to be the remnant of a supernova. It appears optically as a faint nebula.

celestial axis see **axis**

celestial bodies, age of As regards the Earth, geological estimates and determinations of the amounts of radioactive materials and their decay products in rocks give the minimum age of our planet as between 2000 and 4000 million years. Radioactive dating of meteorites gives values of 1000 to 5000 million years for their age. The age of the Sun and planets is currently assumed to be 4600 million years.

The radiation of energy (during formation of helium from hydrogen) enables the maximum time during which certain spectral types of stars will continue to shine to be calculated. The maximum value varies from about 200 million years to about 10,000 million years. Calculations related to the evolution of stars (i.e., to the amount of hydrogen converted to helium) give an age of about 10,000 to 20,000 million years, which must be the approximate age of the Galaxy, and agree with the figure derived from kinetic considerations of star clusters.

The **red shift** exhibited by galaxies is interpreted as a Doppler effect, so that they are receding from us and each other; the velocity of recession is proportional to their distance from us. The rate of expansion of the universe is therefore constant, or nearly so, and it is possible to calculate when the expansion started. Calculations based on observational measurements give the age of the universe as between approximately 10,000 and 20,000 million years.

celestial bodies, temperature of The temperatures of such bodies can be ascertained only from the radiation received from them. In the case of planets and satellites the temperatures are obtained from measurements with a **thermocouple** or **bolometer**. The temperature of the Sun and other stars is obtained from the radiation received from the observable outer layers, i.e. the stellar atmosphere.

Stellar temperatures are quoted in various ways. The EFFECTIVE TEMPERATURE of a star is defined as the temperature which a black body radiating the same total amount of energy per unit time per unit area would have. The **colour temperature** is derived from the distribution of energy over the spectrum. The RADIATION TEMPERATURE of a star is defined as the temperature which a black body would have radiating the same total amount of energy per unit time per unit area over the same spectral range. The IONIZATION TEMPERATURE is determined from the strength of emission lines in the spectrum. Similarly, the EXCITATION TEMPERATURE is determined from the strength of two absorption lines produced in the spectrum of the same element at different levels of excitation. The temperatures of stars are given also by their spectral classification (see **stars, spectral classification of**) and can range from 3000 K to maybe 50,000 K. Temperatures in the interior of stars could be of the order of 10^7 K.

celestial equator The great circle of the celestial sphere in which the Earth's equator meets the sphere and divides it into the northern and southern hemispheres.

celestial horizon The great circle cut on the celestial sphere by a plane whose poles are an observer's zenith and nadir.

celestial latitude see **co-ordinates**

celestial longitude see **co-ordinates**

celestial mechanics That branch of astronomy which deals with the motions of heavenly bodies. It is concerned with the analysis of observations of the movements of bodies under the influence of their mutual attractions, for the purpose of predicting their future motions and the calculation of ephemerides.

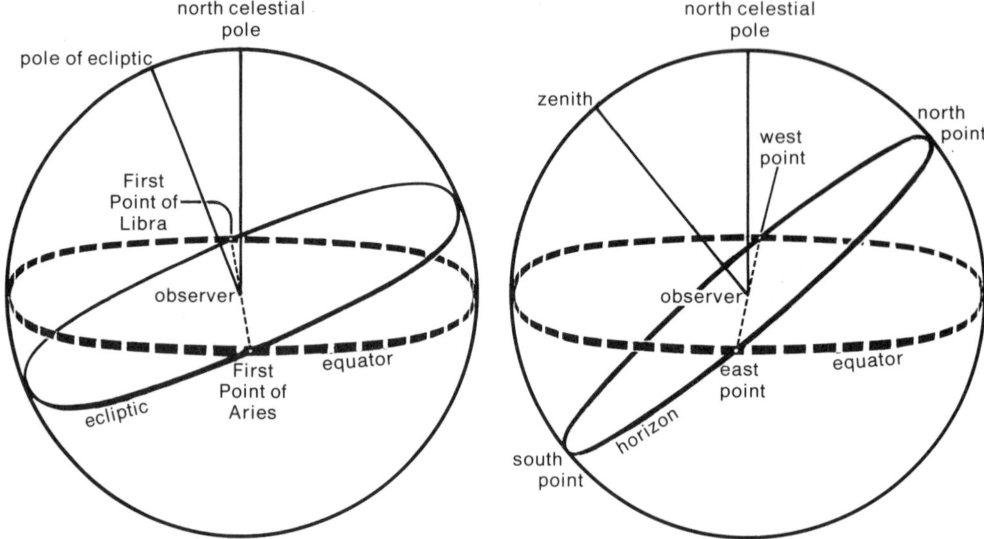

Two aspects of the celestial sphere.

celestial meridian The great circle passing through the zenith, the nadir, and the celestial poles, cutting the horizon in the north and south points.

celestial poles see **pole**

celestial sphere The arbitrary sphere, assumed to have infinite radius and having the Earth as centre, on which to a terrestrial observer all the celestial bodies appear at a uniform distance, though in truth they are at very different distances. Against this background the celestial bodies appear to move from east to west on account of the Earth's rotation. The Earth's axis points towards the CELESTIAL POLES; the north pole is near **Polaris** in the constellation Ursa Minor. The celestial sphere is divided into two halves by the **celestial equator**.

Centaurus A A strong radio source of immense area that lies in the southern constellation Centaurus.

Centaurus X-3 A pulsating **X-ray binary** discovered in 1971 by the Uhuru X-ray satellite. Its optical component was identified in 1974 as an O-type supergiant by W. Krzemiński, after whom it has been named.

centre of mass That point in a body, or a system of bodies, at which the total mass of the body, or system, may be regarded as concentrated. It is the point from which gravitational attraction appears to act. In a uniform gravitational field a body's centre of mass coincides with its CENTRE OF GRAVITY.

centrifugal force The mechanism which impels a revolving body from the centre to the circumference of its orbit. It results from the inertia of the body and is a reaction to the centripetal force acting on the body.

centripetal force The mechanism which tends to make a body move towards the centre of its orbit. The centrifugal and centripetal forces maintain the planets in revolution around the Sun. If the centrifugal force failed, they would fall into the Sun; if the centripetal force failed, they would fly off into space.

Cepheid A member of an important class of variable periodic pulsating stars. They are so called because the variability of δ Cephei, visible to the naked eye, is a typical example; it was first observed by J. Goodricke in 1784. Cepheids, which are supergiant stars, are divided into two classes: type I, or CLASSICAL CEPHEIDS, are younger and more massive than type II, or W VIRGINIS STARS. The periods of Cepheids are in the range 1–50 days; classical Cepheids have a somewhat shorter period on average than W Virginis stars.

Cepheids became important for astronomy when H. S. Leavitt discovered in 1912 a simple relationship between the period of light variation and the absolute magnitude of a Cepheid. This relationship, the **period-luminosity law**, enables the distances of stars to be ascertained. The detection of Cepheids in the Andromeda Galaxy by E. E. Hubble in 1923 enabled the distance of that galaxy to be determined and demonstrated that it is far beyond our Galaxy. As Cepheids have been observed in other galaxies, it has become possible to determine the distances of those remote systems.

Ceres The first **minor planet** to be discovered. It was found as a result of a search for a planet between Mars and Jupiter (see **Bode, Johann**). Ceres, the largest of the minor planets, was discovered by G. Piazzi on January 1, 1801.

Challis, James (Braintree, Essex, December 12, 1803 – Cambridge, December 3, 1882) Professor of Astronomy and Director of the Observatory at Cambridge. He searched for an unknown planet (Neptune) on the basis of computations made by J. C. Adams (see **Adams, John**) and unknowingly observed it on several occasions. He could have had the credit of being the first to identify this planet had he been more assiduous.

Chandler period Observations made by the American astronomer S. C. Chandler (1846–1913) showed that the altitude of the celestial pole varies within a period of about 415–433 days, with a peak at 428 days. This results from a small variation in the position of the geographical poles. The movement of up to 15 m from the mean position affects the altitude by a maximum of 0″·35, which varies the geographical latitude by the same amount. See **latitude, variation of**.

Chandrasekhar limit The limiting mass for the formation of a **white dwarf**. If a star has a mass which exceeds this limit, i.e. is more than 1·4 solar masses, it will collapse to become a **neutron star** or possibly a **black hole**.

Chandrasekhar-Schönberg limit The limiting mass for an isothermal stellar core necessary to maintain the high temperature and pressure required for the conversion of hydrogen to helium just outside the isothermal core.

Chandrasekhar, Subrahmanyan (Lahore, India, October 19, 1910–) A naturalized American who is a theoretical astrophysicist and has made significant contributions to theories of stellar evolution, stellar dynamics, stellar atmospheres, and relativity. His theory of white dwarf stars postulates a limit to their mass. See **Chandrasekhar limit**.

Charon Name proposed for the satellite of Pluto.

Chiron A very faint object (magnitude 18) which was discovered on November 1, 1977 by C. T. Kowal, and has since been found on much earlier photographic plates. Although it has been given the minor planet number 1977 UB, its characteristics are unusual for such a body. Its orbit lies to a great extent between those of Saturn and Uranus, and its aphelion distance is 18·5 A.U. Whether or not it is a **planetesimal**, an escaped satellite of Saturn, or merely a peculiar minor planet is as yet an unresolved problem.

chondrite A stony **meteorite** composed mainly of the silicates of iron, calcium, aluminium, magnesium, and sodium. Spherical grains called **chondrules** and flakes of iron-nickel are embedded in chondrites. The uncommon CARBONACEOUS CHONDRITE is fragile and porous, besides being very dark in colour from the presence of some bitumen-like material.

chondrules Spherical grains of similar composition to that of **chondrites**, in which type of meteorite they are dispersed, giving a structure not found in any terrestrial mineral. Chondrules were evidently formed at the same time as the planets and so represent primordial matter.

Chrétien telescope see **Ritchey-Chrétien telescope**

Christie, William Henry Mahoney (Woolwich, October 1, 1845 – at sea on S.S. Morea on way to Morocco, January 22, 1922) Astronomer Royal and Director of Greenwich Observatory: he succeeded G. B. Airy (see **Airy, George**) in 1881 and continued to improve the observational facilities there. He founded in 1877 the astronomical periodical *The Observatory*, which still appears.

chromatic aberration see **aberration**

chromosphere The transition layer between the corona and the photosphere, which is the visible disc of the Sun. The chromosphere is normally invisible because of the glare of the photosphere shining through it. It is briefly visible during a total eclipse, when its spectrum (see **flash spectrum**) can be secured. With a special instrument, the **coronograph**, it can be observed telescopically at any time. See **Sun**.

chronograph An instrument for making a permanent record of the exact time at which events occur, as well as the duration of a period of time. Data-processing methods utilizing computers have now replaced some applications of the chronograph, for example, in transit-circle observations.

chronological era A fixed point in time from which a series of years is reckoned. It is also the period of years comprised between two fixed points in time. Various eras have been distinguished, e.g. Byzantine, Christian, Julian, and Roman.

chronometer A precision clock used mainly for accurate timekeeping at sea, where it is necessary in connection with the determination of longitude.

circumpolar stars Those stars which at a given place of observation never dip below the horizon. Their north polar distance (south polar distance for the southern hemisphere) does not exceed the observer's latitude. They are important for observers in connection with the adjustment of instruments.

cislunar On this side of the Moon, i.e. between the Moon and Earth.

civil twilight see **twilight**

Clairaut, Alexis Claude (Paris, May 7, 1713 – Paris, May 17, 1765) A French mathematician and astronomer, distinguished for his work in celestial mechanics. He determined the Earth's flattening, formulated the equilibrium conditions of the rotating Earth (Clairaut's Theorem), predicted the return of Halley's Comet, obtained the mass of Venus, and estimated the perturbing influence of planets and comets.

Clark, Alvan (Ashfield, Massachusetts, March 8, 1804 – Cambridge, Massachusetts, August 19, 1887) An American portrait painter who turned astronomer and optician. His firm, Alvan Clark & Sons, made telescopes, and became renowned for the excellence of its lenses: five times the firm surpassed the world's record for the existing size of lens. The firm made, amongst others, the 30-inch lens for the Pulkovo Observatory and the 40-inch lens for the Yerkes Observatory, which is still the largest lens in a refractor. With his son (A. C. Clark) he observed the companion star of Sirius (1861).

classical Cepheid see **Cepheid**

C line J. von Fraunhofer's designation for one of the absorption lines in the solar spectrum. It is in the red part of the spectrum and corresponds to the Hα hydrogen line at 6563 Å. See also **Fraunhofer lines**.

clock An instrument other than a watch or a chronometer designed for measuring intervals of time. Astronomical clocks are precision instruments maintaining a constant rate. They need not indicate the right time, but it must be possible to determine the **clock error** in the time shown by the clocks. Free-pendulum clocks, quartz-crystal clocks, and atomic clocks are basic requirements in an observatory.

clock drive The mechanism, either mechanical or electrical, used to turn an equatorially mounted telescope around the polar axis in order to counteract the Earth's diurnal motion, and so maintain the telescope trained on the object under observation.

clock error The amount by which the time shown by a clock differs from the right time. The error is taken as positive if the clock is slow and negative if the clock is fast, with respect to the right time.

clock rate The amount of change in the **clock error** in 24 hours: it is reckoned positively when the clock is losing and negatively when the clock is gaining. A good astronomical clock should maintain a constant rate, losing or gaining the least possible amount in 24 hours.

clock stars Certain bright stars whose positions and proper motions in Right Ascension are so accurately known that they can be assumed and their times of transit used for the determination of time.

cluster An assembly of stars moving as a whole through space. Star clusters are classified as globular, open, or moving. A GLOBULAR CLUSTER is one in which an extremely large number of stars is closely packed so as to form roughly a sphere, their concentration decreasing from the centre. An outstanding example is M13 in the constellation Hercules. Globular clusters contain very old (Population II) stars and generally lie in the halo of our Galaxy. An OPEN or GALACTIC CLUSTER contains relatively few stars distributed over a large space. An example is M45, the Pleiades, in the constellation Taurus. Open clusters contain young (Population I) stars and occur in or near the central plane of the Galaxy. A MOVING CLUSTER is a group of stars (not necessarily comprising a constellation) whose motions and directions in space are the same. It is similar to an open cluster, but the separation between stars is much greater. An example is the Ursa Major cluster which includes β, γ, δ, ε, ζ UMa, Sirius, and other stars.

cluster of galaxies see **Galaxy**

Coalsack The prominent dark nebula adjacent to the constellation Crux in the southern sky.

coelostat An improved form of **heliostat** and **siderostat**. It comprises two mirrors. The first mirror is mounted equatorially and is clock-driven at one half the rate of the Earth's diurnal rotation, whereby not only one single star but the whole field is reduced to rest. The beam of light received by the first mirror is reflected on to a second (fixed) mirror, which in turn directs the beam into the required instrument for observation. A coelostat is needed whenever the image has to be examined by heavy, cumbrous or fixed equipment, for example, a spectroheliograph.

The term 'coelostat' has also been applied to tower telescopes designed for research in solar physics. See illus. on page 36.

co-latitude The complement (90–α)° of the latitude, α, of a body.

collapsar Another name for **black hole**.

The globular cluster M13 in the constellation Hercules.

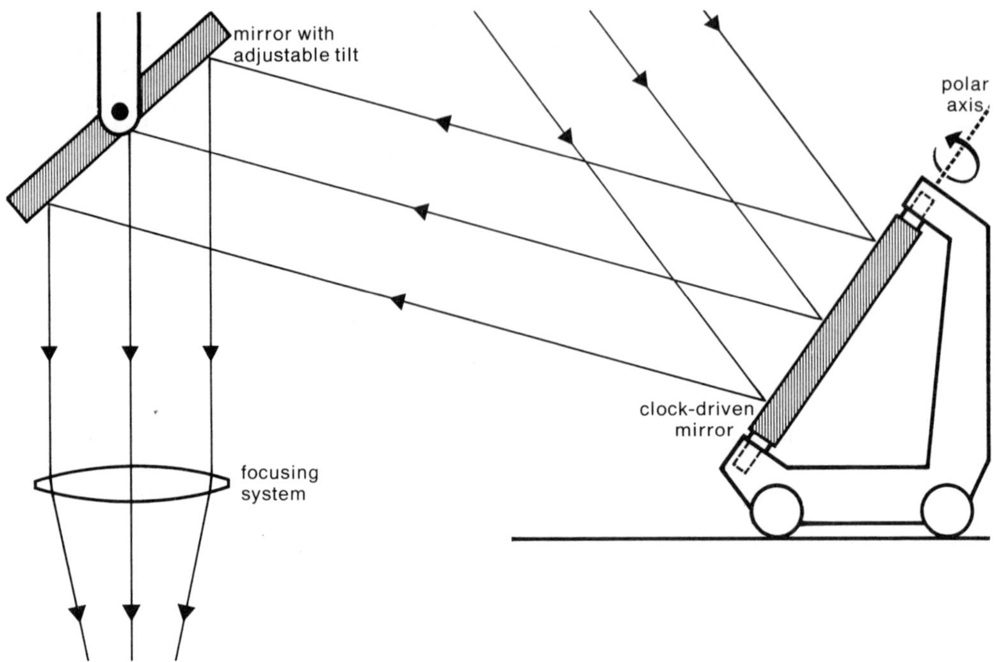

The reflection of light in a coelostat.

collimation In optics, the procedure of aligning an optical system or of rendering a beam of light parallel.

collimation error The amount (angle) by which an object viewed through an optical instrument departs from the point which it should occupy.

colour The sensation experienced by the brain caused by the action of light on the retina of the eye. A PRIMARY COLOUR is one that cannot be resolved into other colours.

colour filter A sheet of transparent material, such as gelatine, glass, or plastic, which transmits light of a restricted band of wavelengths.

colour index (*abbrev.:* C.I.) The difference between the photographic and photovisual magnitudes of a celestial object, or between the photoelectric magnitudes of an object measured at two specified wavelengths. The photographic magnitude is usually measured at wavelength 4250 Å and the photovisual magnitude at 5280 Å. The photoelectric C.I. is usually equal to the value $B-V$ or sometimes $U-B$ (see **UBV system**). The C.I. has been devised so that a star of spectral type A0 has a C.I. of zero. There is a fairly uniform relationship between the C.I. of a star and its spectral type, which is closely related to its temperature. The C.I. of cool red stars is positive while that of hot blue stars is negative.

colour temperature A temperature which is deduced from the distribution of energy over the spectrum of a body. It is the temperature of a black body whose radiation is of the same spectral intensity as that of the object under observation.

coma 1. A defect in an optical system which results in the image of a point appearing as a blurred pear-shaped patch with a flared appearance resembling a comet. It is caused by incident light striking a lens obliquely: the defect increases with distance from the optical axis. **2.** The faint hazy luminous cloud of gas and dust surrounding the **nucleus** of a comet. It does not include the tail of the comet, when one is present.

comes (*plural:* comites) The fainter companion star of a binary.

comet A luminous celestial body consisting of a solid icy **nucleus** surrounded by a luminous **coma** of gas and dust, with or without a luminous **tail**. The whole moves under the gravitational influence of the Sun, first towards that body, then around it, and finally away from it. In earlier times the appearance occasionally of a comet evoked much apprehension, for it was regarded as the portent of disastrous events. Since the application of telescopes and photography it is now known that comets are far more numerous than naked-eye observation suggests. Several new ones are now discovered every year. It is customary for a new comet, after its orbit has been definitely confirmed, to bear the name of its discoverer. The number of known comets has increased to such an extent as to justify the compilation of a *Catalogue of Cometary Orbits Equinox 1950.0* (B.A.A., 1961) and a *Supplementary Catalogue of Cometary Orbits Equinox 1950.0* (B.A.A., 1966).

Comets consist of extended dispersed matter with a relatively small mass. Their orbits are generally ellipses, but in some cases are parabolae or hyperbolae. Short-period comets are those whose orbital period is less than 200 years. Long-period comets have a period greater than 200 years. Some periodic comets have made one apparition only: others have disintegrated after being observed at several returns.

Comet Crommelin The comets 1818 I, 1873 VII and 1928 III were proved by computation by A. C. C. Crommelin to be reappearances of the same short-lived comet and not three separate comets.

commensurable A mathematical term applied to two magnitudes having a common measure. If the orbital periods of two bodies in the solar system can be expressed as a ratio of small whole numbers, then the periods are commensurable. **Bode's Law** is an early example. As regards the satellites of the solar system it has been found that stable orbits tend to form at commensurable distances. The extent to which this occurs indicates that the commensurabilities are not accidental.

Common, Andrew Ainslie (Newcastle-upon-Tyne, August 7, 1841 – Ealing, June 2, 1903) An English astronomer who was a pioneer of astronomical photography. He was the first to apply photography to the study of nebulae, and successfully photographed the Orion Nebula.

commutation, angle of The angular distance between the Sun's true position from the Earth and the position of a planet measured in the plane of the ecliptic.

comparator or **blink microscope** An instrument in which the images of the same star field on two successive photographic plates can be rapidly alternated in the field of view of the eyepiece. If any object has changed its position during the interval between exposure of the two plates, then it will oscillate and so be detected. Variable stars reveal themselves as a pulsation, because they give images of different sizes depending on their brightness. In the STEREO COMPARATOR the two exposures are viewed simultaneously with binocular vision, thereby producing a stereoscopic impression of the object causing it.

computer A person who makes calculations, specifically one employed in an observatory. The work of human computators has been facilitated by the invention of various kinds of calculating machines, especially in recent years by the introduction of the electronic computer. This can perform a systematic sequence of operations on data, including complicated arithmetical and logical operations, at very great speed, and can store information in its memory.

Comrie, Leslie John (Pukehohe, New Zealand, August 15, 1893 – London, December 11, 1950) A computator and compiler of mathematical tables who developed new techniques for astronomical computation.

concavo-convex lens A lens that is bounded by one convex and one concave surface, the latter having the greater radius of curvature. The lens is converging.

conjunction An alignment of three celestial

bodies. Two bodies are in ECLIPTIC CONJUNCTION with a third when they have the same longitude measured on the ecliptic of the third. Two celestial bodies are in EQUATORIAL CONJUNCTION when they have the same Right Ascension on the equator of the third. In the case of Mercury and Venus these planets are in **inferior conjunction** when they are at the point of minimum angular distance from the Sun on passing between Earth and Sun. These planets are in **superior conjunction** when they are behind the Sun. CONJUNCTION OF A SUPERIOR PLANET occurs when it, Sun, and Earth are in line, the two planets being on opposite sides of the Sun. The Moon is in conjunction at New Moon. See also **aspect**.

constellation A region of the sky containing certain stars once grouped together for convenient identification within the confines of an imaginary figure traced on the sky. The original number of 48 given in Ptolemy's **Almagest** has grown to 88, which have now been assigned definite boundaries on the celestial sphere by the I.A.U. (E. Delporte, *Atlas Céleste*, Cambridge, 1930). Noteworthy stars within a constellation are designated by Greek letters followed by the name of the constellation in the genitive case, as with α Lyrae. Many stars, particularly those which have been known for centuries, have received distinctive names in addition. One example is Sirius, also known as the Dog Star.

Constellation names and their abbreviations

Andromeda	And	Carina	Car
Antlia	Ant	Cassiopeia	Cas
Apus	Aps	Centaurus	Cen
Aquarius	Aqr	Cepheus	Cep
Aquila	Aql	Cetus	Cet
Ara	Ara	Chameleon	Cha
Aries	Ari	Circinus	Cir
Auriga	Aur	Columba	Col
Boötes	Boo	Coma	
Caelum	Cae	Berenices	Com
Camelopardalis	Cam	Corona	
Cancer	Cnc	Australis	CrA
Canes Venatici	CVn	Corona	
Canis Major	CMa	Borealis	CrB
Canis Minor	CMi	Corvus	Crv
Capricornus	Cap	Crater	Crt
Crux	Cru	Orion	Ori
Cygnus	Cyg	Pavo	Pav
Delphinus	Del	Pegasus	Peg
Dorado	Dor	Perseus	Per
Draco	Dra	Phoenix	Phe
Equuleus	Equ	Pictor	Pic
Eridanus	Eri	Pisces	Psc
Fornax	For	Piscis Austrinus	PsA
Gemini	Gem	Puppis	Pup
Grus	Gru	Pyxis	Pyx
Hercules	Her	Reticulum	Ret
Horologium	Hor	Sagitta	Sge
Hydra	Hya	Sagittarius	Sgr
Hydrus	Hyi	Scorpius	Sco
Indus	Ind	Sculptor	Scl
Lacerta	Lac	Scutum	Sct
Leo	Leo	Serpens	Ser
Leo Minor	LMi	Sextans	Sex
Lepus	Lep	Taurus	Tau
Libra	Lib	Telescopium	Tel
Lupus	Lup	Triangulum	Tri
Lynx	Lyn	Triangulum	
Lyra	Lyr	Australe	TrA
Mensa	Men	Tucana	Tuc
Microscopium	Mic	Ursa Major	UMa
Monoceros	Mon	Ursa Minor	UMi
Musca	Mus	Vela	Vel
Norma	Nor	Virgo	Vir
Octans	Oct	Volans	Vol
Ophiuchus	Oph	Vulpecula	Vul

continuous spectrum The unbroken sequence of colours, merging one into the other, produced when light is decomposed by refraction through a prism. Substances heated to incandescence give a continuous spectrum. The visual spectrum is continued at both ends by further wavelengths of electromagnetic radiation, detectable by chemical, photographic, or other effects.

continuum A continuous spectrum, i.e. an unbroken band of wavelengths.

converging lens or **positive lens** A lens causing a parallel beam of light to converge at the focus of the lens; a diverging beam has its divergence reduced.

convexo-concave lens A lens that is bounded by one convex and one concave surface, the latter having the smaller radius of curvature. The lens is diverging.

Cookson, Bryan (April 23, 1874 – London, March 18, 1909) An English astronomer who was the designer of the floating zenith tube, used by him to redetermine the constant of aberration. The instrument was used for many years also for determining the variation of latitude.

Co-ordinated Universal Time see **Universal Time**

co-ordinates Two or more magnitudes which can be used to define the position of a point on a plane or on a sphere. Several systems of co-ordinates are used in astronomy for fixing the position of celestial bodies. In the system of CARTESIAN CO-ORDINATES the numerical values on two axes at right angles define the position of a point on a plane. The introduction of a third axis at right angles to the other two enables a point to be defined in space.

For defining the position of a body on the celestial sphere there are three systems. In the HORIZONTAL SYSTEM the reference plane is that of the observer's horizon, and the co-ordinates are provided by **azimuth** and **altitude**. In the EQUATORIAL SYSTEM the reference plane is that of the celestial equator, and the co-ordinates are provided by **hour angle** and **polar distance**, or more usually by **Right Ascension** and **Declination**. In the ECLIPTIC SYSTEM the reference plane is that of the ecliptic, and the co-ordinates are celestial latitude and longitude. The CELESTIAL LATITUDE is the angular distance between the object in question and the ecliptic, measured along a secondary: it is measured from 0° to 90°, positively towards the north ecliptic pole and negatively towards the south ecliptic pole. The CELESTIAL LONGITUDE is given by the angle between the secondary to the ecliptic, which contains the object in question, and the First Point of Aries: it is measured from 0° to 90° eastwards from the First Point of Aries.

When dealing with the distribution of stars in the Galaxy, the GALACTIC SYSTEM of co-ordinates is used, in which the plane of reference is the galactic plane (plane of the Milky Way). Its co-ordinates are galactic latitude and longitude. GALACTIC LATITUDE is measured from 0° to 90° from the galactic equator, positively towards the north galactic pole and negatively towards the south galactic pole. GALACTIC LONGITUDE is measured from 0° to 360° along the galactic equator. In 1959 it was decided by the I.A.U. that the position of zero galactic longitude was to be R.A. 17h 42.4m and dec. −28°55′ (1950). See illus. on page 40.

Copernicus The name given to **Orbiting Astronomical Observatory** 3.

Copernicus, Nikolaus (Latinized form of Koppernigk, Niklas; Toruń, February 19, 1473 – Frauenburg, May 24, 1543) A canon of Frauenburg Cathedral who became interested in astronomy, and through his study of planetary motions developed a heliocentric theory of the universe. In this COPERNICAN SYSTEM, as it is called, the Earth is not the centre of the universe: instead, the planets are in orbit around the Sun, and the apparent diurnal motion of the sky is the result of the Earth's rotation on its axis. An account of his work was published under the title of *De Revolutionibus Orbium Coelestium Libri VI*, (Nürnberg, 1543). It provides the basis for later developments in celestial mechanics.

A 17th-century engraving of Copernicus.

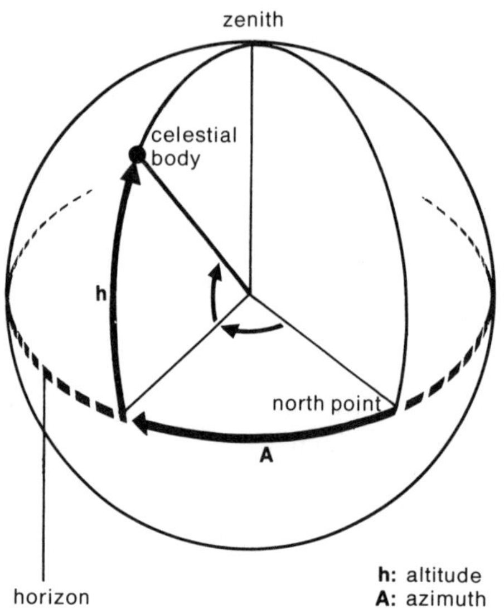

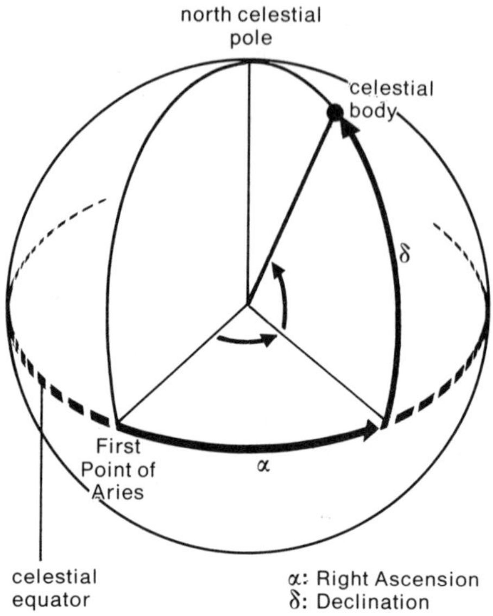

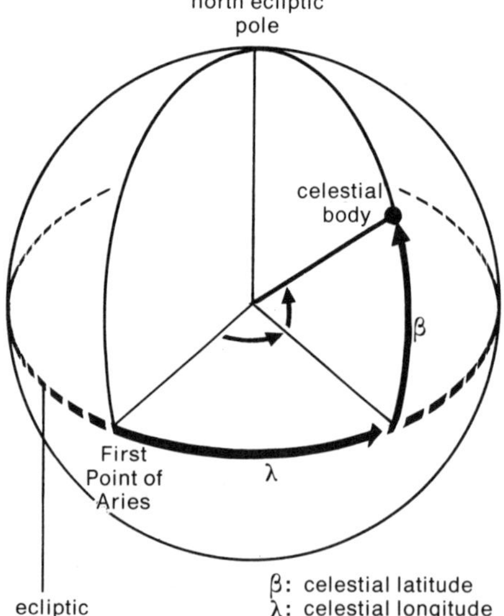

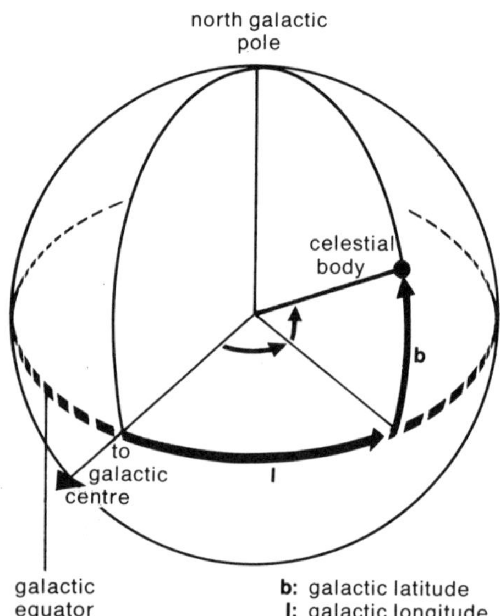

Co-ordinate systems. Top: horizontal system. Top right: equatorial system. Above: ecliptic system. Above right: galactic system.

Coriolis force An effect of the Earth's rotation. J. Hadley (1622–1744), a mathematical instrument maker, was the first to recognize the effect of the Earth's rotation on the movement of air currents. The velocity of rotation of a point on the Earth's surface is lower the further removed it is from the equator. Air streaming from one latitude to another retains its original rotational movement, and so obtains a relative velocity with respect to the new latitude. A general deduction of the displacing force of the Earth's rotation was made by G. G. de Coriolis (1792–1843). A Coriolis force acts on all bodies moving relative to the Earth, and provides an explanation of a number of phenomena, such as the eastward displacement of a free falling body, the displacement of a pendulum in the Foucault experiment, and movement of the atmosphere.

corona The outermost layer of the Sun's atmosphere, extending many millions of kilometres into interplanetary space. It is seen in profile beyond the Sun's limb during a total solar eclipse. Its overall shape changes during the 11-year solar cycle. The temperature of the corona is about 2,000,000 K at 75,000 km above the photosphere. See also **Sun**.

coronal lines Emission lines in the spectrum of the Sun's corona. The strongest is a green line at 5303 Å produced by Fe XIV and a red line at 6374 Å. It was not until 1941 that an interpretation of these and other lines was provided by the Swedish physicist B. Edlén (1906–) who showed that they were produced by atoms in a highly ionized state. The presence of the green line from Fe XIV, the red line from Fe X, and the yellow line from Ca XV depends on the amount of solar activity, and indicates that the temperature of the corona can vary between 2 and 4 million K.

coronium A hypothetical element which was thought to be a constituent of the Sun's corona. Spectroscopic analysis of the radiation from the corona showed bright lines not identifiable with those from any known chemical element. It was established (1942) by B. Edlén that iron, calcium, and nickel when very highly ionized would emit radiation of wavelengths agreeing with those of coronium. See also **coronal lines**.

coronograph (frequently misspelt *coronagraph*) An instrument devised by B. F. Lyot (see **Lyot, Bernard**) in 1930. When used in conjunction with a telescope the instrument permits study and photography of the Sun's corona and prominences by producing an artificial eclipse. Before the invention of this instrument such studies were possible only during the brief duration of a total eclipse.

cosmic background radiation see **background radiation; microwave background radiation**

cosmic dust Small solid particles present in dark nebulae in the Galaxy, particularly in the plane of the Milky Way. Such material is responsible for the dimming and reddening of starlight by scattering and absorbing light. The density of a cloud of cosmic dust is extremely low, being about 10^{-25} g cm^{-3} and the temperature is 20 K or lower. See also **interstellar matter**.

cosmic light Part of the constant illumination of the night sky in the absence of moonlight. It comes from extragalactic sources and is less than 1% of the night sky light.

cosmic rays Particles of extremely high energy, travelling at relativistic speeds, which enter the Earth's atmosphere from space. They cause disintegration of the atoms they encounter and produce various fundamental particles, such as protons, electrons, α-particles, and muons, which can be detected at ground level. Some particles of very high energy can be detected by instruments carried aloft by sondes. The origin of cosmic rays is not known with certainty, although supernovae are probable sources. Some are ejected from the Sun during the occurrence of **solar flares**. They come from all directions in space, and are considered to have their origin within rather than outside the Galaxy and to spend millions of years in the vast regions of interstellar space. Their energies lie between 10^7 and 10^{21} electron volts. Many of the particles become directed to the polar regions under the influence of the Earth's magnetic field.

cosmochemistry The study of the chemical composition of celestial bodies, e.g. meteorites, or samples of material either brought back from the Moon or analysed *in situ*.

cosmogony The study and theory of the origin or mode of creation of the universe. Apart from belief in the divine creation of the universe there has been much theorizing on this subject. I. Kant (1724–1804) suggested that the members of the solar system were formed by condensation from a rotating nebula consisting of small particles. P. S. Laplace (see **Laplace, Pierre**) envisaged a rotating primordial nebula which developed into concentric rings of nebular matter that in turn condensed into planets. T. C. Chamberlin (1843–1928) and F. R. Moulton (1872–1952) postulated that the planets were formed by accretion of planetesimals. J. H. Jeans (see **Jeans, James**) and H. Jeffreys (see **Jeffreys, Harold**) suggested that the planets were formed by condensation of matter drawn off the Sun as the result of the close approach of a star to the Sun. It has been proposed that the Sun was originally a binary: the capture of the companion star by a passing star (R. A. Lyttleton), or explosion of one star of the binary (F. Hoyle, 1915–), drew out stellar material which condensed to form the planetary system.

During this century theories have been advanced based on the ejection of charged particles from the Sun (K. Birkeland, 1867–1917), the effect of electromagnetic forces (H. Alfvén, see **Alfvén, Hannes**), the accretion of nebular dust particles to form protoplanets (C. F. von Weizsäcker, 1912–), and the formation of protoplanets by the breaking up of a nebulous disc by gravitational contraction (G. P. Kuiper, 1905–). Whereas early theories postulate high-temperature conditions for the formation of the planets, present views favour the accretion of small particles in the cold.

cosmological distance A distance derived by accepting the validity of the **Hubble Law**.

cosmological principle A basic principle of cosmology which states that the universe is homogeneous and isotropic throughout. The universe will appear the same at any one moment to any observer at any point in it. An extension, known as the PERFECT COSMOLOGICAL PRINCIPLE, states that the universe is homogeneous and isotropic throughout time. It follows that the universe will appear the same *at any time* to any observer at any point in it. This principle is fundamental to the **steady-state theory** of the universe.

cosmological red shift The **red shift** produced in spectra by the expansion of the universe.

cosmology The study and theory of the secondary causes by which the present order of the universe as a whole has evolved and is maintained. At the present time cosmological theories fall into two groups: relativistic (see **big bang theory**) and **steady-state theory**.

Cosmos The name given to a series of artificial Earth satellites launched from the USSR. Cosmos 1 was launched on March 16, 1962 and Cosmos 1000 on March 31, 1978.

coudé (Adjective from French *couder*, 'to form like an elbow'. It is not the name of a person, as is sometimes thought.) A term applied to a particular optical arrangement whereby the beam of light received by a telescope is reflected by secondary mirrors so that it emerges along its polar axis. The final image is brought to focus in a fixed point on the polar axis, known as the COUDÉ FOCUS.

Couder telescope A modified form of **Schwarzschild telescope** designed by A. Couder (1897–1979), in which astigmatism is removed. In spite of its optical advantages the instrument has not become popular since it is cumbersome and requires the use of special photographic plates or a field-flattener because the field surface is curved.

coudé telescope A telescope using a **coudé** optical system whereby light from the object under observation is passed down the polar axis to the eyepiece or heavy auxiliary equipment. The coudé focus is used particularly for spectroscopy. Large modern reflectors are designed so that they can be operated at the focus of various optical systems, when they receive a descriptive identification, such as coudé–Newtonian–Cassegrain. See illus. opposite.

Cowell, Philip Herbert (Alipore, India,

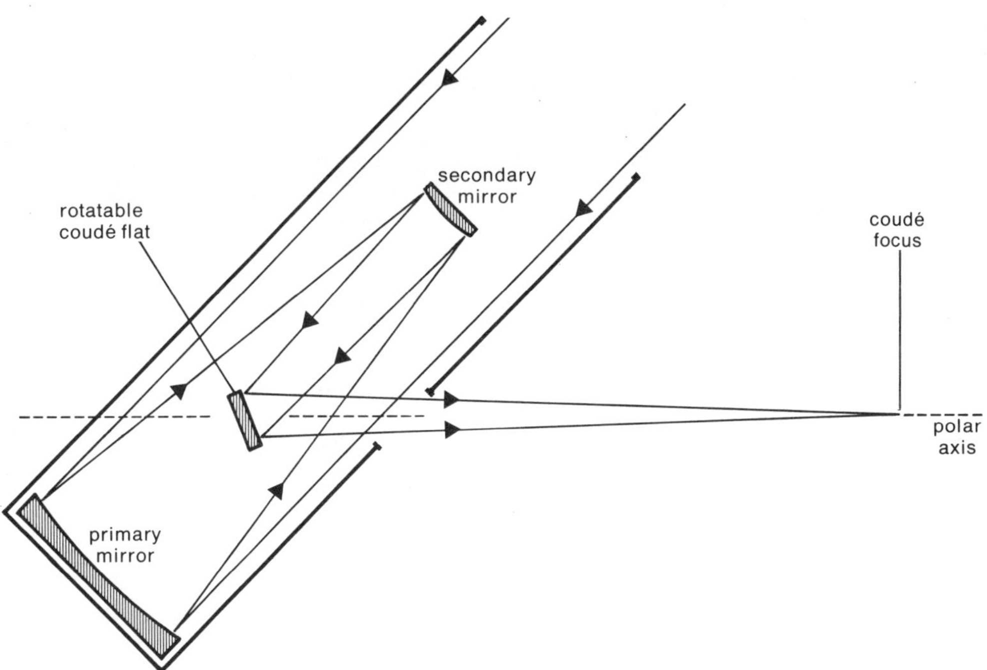

August 7, 1870 – Aldeburgh, Suffolk, June 6, 1949) An English astronomer who discovered that the terrestrial day is lengthening by about one second in 100,000 years. With A. C. D. Crommelin (see **Crommelin, Andrew**) he made a successful prediction of the return of Halley's Comet in 1910.

Crab Nebula The luminous emission nebula M1 in the constellation Taurus. It was discovered by J. Bevis (1695–1771) in 1731 and independently by C. Messier in 1758. It is the remnant of a supernova, the appearance of which was noted by Chinese astronomers in 1054. It is still expanding at about 1100 km s^{-1}. It is an intense source of radio emission, as deduced from spectroscopic examination, which reveals both continuous and emission spectra; the former is attributed to **synchrotron radiation** resulting from the movement of high-energy electrons in a strong magnetic field. The nebula is also a source of X-rays produced by synchrotron emission. Inside the Crab Nebula lies the young CRAB PULSAR, which is a major source of energy for the nebula and is most likely the remaining core of the supernova of 1054. See **pulsar**.

Light path in a coudé telescope.

The Crab Nebula – the remains of a supernova explosion observed by the Chinese in 1054.

crater A prominent circular formation on the Moon, Mars, Mercury and many other bodies in the solar system. Craters are of many sizes up to 100 km or more in diameter, though the term is usually applied to formations having a diameter of 15–60 km. Many of these features have been given names, which are used on maps of the bodies in question. The floor of a crater is below the level of the outer lunar, planetary or satellite surface.

crêpe ring see **Saturn**

Crommelin, Andrew Claude de la Cherois (Cushendun, Ireland, February 6, 1865 – London, September 20, 1939) An English astronomer noted for his work on cometary orbits (see **Comet Crommelin**) and his collaboration with P. H. Cowell (see **Cowell, Philip**) in predicting, correct to within three days, the return of Halley's Comet in 1910.

cross-bar micrometer see **micrometer**

crossed lens A lens that is bounded by two convex surfaces, the first having a radius of curvature about six times that of the second. Spherical **aberration** is reduced to a minimum for a parallel beam of light.

Crux The Southern Cross, the smallest of all constellations. It is conspicuous in the southern sky, where its four brightest stars form a cross. This constellation, symbolized by five mullets for its five brightest stars, is frequently used in Antipodean coats of arms.

C stars see **carbon stars**

culmination The moment of transit of a celestial body across the observer's meridian. It then reaches its greatest altitude above, or least altitude below, the observer's horizon. In the case of circumpolar stars both culminations are observable. INFERIOR or LOWER CULMINATION occurs when a star transits between the pole and the horizon: the star's hour angle is then exactly 12^h. SUPERIOR or UPPER CULMINATION occurs when a star transits between the pole and the zenith: the star's hour angle is then exactly 0^h.

curvature of a field In certain optical systems, the gathering of the rays of light to form a focused image in a curved field, for example, on a curved surface.

curvature of space One of the results of Einstein's Theory of General Relativity is that the geometry of space and time is modified by the presence of matter. This curvature affects the motion of bodies. Euclidian geometry deals with the geometry of a flat surface, and the sum of the three angles in all triangles is 180°. In Riemannian geometry the sum of the angles in all triangles is *greater* than 180°; this is the geometry of a surface with positive curvature, as in the case of a sphere. In Gaussian geometry the sum of the angles in all triangles is *less* than 180°; this is the geometry of a surface with negative curvature. The astronomer K. Schwarzschild (see **Schwarzschild, Karl**) tried to determine the geometry (curvature) of space by measuring the angles of an enormous triangle formed by the intersection of the rays of light from a star with the Earth at widely separate points in its orbit. The only conclusion he could make from his experiments was to the effect that if space is indeed curved, then its radius of curvature must be extremely large.

cusp One of the points or horns of the crescent phase of the Moon, or of an inferior planet.

cusp caps The bright areas sometimes seen at the cusps of Venus when in the crescent phase.

cyanogen (CN) absorption bands Bands produced by molecular cyanogen in the spectra of G0 and later stars, as well as some comets. Cyanogen absorption is greater in the spectra from giant stars than from dwarf stars of the same spectral type.

cycle of the indiction A cycle of 15 years. It is not of astronomical origin but probably originated in the taking every 15 years of a census in Egypt for fiscal purposes. Its use has been adopted in other countries to designate years without special reference to matters of

Craters on the far side of the Moon photographed by Lunar Orbiter 3.

finance. The Papal Indiction was reckoned from January 1, 313. To find the indiction for any year of the Christian era add 3 to the year, and divide the sum by 15; the remainder, if other than zero, is the indiction; if the remainder is zero, then the indiction is 15.

Cygnus A One of the strongest sources of radio emission in the sky, situated in the constellation Cygnus, and the first to be detected. It was formerly thought to have been caused by the collision of two galaxies. It is also a source of X-ray emission.

Cygnus loop The remnant of a supernova (NGC 6992) consisting of a large loop of gas ejected from a star. It has a temperature of 2×10^6 K, and is the X-ray source Cygnus X-5.

Cygnus X-1 A strong source of X-ray emission in the constellation Cygnus. It was discovered by the satellite Uhuru which was devoted entirely to the study of cosmic X-ray sources. The visible star (HDE 226868), a supergiant, is in orbit about an invisible primary of at least six solar masses in a period of 5·59 days. It has been suggested that this massive invisible object may be a **black hole**. The constellation also contains the sources X-2, X-3, X-4, and X-5.

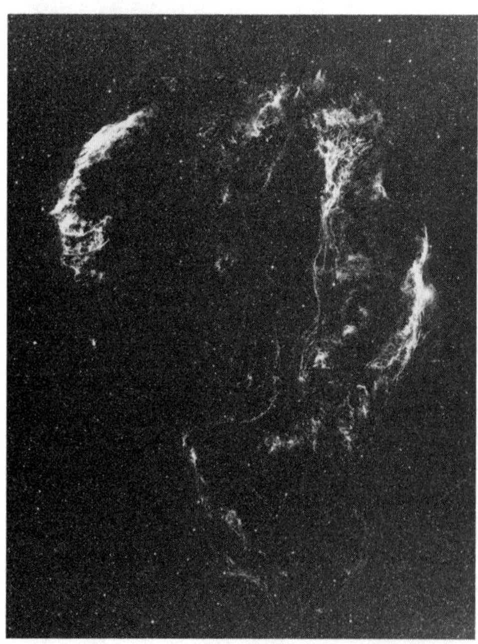

The Cygnus loop supernova remnant.

Cynthian Pertaining to the Moon. The word is derived from Cynthia, another name of Diana, the Roman goddess of the Moon, from Mount Cynthus where she was born.

Cytherean Pertaining to Venus. The word is derived from Cythera (now Cerigo), the most southerly of the Ionian Islands, sacred to the goddess Venus, who, it was supposed, rose from the sea near its coast.

dark nebula see **nebula**

Darwin, George Howard (Down, Kent, July 9, 1845 – Cambridge, December 7, 1912) The second son of the naturalist Charles Darwin, he was an astronomer who investigated the effect of tides and tidal friction of the Earth-Moon system and the solar system. He developed the theory that the Moon was thrown off from the Earth.

Dawes' limit see **resolving power**

Dawes, William Rutter (London, March 19, 1799 – Haddenham, February 15, 1868) An English amateur astronomer known for his observations of double stars, planets (in particular Saturn and Mars), and the Sun. He discovered the crepe ring of Saturn independently of W. C. Bond (see **Bond, William**). He also invented a wedge photometer.

day The time taken by the Earth to rotate once on its axis. This time depends on the way the calculation is made. An APPARENT SOLAR DAY is the interval between two successive transits of the Sun across the meridian, and is the time given by a sundial. This time is variable because the Sun's apparent motion throughout the year is not uniform. An average for the year is therefore taken. This provides the MEAN SOLAR DAY as measured by a MEAN SUN, which is an imaginary point moving around the celestial equator at the average (uniform) rate of motion of the true Sun. A SIDEREAL DAY is the interval between two successive transits

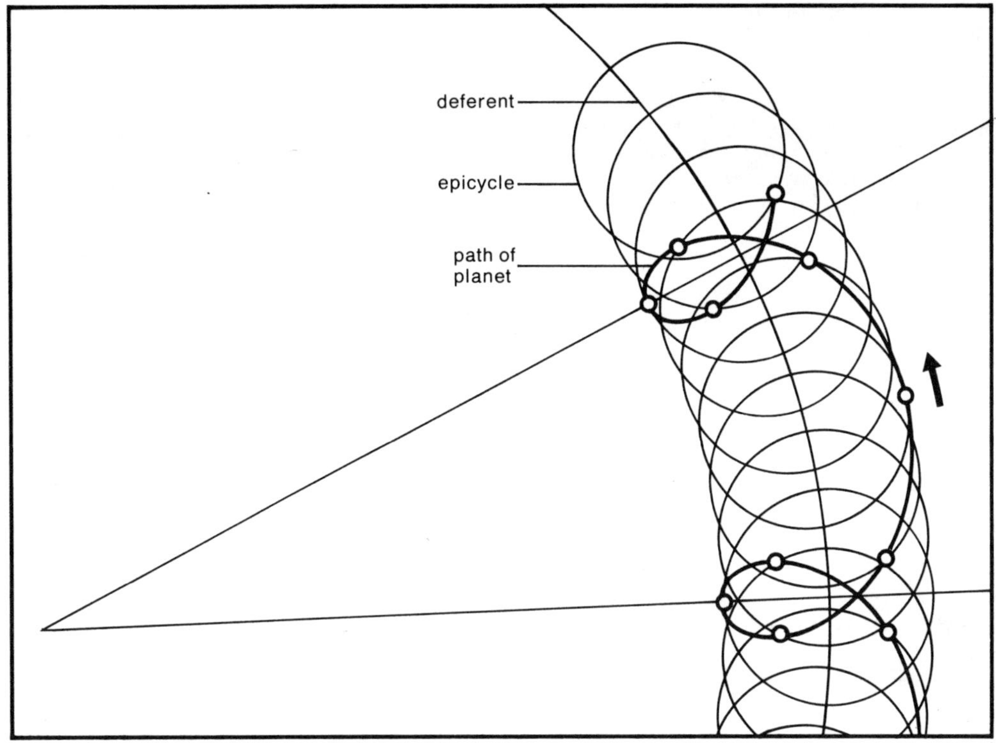

Planetary path in the Ptolemaic system.

of the First Point of Aries (see **Aries, First Point of**) across the meridian. This day is slightly shorter by four minutes than the mean solar day because of the Sun's apparent movement eastwards by about 1° each day.

> 24^h of mean solar time = 24^h 03^m $56^s{\cdot}55536$ of mean sidereal time
> 24^h of mean sidereal time = 23^h 56^m $04^s{\cdot}09054$ of mean solar time

The ASTRONOMICAL DAY, previous to 1925, meant the mean solar day beginning at noon, 12 hours after the midnight at the beginning of the same civil date. See **time**.

Declination (*abbrev.:* dec.; *symbol:* δ) One of the two co-ordinates in the equatorial system. It is the angular distance of a celestial object north or south of the celestial equator. It is reckoned positively from 0° to 90° from the equator to the north pole, and negatively from 0° to 90° from the equator to the south pole.

deferent In the Ptolemaic geocentric world system the observed apparent motions of the Sun, Moon, and planets are reproduced by superimposed circular motions. The deferent, a circle, is selected to be the path for the uniform motion of the mean planet; the true planet moves on the circumference of the epicycle, a second circle whose centre moves on the circumference of the deferent. By suitably choosing the radii of deferent and epicycle, as well as the orbital times on them, the epicycloid path traced by the true planet corresponds quite closely to the observed motion from Earth, which is situated eccentrically within the deferent.

degenerate matter A state of matter existing in stars in the final stage of their evolution, when the hydrogen in their core has been exhausted, leaving a surrounding tenuous atmosphere of hydrogen. The matter is completely ionized. Atomic nuclei and electrons are packed so closely together that ultra-high densities prevail. The laws of classical physics no longer apply. Pressure ceases to be dependent on temperature and is a function only of density.

degree (of arc) A unit of angular measure equal to the 360th part of the circumference of a circle. A star is so many degrees above the horizon as there are degrees in the angle subtended by the arc between the star and the horizon.

Deimos Satellite II of Mars. See also **satellite**.

De La Rue, Warren (Guernsey, January 18, 1815 – London, April 19, 1889) An English astronomer who was a pioneer of astronomical photography, especially of the Moon and the Sun. He designed the first photoheliograph. He observed the solar eclipse of 1860 in Spain, and demonstrated that prominences are of solar and not lunar origin.

Delta Aquarids One of the meteor showers which occur at the end of July and the beginning of August. It has a double radiant in Aquarius near δ Aqr.

Demeter Name proposed for Satellite X of Jupiter but not adopted.

Deneb (α Cyg) A remote luminous white supergiant which is the brightest star in the constellation Cygnus.

Denning, William Frederick (Radstock, November 25, 1848 – Bristol, June 9, 1931) An English astronomer who was a distinguished systematic observer. He discovered comets, novae (Nova Cygni on August 20, 1920), and the meteor shower associated with Comet Pons-Winnecke. He made deductions as to the temperature of the Earth's atmosphere at high altitudes.

density function The total number of stars in a given volume of space.

descending node see **node**

De Sitter, Willem (Sneek, May 6, 1872 – Leiden, November 20, 1934) A Dutch astronomer who was director of the observatory at Leiden. He applied Einstein's theory of relativity to the origin and expansion of the universe, the radius of which he computed as 200 million light-years. He also proposed models for the universe.

diagonal An optical device for deflecting the path of a beam of light, usually through a right angle, either to facilitate access to an eyepiece or to reduce the intensity of the light. It consists of an optical flat, or of a right-angle prism, set usually at an angle of 45° to the path of the light. The SOLAR DIAGONAL is used for solar observations. It allows a large amount of light and heat to be transmitted, whilst a small amount is reflected, thereby giving an image of reduced intensity. In the STAR DIAGONAL the surface of the optical flat, or of the prism face, is silvered.

dichotomy That phase of the Moon, or of an inferior planet, when it appears only half-illuminated and the terminator is a straight line.

differential rotation The difference in speed of rotation shown by two or more parts of a body, as, for example, the Sun, which does not rotate as a solid body. R. C. Carrington (see **Carrington, Richard**) discovered that the Sun's equatorial rotational period is shorter than the polar.

diffraction The modification which light undergoes when it passes by the edge of an obstacle and is deflected from its direct path, so that the light waves spread sideways behind the obstacle. When starlight passes through the aperture of a telescope, the effect of diffraction upon the image of a sharply focused star is to make it appear as a central disc of light surrounded by a number of concentric rings of light of diminishing intensity with dark spaces in between. These alternate bright and dark rings are DIFFRACTION RINGS. The bright central disc is called the AIRY DISC. Obstructions in the path of a beam of light in a telescope can also produce diffraction patterns on images of stars.

diffraction grating A plane or curved surface upon which a very large number of equidistant lines have been ruled very closely together. When light falls upon them it is diffracted and thereby produces a spectrum. Gratings are usually of metal, and so function by reflection; they are sometimes made of glass, when they function by transmission. Gratings are ruled with 15,000 lines or more to the inch. Whereas a prism produces a single spectrum, a grating gives several on either side of the incident ray of light. The spectra which are least deviated from the incident light are first-order spectra, while further from the undeviated beam lie the fainter second, third and higher orders. The dispersion of the spectrum increases with its order, which in turn depends on the displacement in wavelengths between in-phase adjacent rays.

diffraction rings see **diffraction**

Dione Satellite IV of **Saturn**. See also **satellite**.

Dipper Short for 'Big Dipper', which is the common name in the United States for that part of the constellation **Ursa Major** elsewhere called the **Plough**.

direct motion The orbital motion of a celestial body, observed from the north pole of the ecliptic, when it is from the west, through south to east. In the solar system the motion of all planets and most satellites and comets is direct. Motion in the reverse direction, as in the case of some satellites and comets, is called **retrograde motion**.

disc meter An instrument for measuring the diameter of a planetary disc by means of an artificial disc placed in the field of the telescope. This artificial disc is made to match exactly in respect of size and brightness the disc of the planet as seen in the telescope.

Discoverer The name given to a series of American Earth satellites used for scientific and military research and first launched in February 1959.

dispersion The decomposition of light into its primary colours on passing through a prism, or something similar. It results from the fact that the component colours of white light are refracted to different extents: red light is refracted least and violet light is refracted most. The term has also a more specific meaning, namely, the angle of separation of two selected rays in the spectrum.

distance The distance between bodies in the solar system is expressed in **astronomical units**. Stellar distances are expressed in **light-years** or in **parsecs**. The angular separation between two stars is called the ANGULAR DISTANCE, which is used with the **position angle** in measuring the positions of binary stars.

distance modulus The difference between the apparent magnitude (m) and the absolute magnitude (M) of a star. It is a measure of the star's distance (r) in parsecs:

$$(m - M) = 5 \log r - 5$$

distortion A defect in an optical system thereby spoiling the definition of an image produced by an objective lens covering a wide field. The distortion results from displacement of the image perpendicular to the axis of the lens. In the case of positive distortion the outer parts of the image are closer to the axis, and straight lines are curved outwards, so that the image of a square will exhibit 'barrel distortion'. In negative distortion the reverse occurs, so that the image of a square will exhibit 'pin-cushion distortion'.

diurnal aberration see **aberration**

diurnal inequality see **tides**

diurnal motion The apparent motion of celestial bodies, as seen from Earth, from east to west. It is a result of the Earth's axial rotation from west to east.

diurnal parallax see **geocentric parallax**

diverging lens or **negative lens** A lens by which any beam of light is rendered divergent.

D layer A low layer of the Earth's **ionosphere**.

D lines Two very close absorption lines in

the Fraunhofer solar spectrum produced by neutral sodium: they are at 5891 Å and 5890 Å. See also **Fraunhofer lines**.

Dolland, John (London, June 10, 1706 – London, November 30, 1761) A silkweaver who turned optical instrument maker, joining his son Peter's optician's business in 1752. Finding that the deviation of light rays could be obtained without dispersion, he used crown and flint glasses in combination to make the achromatic lens, the principle of which had been discovered earlier by Chester Moor Hall (1703–71). He produced an improved heliometer.

Dollond, Peter (London, 1730 – London, July 2, 1820) An optical instrument maker who improved the achromatic telescope developed by his father John. He deduced the Sun's distance from Earth using the value for parallax obtained by J. Short (1710–68) at the transit of Venus. In 1804, he took his nephew George Dollond (formerly Huggins, 1774–1852) into his business.

Donati, Giovanni Battista (Pisa, December 16, 1826 – Florence, September 20, 1873) An Italian astronomer who became Director of the observatory at Pisa. He discovered six comets; that most brilliant one of June 2, 1858 bears his name. He was the first to observe the spectrum of a comet, and to discover the gaseous nature of comets. He concluded that the scintillation of stars was caused by short-period variations in the refraction of light by the Earth's atmosphere. He found the relationship between aurorae and the Sun.

Doppler broadening see **spectrum**

Doppler effect The phenomenon first described by the Austrian physicist C. J. Doppler (1803–53). The DOPPLER PRINCIPLE states that the frequency of a source of waves (sound, light, etc.) as recorded by an observer is greater when the source moves towards him than when it is at rest relative to him, and that the frequency is less when the source is moving away from him. A clear explanation of, and deductions from, the effect were given by A. H. L. Fizeau (1819–96). The effect has proved of great value in astronomy with respect to the displacement of spectral lines. If a star is moving towards an observer, then he will receive a greater number of wavelengths of light per second; the frequency of the waves therefore increases and the wavelength will be shifted to a lower value towards the violet end of the spectrum. If the star is moving away from the observer, then the shift will be towards the red end of the spectrum. This DOPPLER SHIFT, as it is called, makes it possible to detect and measure relative motion in the line of sight, radial velocity, and rotation of celestial objects. See also **red shift**.

double star A pair of stars that can fall into either of two categories: a pair which are close together in space and constitute a physical unit as a gravitational system known as a **binary**; a pair which are quite distant from each other, but as a result of being on adjacent lines of sight appear close together in the field of a telescope.

doublet lens One that has two component lenses, which may either be cemented together or separated by an air-gap. This construction is used to reduce chromatic **aberration** to a minimum.

draconic month see **month**

Draper classification A classification of stellar spectra developed by A. J. Cannon at the Harvard College Observatory. The work was financed by the Henry Draper Memorial Foundation and published over the years 1918 to 1936. See **stars, spectral classification of**.

Draper, Henry (Prince Edward County, Virginia, March 7, 1837 – New York, November 20, 1882) An American natural philosopher and astronomer who was noted for his work in stellar spectroscopy. In 1872 he obtained the first successful stellar spectrogram showing absorption lines. He is commemorated by the *Henry Draper Catalogue* of stellar spectral types.

Dreyer, Johan Ludvig Emil (Copenhagen, February 13, 1852 – Oxford, September 14, 1926) A Danish astronomer who worked for many years in Ireland at the Earl of Rosse's observatory at Birr Castle, also at the Dublin University Observatory and at the Armagh Observatory, where he was director. His work, *New General Catalogue of Nebulae and Clusters of Stars* (London, 1888), with its two supplementary *Index Catalogues* (London, 1895, 1908), is standard: the NGC numbers are still used. He edited *The Scientific Papers of Sir W. Herschel*, 2 vols. (London, 1912).

Dunham Jr., Theodore (New York, December 17, 1897–) An American astronomer distinguished for his work in stellar spectrography, on planetary atmospheres and on material in interstellar space.

dwarf A star of relatively small diameter and low absolute brightness, lying in the main sequence of the **Hertzsprung-Russell diagram**. These stars are mainly red dwarfs at the lower end of the main sequence. The Sun is a yellow dwarf. Compare **white dwarf**.

dwarf galaxy A galaxy that is much smaller than a normal galaxy and is of low luminosity. Such galaxies are usually elliptical or irregular. Many have been discovered in the Local Group of galaxies.

dwarf nova A member of a class of irregular variable stars whose light-curves resemble those of novae. The luminosity remains steady for long periods, then rapidly increases in magnitude, and finally slowly returns to its normal minimum. U Geminorum stars and Z Camelopardalis stars are the main groups of stars of this type. Dwarf novae are thought to be close binary stars in which one component is a white dwarf.

dwarf variable An intrinsic **variable star** whose spectral-luminosity classification places it on the main sequence of the **Hertzsprung-Russell diagram**. Flare stars and U Geminorum stars are examples.

dynamical parallax Formerly known as HYPOTHETICAL PARALLAX, it is the parallax of a binary system determined from a relationship between the known masses of the components, the semi-axis major of the relative orbit, and the period of revolution.

Dyson, Frank Watson (Measham, near Ashby de la Zouch, January 8, 1868 – at sea on a voyage from Australia, May 25, 1939) Astronomer Royal for Scotland (1906–10) and Astronomer Royal (1910–33). He proved that J. C. Kapteyn's hypothesis of star-streaming was confirmed by stars of large proper motion. He made extensive studies of the spectra of the Sun's chromosphere and corona, and organized solar eclipse expeditions which verified Einstein's theory of the effect of gravitation on light.

Early Bird The first commercial satellite for telephony, telegraphy, and television, launched April 6, 1965, for the Communications Satellite Corporation (COMSAT).

early-type stars Hot stars whose spectral type places them at the beginning of the Harvard-Draper classification, i.e. types O, B, A, and early F.

Earth The third planet of the solar system, reckoned from the Sun, whose orbit lies between Venus and Mars. Earth appears blue when viewed from space, as predicted by A. L. Danjon (1890–1967) from colorimetric examination of **earthshine** and confirmed by the first voyagers in space. Our planet has one natural satellite, the **Moon**, but now has artificial satellites as well as much space junk circulating around it. Earth has an **atmosphere**. The main data relating to Earth are given in the table.

Globe	
Diameter (equatorial)	12,756 km
Diameter (polar)	12,714 km
Density (water = 1)	5·52 g cm^{-3}
Mass	$5{\cdot}977 \times 10^{24}$ kg
Volume	$1{\cdot}083 \times 10^{12}$ km^3
Sidereal period of axial rotation	$23^h\ 56^m\ 04^s$
Escape velocity	11·18 km s^{-1}
Albedo	0·36
Inclination of equator to orbit	23° 27′
Surface temperature	394 K (maximum)

Orbit	
Semi-axis major	1 A.U. = $149{\cdot}60 \times 10^6$ km
Eccentricity	0·0167
Inclination to ecliptic	0 (by definition)
Sidereal period of revolution	365·256 d
Mean orbital velocity	29·79 km s^{-1}

View of North and South America and West Africa taken from the satellite GOES-1.

earth-grazer A minor planet which in its orbit passes relatively close to the Earth. On October 28, 1937 Hermes came within 800,000 km of the Earth.

earthshine Also known as ASHEN LIGHT, EARTH LIGHT, and popularly as 'The Old Moon in the New Moon's arms'. Light reflected from the Earth upon that part of the Moon which is in the Sun's shadow, i.e. during the first days of a lunation or when the Moon has waned considerably. The Moon appears as a thin brightly illuminated crescent

with a pale greyish light over the remainder of its disc.

east point That point on the celestial sphere, due east of the observer, where the celestial equator intersects the horizon.

eccentric anomaly see **anomaly**

eccentricity One of the elements of an orbit of a planet, satellite, etc. It indicates how much an elliptical orbit departs from a circle. It is designated e and is given by the ratio c/a, where c is the distance between the centre and a focus of the ellipse and a is the semi-axis major.

eclipse The partial or total obscuration of the light coming from a celestial body, caused either by the passage of another body passing between it and the observer, as in a **solar eclipse**, or by the intervention of another body between it and the source of light which it reflects, as in a **lunar eclipse**. The eclipse of a star by the Moon or by a planet is called an **occultation**. Eclipses of the Sun and Moon can be calculated backwards or forwards; in the former case the calculation provides a means of verifying historical dates. Particulars of eclipses are given in the annual volumes of *The Astronomical Ephemeris* and in *Whitaker's Almanack*. Technical data on all eclipses from 1207 B.C. to A.D. 2161 are given in T. von Oppolzer, *Canon der Finsternisse* (Vienna, 1887).

eclipse year see **year**

eclipsing binary see **binary**

ecliptic The great circle traced on the celestial sphere by the Sun's annual path relative to the stars. It intersects the celestial equator in the First Point of Aries (see **Aries, First Point of**) and the First Point of Libra (see **Libra, First Point of**). The ecliptic is inclined to the celestial equator by 23½° approximately.

ecliptic conjunction see **conjunction**

ecliptic co-ordinates see **co-ordinates**

ecliptic limits The limiting distances of the Moon from its nodes beyond which eclipses cannot occur, because of the inclination of the ecliptic to the Moon's orbital plane.

ecliptic poles see **pole**

E component Part of the solar corona whose spectrum consists of emission lines.

ecosphere The space surrounding the Earth, another planet, a satellite, or a star in which life can be maintained.

Eddington, Arthur Stanley (Kendal, December 28, 1882 – Cambridge, November 22, 1944) An English astronomer and physicist who was director of the observatory at Cambridge. His main work dealt with astrophysics. He was a pioneer of theories of stellar motions and the internal constitution of stars. He discovered the **mass-luminosity relationship** of stars and upheld Einstein's Theory of Relativity which he tried to unify with the Quantum Theory.

Eddington limit According to A. S. Eddington stars should occur in the range of mass between 10^{32} and 10^{34} g. If the mass is smaller, then the gravitation is too large to allow effective radiation; if the mass is larger, then the radiation pressure is so great that it makes the star explosive. There is a limit beyond which the radiation pressure on matter is greater than the gravitational force holding the star together. For a star of one solar mass the Eddington limit is 10^{31} watts.

effective temperature see **celestial bodies, temperature of**

Einstein, Albert (Ulm, Germany, March 14, 1879 – Princeton, New Jersey, April 18, 1955) A German-born theoretical physicist whose contributions have made a great impact on 20th-century science. He is noted especially for his explanation of the photoelectric effect (1905), which was based on the concepts of Quantum Theory; the Special Theory of Relativity (1905), which led to the relation between mass and energy; the General Theory of Relativity (1916). He attempted to form a Unified Field Theory to unify mechanics and electrodynamics. His theories of relativity have been of great significance for astronomy and cosmology, for example, in providing a solution to the problem of the advance of the perihelion of Mercury, the curvature of light by a gravitational field, and the displacement (red shift) of spectral lines by a gravitational field.

Einstein displacement A small displacement in the apparent position of a star when the light from it passes close to a massive body, such as the Sun, on its way to the Earth. This is a prediction of Einstein's General Theory of Relativity and results from the influence of the massive body's gravitational field. The displacement by the Sun amounts theoretically to 1·75 seconds of arc for light grazing the solar disc. The first measurements to test this prediction were made during the total solar eclipse on May 29, 1919 by photographing stars near the Sun and comparing their positions on plates taken months earlier or later. The results seemed to verify the prediction, but analysis of the errors involved in such difficult observations made the result uncertain. Since then radio waves instead of light waves have been used in the experiment and given results in agreement with theory.

Elara Satellite VII of Jupiter. See also **satellite**.

E layer Part of the Earth's **ionosphere**.

electromagnetic radiation Radiation ranging from radio waves through light to high-energy X-rays and gamma rays. It is a flow of energy that can be considered either as a wave motion, each wave having a specific frequency and related wavelength, or as a stream of particles called **photons**. The photons are tiny packets of energy, their energy being related to the frequency and wavelength of the radiation. Electromagnetic radiation can travel through a vacuum (unlike sound waves), both waves and particles then moving at a constant speed, known as the speed of light and denoted *c*. The value of *c* is 299,792·5 km s^{-1}. Each type of electromagnetic radiation has a different but continuous range of frequencies (and wavelengths), the various ranges overlapping, as shown in the table. There is no known upper or lower limit to the frequency.

Almost all our information about the universe and its contents has been learnt from studies of the electromagnetic radiation that reaches us from celestial objects. Much of the radiation that falls on the Earth is absorbed by its atmosphere, but there are two regions of the electromagnetic spectrum which can reach the surface of the Earth. They are known as the **optical window** and the **radio window**. Other types of cosmic radiation, including X-rays and γ-rays, can now be detected and analysed by instruments carried above the atmosphere in satellites or rockets.

Type of radiation	*Wavelength range (approx.) in metres*
gamma rays	$<10^{-10}$
X-rays	10^{-8}–10^{-12}
ultraviolet	4×10^{-7}–10^{-8}
light	8×10^{-7}–4×10^{-7}
infrared	10^{-3}–8×10^{-7}
radar, radio	$>10^{-3}$

electron An elementary particle which is a constituent of all atoms. Its rest mass is $9{\cdot}109\times10^{-31}$ kg; the electron's negative electric charge is $1{\cdot}602\times10^{-19}$ coulomb; its radius is $2{\cdot}81777\times10^{-15}$ m.

electron density A figure which indicates the number of electrons in a given volume of a substance. It is used in astronomy in connection with gaseous media such as planetary atmospheres.

electron shell An atom of an element is regarded as consisting of a nucleus and one or more outer shells in which a definite number of electrons move in orbit around the nucleus. The number of these orbiting electrons in a neutral atom is equal to the number of protons in the nucleus, and is called the ATOMIC NUMBER. The arrangement of these electrons in the shells is different for each element. There are seven electron shells, designated consecutively by the letters K to Q, each of which can accommodate only a certain maximum number of electrons.

electron volt (*abbrev.*: eV) A unit of energy

used in atomic physics. One electron volt is the kinetic energy acquired by a particle carrying a charge *e* passing through a potential difference of one volt. 1 eV is equal to $1{\cdot}6021 \times 10^{-12}$ ergs or $1{\cdot}6021 \times 10^{-19}$ J, and corresponds to a frequency of $2{\cdot}42 \times 10^{14}$ Hz or a wavelength of 12,398 Å. An electron with a kinetic energy of 1 eV has a velocity of about 580 km s^{-1}.

elements, chemical Elements are chemically homogeneous substances which cannot be decomposed by chemical means. In the physical sense elements are not homogeneous because they contain several kinds of atoms (isotopes) which can be isolated. The number of known elements is now 104 of which 88 occur in nature and 16 have been prepared artificially.

elements, cosmic abundance of It has been found from spectroscopic studies of the atmospheres of the Sun and stars that (with the exception of hydrogen and helium) the relative distribution of the chemical elements in the universe as a whole is in general much the same as found in the Sun. The relative abundance of certain elements, referred to that of hydrogen expressed as log 12, is as follows: hydrogen 12, helium 11·2, oxygen 8·8, neon 8·7, nitrogen 8·4, carbon 8·2, magnesium 7·8, silicon 7·7, iron 6·5. Uncommon types of stars, e.g., Wolf-Rayet, show anomalous abundances of certain elements. Population I stars have a greater proportion of heavy elements than Population II stars.

Elger, Thomas Gwyn Empy (Bedford, 1838 – Bedford, January 12, 1897) A notable English amateur astronomer who was the first director of the Lunar Section of the British Astronomical Association. He produced an excellent map of the Moon.

E line J. von Fraunhofer's designation for one of the absorption lines in the solar spectrum. It is in the green part of the visible spectrum at 5270 Å and is produced by iron. See also **Fraunhofer lines**.

ellipse A conic section obtained by cutting a right circular cone obliquely by a plane through the axis. The ellipse can be defined in several ways; one is that the sum of the distances of any point on the curve from the two foci of the ellipse is constant, and equal in length to its major axis. The shape of the ellipse is governed by its **eccentricity**, which is always less than unity.

The ellipse is important in astronomy because the orbits of planets, satellites, companion stars, comets, etc., are all of that form to a first approximation. It is also basic to **Kepler's Laws**.

elliptical galaxy see **galaxy**

Ellison, Mervyn Archdall (Fethard-on-Sea, Southern Ireland, May 5, 1909 – Dunsink, September 12, 1963) An amateur astronomer distinguished for his work as a solar observer. He became a professional astronomer at the Royal Observatory, Edinburgh, and was later appointed Professor at Dunsink Institute for Advanced Studies and Director of the Dunsink Observatory. He was the Solar Reporter for the **International Geophysical Year**.

elongation Of a planet or of the Moon, the difference between its celestial longitude and that of the Sun. See **aspect**.

emersion The reappearance of a celestial body after it has been eclipsed or occulted.

emission nebula see **nebula**

emission spectrum The bright line spectrum produced by incandescent substances. See **spectrum**.

Enceladus Satellite II of **Saturn**. See also **satellite**.

Encke, Johann Franz (Hamburg, September 23, 1791 – Spandau, August 26, 1865) A German mathematician and astronomer who became director of the observatory at Berlin, where he supervised the erection of a new observatory. He calculated the solar parallax from past transits of Venus. He computed the orbit of the comet which now bears his name. He developed a method of determining elliptic orbits from three observations, and made investigations on planetary orbits and perturbations.

Encke's Comet A comet which was discovered by J. L. Pons on November 26, 1818. Encke computed the orbit and period of this body, and proved that it was identical with three comets that had been previously observed. It was the second periodic comet to be discovered, and has the shortest known period: 3·3 years. Regular observation since 1818 has revealed a reduction in period of 2·5 days in 100 years, thereby indicating that the orbit is shrinking. The comet has a small nucleus, a tenuous head consisting mainly of gas, and a tail.

Encke's division A dark marking just beyond the middle of Ring A of Saturn, probably a region with a sparser distribution of particles. Unlike the Cassini division it is not a true gap in the ring system.

energy level The energy of one of the possible stationary states of an atomic system, such as an atom, molecule or nucleus. According to the quantum theory such a system can exist in only one of a number of discrete energy levels. An atomic system normally exists at its lowest level of energy (the ground state). On excitation higher energy levels are attained. Passage from one energy level to another is accompanied by the absorption or emission of quanta of energy, which are responsible for line or band spectra.

English yoke mounting see **telescope, mounting of**

enthalpy A term used by H. Kamerlingh Onnes (1853–1926) and equivalent to the 'heat function' of J. W. Gibbs (1839–1903). It is the heat content of a substance at constant pressure, and is defined thermodynamically as

$$H = U + pV$$

where H is the heat content, U is the internal energy, p is the pressure, and V is the volume.

entropy A measure of the amount of thermal energy in a system which is unavoidably not available for conversion into mechanical work. It may be expressed as a quantity of heat divided by the absolute temperature at which that heat is developed. It expresses also the degree of disorder of a system. As all natural processes are irreversible, the entropy of such processes increases, tending towards a maximum, and the amount of available energy in the system decreases. The significance of this for the universe, if indeed it be a closed system, is that its entropy is increasing and tending towards a maximum, at which point all temperature differences will disappear and the universe's available energy will be exhausted. This state has been called 'heat-death' by J. H. Jeans.

envelope star see **shell star**

epact The age of the Moon, diminished by one day, on January 1 in the Gregorian ecclesiastical lunar calendar. In the Gregorian calendar Easter is determined by the epact and the dominical letter. A table for the epact is given in the *Explanatory Supplement to the Astronomical Ephemeris*.

ephemeris (*plural:* ephemerides) An astronomical calendar giving the calculated future positions of celestial bodies for a number of successive days. The governments of most countries publish an authoritative ephemeris. For example, *The Astronomical Ephemeris*, produced jointly by the Nautical Almanac Offices of the United Kingdom and the United States is published in the United Kingdom by HMSO and is issued in the United States by the Nautical Almanac Office of the US Naval Observatory. The *Astronomicheskii Ezhegodnik* is produced in the USSR.

Ephemeris Time (*abbrev.:* E.T.) A system of time measurement that depends on the orbital motions of the Earth, Moon, and planets in the solar system, and is used only in computing. It is defined by the I.A.U. as follows: 'Ephemeris Time is reckoned from the instant, near the beginning of the calendar year A.D. 1900, when the geometric mean longitude of the Sun was 279° 41′ 48″·04, at which instant the measure of Ephemeris Time was 1900 January $0^{d}\,12^{h}$ precisely.' The ephemeris second had previously been adopted as the fundamental invariable unit of time, until superseded by the atomic second.

It is defined as the fraction 1/31 556 925·9747 of the tropical year for 1900 January 0 at 12^h Ephemeris Time.

epoch A point in time used as a fixed reference for comparison of astronomical data in star catalogues, etc. Observations made over any considerable period of time are reduced to a common epoch so that they are comparable. The starting point in the calculation of ephemerides is at a certain epoch. The maximum or minimum magnitudes of variable stars are referred to some epoch. It is also one of the elements of orbital data, where it is the time of perihelion passage.

equant A notion introduced by **Ptolemy** in his geocentric world system, being an imaginary circle placed in the plane of the **deferent** to adjust and reconcile the planetary motions with the hypothesis of uniform velocity.

equation of position see **visual photometry**

equation of the centre An inequality in the Moon's motion, caused by its orbital eccentricity. It is the difference between the true anomaly and the mean anomaly.

equation of time The difference between mean time and apparent solar time caused by the combined effect of the eccentricity of the Earth's orbit and the obliquity of the ecliptic.

equatorial conjunction see **conjunction**

equatorial co-ordinates see **co-ordinates**

equatorial horizontal parallax The **geocentric parallax** of a celestial body when viewed on the horizon.

equatorial telescope A telescope whose polar axis is mounted parallel to the Earth's axis, and therefore points to the celestial pole, and whose declination axis is at right angles to the polar axis. See also **telescope, mounting of**.

equinoctial colure The declination circle passing through the north and south celestial poles and the equinoctial points.

equinoctial points see **equinox**

equinox Either of the two moments during the year when the astronomical day and night are equal throughout the world, except in so far as that equality is modified by atmospheric refraction at the time of sunrise and sunset. The Sun rises due east and sets due west on these days. The VERNAL EQUINOX occurs when the Sun crosses the equator from south to north on March 20 or 21 in the northern hemisphere. It is thus one of the points of intersection of the ecliptic with the celestial equator. This point is often called the First Point of Aries. The other equinox is the AUTUMNAL EQUINOX, which occurs on September 22 or 23 when the Sun crosses the equator from north to south. This point is often called the First Point of Libra. The points of intersection of the ecliptic with the celestial equator are known also as EQUINOCTIAL POINTS.

Eratosthenes (Cyrene, *c.* 273 B.C. – Alexandria, *c.* 192 B.C.) A Greek geographer, mathematician, and astronomer. He made the first scientific calculation of the Earth's circumference, and obtained a value corresponding to a diameter within 80 km of the polar value. He is credited with the compilation of a book on stars and also a star catalogue.

Eros Minor planet No. 433, discovered by G. Witt in 1898. It is an irregularly shaped elongated object, whose orbit comes within that of Mars. Its occasional close approach to the Earth has several times provided opportunities to redetermine the **solar parallax**.

escape velocity The minimum velocity that a body, such as a rocket or a space probe, must attain in order to overcome the gravitational attraction of the primary body and go into space. The escape velocity depends on the mass of the primary body and its diameter. The escape velocity at the surface of some of the bodies belonging to the solar system is as follows: Mercury 4·2; Venus 10·3; Earth 11·2; Moon 2·4; Mars 5·0; Jupiter 60; Saturn 37; all values are in km per second.

Eta Aquarids A meteor shower that occurs

in the first week of May. Its radiant is in Aquarius near η Aqr. This shower is associated with Halley's Comet.

Eudoxos of Cnidos (*c.* 408 B.C. – *c.* 355 B.C.) A Greek mathematician and astronomer. His theory of 27 homocentric spheres was the first attempt to account for irregularities in the motion of the Sun, Moon, and planets on a quantitative basis. According to Pliny he fixed the length of the year at 365·25 days, and according to Vitruvius he invented the sundial.

Euler, Leonhard (Basel, April 15, 1707 – St Petersburg, September 18, 1783) A Swiss mathematician who made outstanding contributions to all branches of natural philosophy by his mathematical work. His contributions to astronomy relate to lunar theory, planetary perturbations, and dynamics. His work on optical systems was important for the technical development of telescopes and microscopes.

Europa Satellite II of Jupiter. See also **satellite**.

European Space Agency (*abbrev.:* ESA) A co-operative astronomical association of certain European countries to promote space research, formed in 1975 from the merging of ESRO (European Space Research Organization) and ELDO (European Launcher Development Organization).

evection A periodical inequality in the motion of the Moon. It is an oscillation of that body about its position as given by the **equation of the centre**, and is caused by the Sun's perturbative effects. Evection was discovered by Ptolemy.

Evening Star The planet Venus when it appears in the west after sunset.

event horizon The boundary surface of a **black hole** from which nothing can escape. Hence, outside observers cannot obtain information about its interior.

Evershed effect A horizontal flow of gases within a sunspot, radially outwards from the centre to the edge of the penumbral region where it ceases. The effect was discovered in 1908 by J. Evershed and is revealed by an examination of the spectrum of a sunspot for Doppler shift.

Evershed, John (Gomshall, Surrey, February 26, 1864 – Ewhurst, Surrey, November 17, 1956) An English astronomer who became Director of the Kodaikanal Observatory, India. He is known for his solar observations, and his discovery of the radial movement in sunspots. He designed a spectroheliograph.

excitation The addition of energy to an atom causing a transfer of one or more electrons to a higher shell. It is brought about either by collisions between particles or by absorption of photons. In each case the energy absorbed is equal to the EXCITATION ENERGY, which is the difference in energy between the ground state and the excited state of the atom. After less than 10^{-8} s the electron falls back to its original lower level of energy, and emits light or other radiation. The physical conditions (high temperature, etc.) in stars cause a high incidence of atoms in the state of excitation.

excitation temperature see **celestial bodies, temperature of**

exit pupil The part of a pencil of light rays, emerging from an optical instrument, which has the smallest cross-sectional area in which are contained all rays from all parts of the field. It is therefore the image of the objective lens or primary mirror of a telescope formed on the side of the eyepiece nearest the eye.

exobiology The study of organized life beyond the Earth. See also **extraterrestrial life**.

exosphere see **atmosphere**

expanding universe The concept of an expanding universe arises from the fact that observation of extragalactic objects reveals a red shift which, if correctly ascribable to the Doppler effect, means that these objects are receding from us at velocities proportional to their distance from us. See **Hubble Law**.

exploding stars or **explosive variables** Cataclysmic variable stars, such as **novae**, whose luminosity increases unpredictably and extremely rapidly.

Explorer The name given to a series of US satellites. The first successful one, Explorer 1, was America's first satellite to enter orbit. It was launched from Cape Canaveral, Florida, on February 1, 1958. Explorers have been used in many scientific experiments on the atmosphere, cosmic radiation, magnetic fields, etc.

extinction The reduction in the observed brightness of a celestial object, such as a planet or a star, caused by absorption, scattering, or diffraction of the light on passing through the Earth's atmosphere. Allowance has to be made for this reduction in all measurements of brightness. See also **interstellar matter**.

extragalactic nebulae An obsolete term for **galaxies**, which were originally seen as cloud-like objects. Its use should be avoided to prevent confusion with gaseous **nebulae**.

extraterrestrial life Whether or not life exists on other celestial bodies in our solar system, or on more remote objects, clearly depends on the definition of life. In so far as our solar system is concerned there cannot be life as we know it on Earth because physical conditions are not the same, and so are unsuitable. Other planets and satellites have either no atmosphere, or have one consisting of toxic gases, such as carbon dioxide, methane, and ammonia; furthermore the temperatures are either excessively high or low, and conditions of humidity are unfavourable. It has been argued on statistical grounds that other planetary systems with worlds such as our Earth might exist somewhere in the universe.

On the assumption that there might be intelligences in space trying to make contact with other lifeforms, an attempt was made in 1960 to pick up possible regular, rhythmic radio signals from space (see **Ozma project**), but without success. More recent attempts have also proved unsuccessful. The detection of pulsars in 1967 did for a while suggest the serious possibility of an extraterrestrial attempt at communication, but the true nature of pulsars was found to have a different cause. The detection of interstellar molecules such as formaldehyde, methanimine, hydrogen cyanide, ammonia, and water make it conceivable that they are basic to the composition of the primeval 'soup' from which life on Earth emerged. The discovery of terrestrial amino acids and of a polynucleotide of DNA in meteorites suggests that biomolecules can be formed in the protoplanetary cloud of gas and dust. F. Hoyle and N. C. Wickramasinghe have recently put forward the theory that micro-organisms enter the Earth's atmosphere from space, are responsible for certain outbreaks of illness and could have played a role in the origin of life.

extrinsic variable see **variable stars**

eye lens The magnifying component nearest to the eye in an eyepiece consisting of two lenses.

eyepiece or **ocular** The system of lenses in an optical instrument nearest to the observer's eye. Its function is to magnify the image formed at the focus by the objective of a refractor, or by the mirror of a reflector. There are numerous designs of eyepiece. They usually have two lenses, the field lens (which points towards the objective) and the eye lens (which is nearer the observer's eye). Eyepieces are described as positive when the image is formed in front of the lenses, as in the Ramsden eyepiece, and negative when it is formed between the lenses, as in the Huygenian eyepiece. Eyepieces are preferably identified by their focal length (e.g. 3/16 inch), though a set of them used permanently with one instrument may conveniently be identified by the magnification (e.g. × 150). The magnification of an eyepiece is given by the ratio of the focal length of the objective or mirror to the focal length of the eyepiece.

faculae A Latin word meaning 'little torches', applied by C. Scheiner (1575–1650) to bright streaks around sunspots. They are visible only within a limited distance from the Sun's east and west limbs. The appearance of faculae is usually closely related to the development of a sunspot or a group of sunspots. They conform to the same 11-year cycle as do other phenomena of solar activity, and are more prominent when the Sun is most active. They are seen some hours before in the place where sunspots will appear; they disappear much more slowly and remain visible even for months after the sunspots have gone. Faculae are considered to be regions of incandescent gas above the photosphere, and at a much higher temperature so that they appear brighter.

falcated Curved like a sickle. The word is applied to the Moon or to an inferior planet when it exhibits the crescent phase.

F component The outer part of the solar corona which gives a continuous spectrum containing absorption lines. 'F' stands for Fraunhofer.

field-flattener A device introduced by C. Piazzi Smyth in 1874 to permit the use of flat photographic plates in telescopes which produce a spherical focal plane. It is usually a thin planoconvex lens of appropriate curvature placed with its plane surface just in front of the photographic plate. It is used also in some designs of **Schmidt camera**.

field lens In an eyepiece consisting of two lenses, the light-gathering component farthest from the eye.

field of view The area visible in an optical instrument at a given setting. In the case of a telescope it depends on the eyepiece that is being used: an increase of magnification usually decreases the field. For a given setting of a telescope the apparent field is expressed by its angular diameter.

filaments Dark thread-like markings, which can be seen on spectroheliograms taken in the light of the K line of calcium. They are flame-like prominences viewed in plan. They appear dark as a result of absorbing light from the bright chromosphere.

filar micrometer see **micrometer**

finder or **guider** A small telescope of low power fixed parallel to the axis of a larger telescope. It has a larger field of view than the main telescope, and serves to locate and facilitate training the main instrument on the object it is desired to observe.

fireball or **bolide** An exceptionally bright meteor: the magnitude can be as much as −4. In its passage through the atmosphere it may leave a luminous trail visible for some minutes.

First Point of Aries see **Aries, First Point of**

First Point of Libra see **Libra, First Point of**

First Quarter One of the Moon's phases when it has almost reached eastern quadrature and is half illuminated as seen from the Earth.

Fitzgerald, George Francis (Dublin, August 3, 1851 – Dublin, February 22, 1901) An Irish physicist who is distinguished for his contributions to the theory of radiation and particularly for the postulate that a body moving through space with a velocity significant with respect to that of light will undergo a change in dimension (contraction) in the direction of its motion. This theory was put forward independently by H. A. Lorentz (see **Lorentz, Hendrik**). The contraction was later shown to be a direct consequence of the Special Theory of Relativity.

fixed stars The name given in ancient times to those celestial bodies which in contrast to the planets appeared not to move, and accordingly were thought to be firmly attached to the surrounding crystal sphere. The planets were called WANDERING STARS. The so-called fixed stars, now known as background stars, do in fact have a proper motion, which is detectable only over a long period of time.

Flamsteed, John (Derby, August 19, 1646 – Greenwich, London, December 31, 1719) An English astronomer who was appointed the first Astronomer Royal by King Charles II for the purpose of obtaining accurate data on the motion of the Moon and positions of stars for use in navigation. He made thousands of observations on stars with greater precision than others before him, besides discovering the mutual influence of Jupiter and Saturn on each other's motion. His ambition to produce his own catalogue of stars was not realized: this great contribution to fundamental astronomy appeared posthumously in 1725 as *Historia Coelestis Britannica*.

Flamsteed numbers A system of stellar identification by numbers allocated to the stars in each constellation in order of Right Ascension. The numbers were allocated by later astronomers to stars in Flamsteed's star catalogue.

flare stars Variable stars, usually red dwarfs, whose luminosity increases unpredictably and irregularly in a very short time (some few minutes) as a result of a sudden intense eruption of energy, and then decreases to its normal value in about 30–60 minutes. They are also known as UV CETI STARS, after the best-known example. The outburst is sometimes accompanied by an increase in radio emission.

flash spectrum The bright line emission spectrum of the solar chromosphere which is observable for a short time at the beginning and end of totality during an eclipse of the Sun. When the photosphere is obscured by the Moon's disc, the absorption spectrum of the chromosphere suddenly changes to this emission spectrum, but lasts only for several seconds (hence the term 'flash').

F layers Two layers, F_1 and F_2, which lie above the E layer of the Earth's **ionosphere**. They are also known as the APPLETON LAYERS after E. V. Appleton (1892–1965), the British physicist who identified them.

F line J. von Fraunhofer's designation for one of the absorption lines in the solar spectrum; it is in the blue-green part of the spectrum and corresponds to the Hβ hydrogen line at 4861 Å. See also **Fraunhofer lines**.

flocculi or **plages** Small features which give a mottled or granulated appearance to the chromosphere of the Sun. Together with **faculae** they are related to the genesis and development of sunspots. They are revealed by spectroheliograms taken by the light of the Hα line and the K line of calcium. They consist of masses of gases composed mainly of calcium (BRIGHT FLOCCULI) and of hydrogen (DARK FLOCCULI), the former being at a higher temperature, the latter at a lower temperature than the surrounding region.

focal length The distance between a lens or curved mirror and its focus.

focal ratio The ratio of the focal length of a lens or mirror to its aperture.

focus The point on the optical axis of a lens or a curved mirror at which the image of a distant point (also on the axis) is formed. It is the point at which rays of light parallel to the axis are made to converge.

Fomalhaut (α PsA) The brightest star in the constellation Piscis Austrinus: it is the most southerly first-magnitude star visible from Great Britain.

forbidden lines Lines observed in the spectra of interstellar gas which were not identifiable with lines in spectra produced under laboratory conditions; they were consequently ascribed to some unknown chemical element. It is now known that they are produced under special conditions which give rise to metastable states of atoms, for example, of oxygen and nitrogen. References in the literature to such lines are made by enclosing them in square brackets, e.g. [O II].

fork mounting see **telescope, mounting of**

Foucault knife-edge test An optical test devised by J. B. L. Foucault (1819–68) to check the optical quality of lenses and mirrors. A knife-edge is placed exactly at the focus of the lens or mirror, which is illuminated by a beam of light parallel to the axis. If the optical element is perfect it should be possible to extinguish the whole beam; a certain amount of light usually passes by the knife-edge because the lens or mirror still retains some surface imperfections which prevent all the light being brought to the focal point.

Foucault pendulum Apparatus first used by J. B. L. Foucault (1819–68) in 1851 to demonstrate the Earth's rotary motion. It consists of a heavy ball suspended by a very long wire mounted so that it swings freely with the minimum amount of friction at the point of support. On being set swinging in a plane in space the Earth's rotation will cause the pendulum to depart slowly from its initial plane of motion.

Franklin-Adams charts Photographic star charts (206 in all) covering both the northern and southern hemispheres. The plates were taken between 1903 and 1913 by John Franklin-Adams (1843–1912) and his assistants at the Royal Observatory, Cape Town, and at his own observatory at Godalming, Surrey.

Fraunhofer, Josef von (Straubing, Bavaria, March 6, 1787 – Munich, June 7, 1826) A German optician who was noted for his skill in making achromatic lenses and optical instruments. He discovered independently of W. H. Wollaston (1766–1828) the dark lines in the solar spectrum, which he studied with improved apparatus. He also studied the diffraction of light through narrow slits, and developed the earliest form of diffraction grating. The spectroscope became a precision instrument as a result of his work.

Fraunhofer lines The dark absorption lines in the solar spectrum, first observed by W. H. Wollaston but carefully studied by J. von Fraunhofer. Fraunhofer designated the most prominent ones by the letters *A* to *K*, distinguishing 574 lines between *B* and *H*; the next most prominent series of lines he designated by the letters *a* to *i*. Over 25,000 lines have now been identified. The most prominent ones at visible wavelengths are caused by the presence of neutral hydrogen (the hydrogen alpha or Hα line), singly ionized calcium (the calcium H and K lines), sodium, and magnesium.

free fall see **zero gravity**

French Revolutionary Calendar see **calendar**

frequency The number of complete vibrations during a suitable unit of time, usually one second. The unit of frequency employed in the SI is the HERTZ (symbol: Hz). ROTATIONAL FREQUENCY is the number of complete revolutions per second (rev/s).

f-spot see **sunspot**

F star see **stars, spectral classification of**

Full Moon One of the Moon's phases in which the Moon is at **opposition**, and its fully illuminated hemisphere is visible from the Earth.

fundamental stars Reference stars whose co-ordinates (positions) and proper motions have been determined with the greatest accuracy. Fundamental stars serve as the points to which positional observations of other bodies can be related. The I.A.U. has adopted 1535 stars as fundamental reference stars: relevant data are contained in the *Fourth Fundamental Catalogue, FK4.*

Gagarin, Yuri Alekseyevich (Gzhatsk, March 9, 1934 – killed in a plane crash, March 17, 1968) A Soviet cosmonaut, the first man to be placed in orbit round the Earth in a space vehicle. This event took place on April 12, 1961. The space vehicle was Vostok 1, which made one complete orbit of the Earth during a flight lasting 108 minutes.

galactic Pertaining to or belonging to the Galaxy, as opposed to EXTRAGALACTIC, i.e. relating to something outside the Galaxy.

galactic cluster see **cluster**

galactic co-ordinates see **co-ordinates**

galactic latitude see **co-ordinates**

galactic light Part of the illumination of the night sky, originating from stars but diffused throughout interstellar space.

galactic longitude see **co-ordinates**

galactic poles see **pole**

galactochemistry The chemistry of objects outside the Galaxy and of interstellar space. Visible, ultraviolet, and microwave spectroscopy have revealed the existence of polyatomic molecules, including organic compounds, which have attracted interest because of their possible biological significance.

galaxy Formerly called an EXTRAGALACTIC NEBULA. A huge assembly of stars, dust, and gas, an example of which is our own Galaxy. Methods of classifying galaxies have been developed by E. Hubble, G. de Vaucouleurs, and others. There are three main groups. ELLIPTICAL GALAXIES are round or elliptical systems, showing gradual decrease in brightness from the centre outwards; they are classified E0 to E7 in increasing degree of ellipticity. LENTICULAR GALAXIES are systems intermediate between elliptical and spiral galaxies. SPIRAL GALAXIES are flattened disc-shaped systems in which young stars, dust, and gas are concentrated in coiled spiral arms; they are classified by S with lower case suffix letters. There are also BARRED SPIRAL GALAXIES distinguished by a bright central bar from which the spiral arms emerge; they are classified by SB with lower case suffix letters. IRREGULAR GALAXIES are systems with no symmetry. See also **Morgan's classification**. Illus. page 64.

Galaxy, the The star system which contains the solar system (the capital G distinguishes it from other galaxies). The term originally applied to the Milky Way, but is now used to designate the MILKY WAY SYSTEM. The Galaxy is a spiral star system about 30 kiloparsecs in diameter (probably similar in appearance to the Andromeda Galaxy) with our Sun and solar system located at the edge of one of the spiral arms about 10 kpc from the centre. Population II stars are found in the core (or nucleus) of the Galaxy, with younger hotter Population I stars occurring in the spiral arms together with interstellar dust and gas. The spiral arms and core form a disc-shaped system with a bulging centre. This is surrounded by the galactic **halo**, which has a diameter even greater than that of the disc and in which occur very old Population II stars.

The CENTRE OF THE GALAXY is that point at which the axis of rotation of the Galaxy intersects with the galactic plane. The radio source Sagittarius A appears to be at the galactic centre. The DIRECTION OF THE GALACTIC CENTRE has been adopted as the zero of galactic longitude, and coincides with that of Sagittarius A. The GALACTIC EQUATOR is the intersection of the galactic plane with the celestial sphere; the Milky Way marks its path across the sky. Study of stellar motion in the Galaxy has revealed that it rotates about an axis which is perpendicular to the centre of the disc. The rotational velocity is not uniform throughout; it varies with the distance from the centre, and at the distance of the Sun is about 250 km s^{-1}.

Knowledge of the structure of the Galaxy became possible only when it could be examined by a telescope. W. Herschel's long investigation on the distribution of stars led him to the fairly true conclusion that it was greatest in the regions of the Milky Way, and that the galactic system was a disc with the Sun close to the centre. Subsequent work in the 20th century by J. C. Kapteyn (see **Kapteyn, Jacobus**) on the statistical analysis of star distribution, by H. Shapley (see **Shapley, Harlow**) on the distribution of globular clusters, and particularly the prediction by H. C. van de Hulst (see **Hulst, Hendrik van de**) that neutral interstellar hydrogen should give radio emission with a wavelength of 21 cm has led to more precise knowledge of the structure of the Galaxy. There is a rotating disc of hydrogen at the centre of the Galaxy. The dark patches have been identified as interstellar dust clouds obscuring the stars behind. The true position of the Sun has been ascertained and the spiral arms have been mapped.

Galaxies do not appear uniformly distributed throughout space. There is a **zone of avoidance** along the galactic equator where very few are seen because of the obscuring dust clouds. Elsewhere there are CLUSTERS OF GALAXIES comprising very large numbers of individual galaxies whose separation is nevertheless great – about 500 pc. The Galaxy, of which our solar system is part, is one member of a cluster known as the **Local Group**, which includes the Andromeda Galaxy and the Magellanic Clouds. The

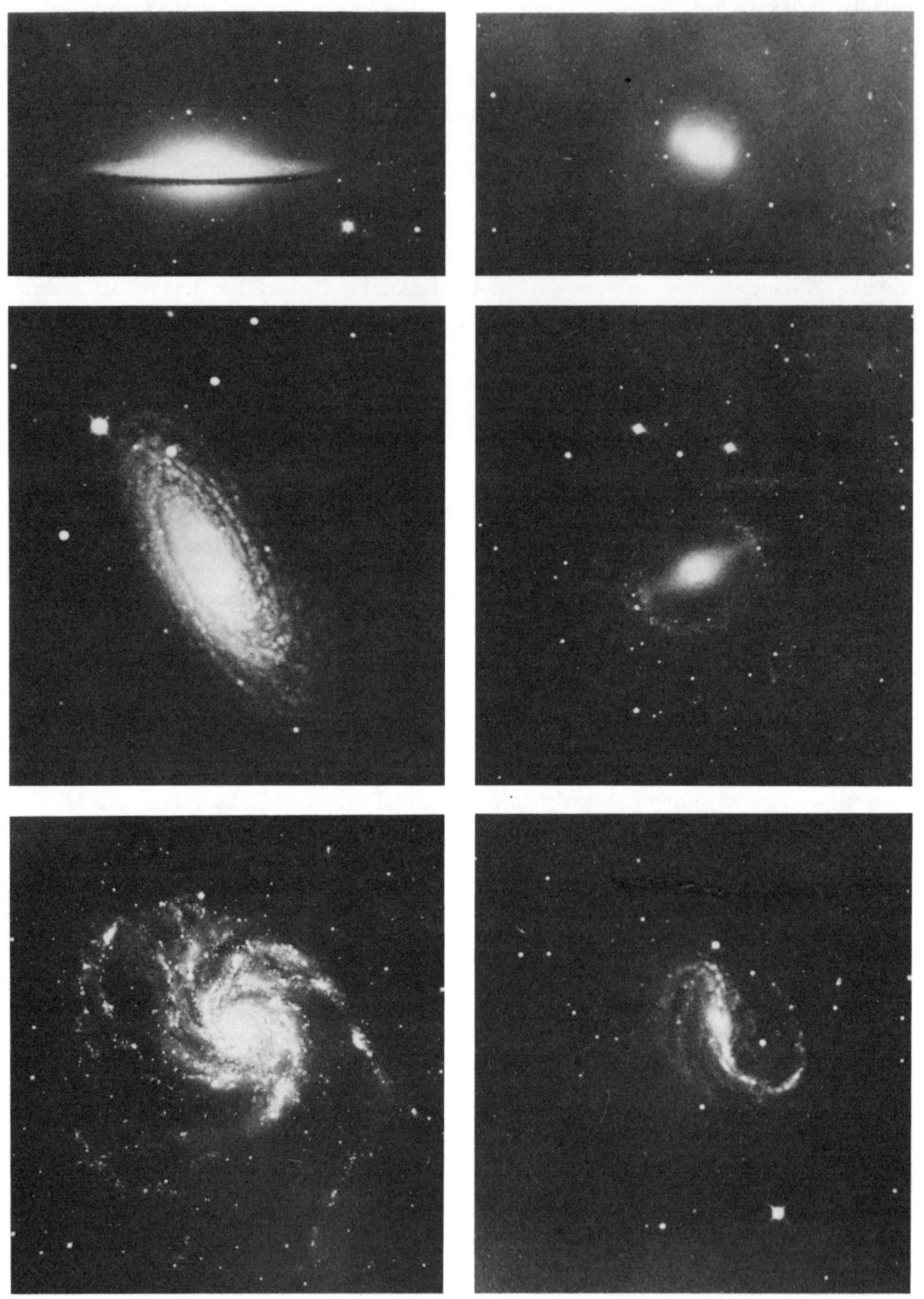

Various examples of both spiral and barred spiral galaxies.

age of the Galaxy is thought to be about 10,000 million years, during which time it would have completed approximately 40 revolutions.

Information relating to the Galaxy is given in the table.

The Galaxy: data

Diameter of disc	25–30 kpc
Thickness of disc at centre	4–5 kpc
Average density	7×10^{-21} kg m^{-3}
Mass of central nucleus	$1{\cdot}8 \times 10^{41}$ kg
Total mass of the Galaxy	about $2{\cdot}9 \times 10^{41}$ kg
Pole of galactic plane (1950·0)	R.A. $12^h 49^m{\cdot}0$ Dec. +27° 24′
Point of zero longitude (1950·0)	R.A. $17^h 42^m{\cdot}4$ Dec. −28° 55′
Galactic longitude of north celestial pole	123°·00
Distance of Sun from centre	10 kpc
Distance of Sun above galactic plane	8 pc
Rotational velocity of Sun	250 km s^{-1}
Period of Sun's revolution about centre	$2{\cdot}5 \times 10^8$ years
Magnitude	−20 to −18

Galilean satellites see **Jupiter; satellite**

Galilean telescope A refracting telescope in which the optical system comprises a convex (positive) objective and concave (negative) eyepiece. It forms an erect image and gives a small field. This type of instrument, first employed by Galileo Galilei to make astronomical observations, is no longer used in astronomy. The principle of the optical system is still used in opera glasses.

Galilei, Galileo (Pisa, February 15, 1564 – Arcetri, January 8, 1642) An Italian natural philosopher, mathematician, and astronomer who discovered the isochronism of the pendulum. He was one of the first to use the telescope for astronomical observations. He discovered surface features of the Moon, and gave a correct explanation of **earthshine** and of the Moon's **libration**. He discovered four satellites of Jupiter, which are now known as the GALILEAN SATELLITES. He observed the phases of Venus, and studied sunspots, from whose behaviour he deduced that the Sun rotates. He discovered that some celestial objects, such as the Milky Way and the Pleiades, are resolvable into many more stars than can be distinguished by the naked eye. His support of the Copernican world system brought him into conflict with the Catholic Church. His works include *Sidereus Nuncius* (Venice, 1610), *Istoria e dimostrazioni intorno alle macchie solari e loro accidenti* (Rome, 1613), *Il Saggiatore* (Rome, 1623), and *Dialogo sopra i due massimi sistemi del mondo* (Florence, 1632).

Galle, Johann Gottfried (Pabsthaus, June 9, 1812 – Potsdam, July 10, 1910) A German astronomer who discovered three comets and compiled a standard reference catalogue of cometary orbits (1894). He was the first to

Original telescopes of Galileo and object glass mounted in an ivory frame.

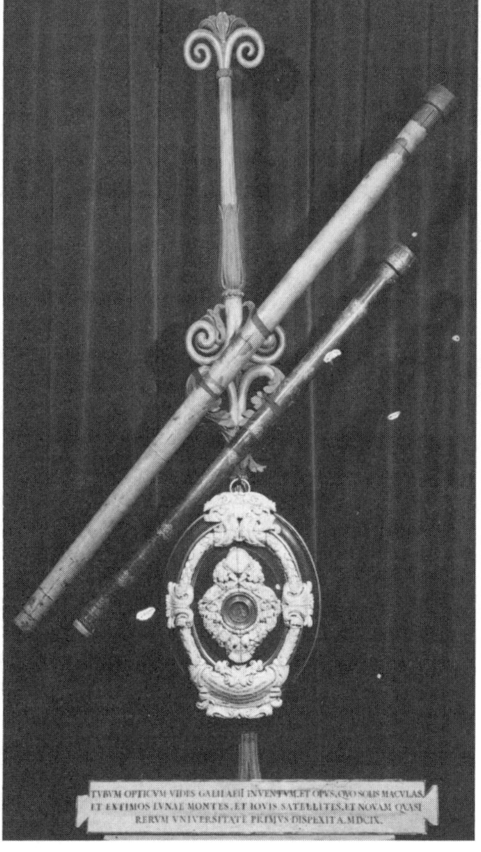

detect the crêpe ring of Saturn (1838) and to observe the planet Neptune (1846), which was located close to the position predicted by U. J. J. Le Verrier. He suggested that the parallax of minor planets be used to ascertain the scale of the solar system.

gamma-ray astronomy That branch of astronomy which deals with the detection, measurement and sources of cosmic **gamma rays**. As this type of radiation of very short wavelength is absorbed by the Earth's atmosphere, all studies have to be made by high-altitude balloons, artificial satellites, space craft, or orbiting astronomical observatories. The U.S. satellite Explorer 48, launched on November 16, 1972, was the first to be devoted exclusively to the study of gamma rays of cosmic origin. The **Crab Nebula** has been identified as a gamma-ray source. Another is in the direction of the Galactic centre, and there are extragalactic sources too.

gamma rays (γ-rays) Very high energy electromagnetic radiation of wavelength about 10^{-8} cm, overlapping the spectral region of X-rays. They are usually considered as photons, quanta of energy moving with the velocity of light. They are emitted as quanta by the atoms of natural and artificial radioactive isotopes during decay. They are not corpuscular like α- and β-rays. A number of satellites have carried instruments to detect and analyse the γ-rays from space. Some come from identifiable sources, such as the Crab Nebula pulsar. Others possibly arise from interactions between cosmic rays and interstellar matter.

Gamow, George (Odessa, March 4, 1904 – Boulder, Colorado, August 19, 1968) Russian physicist, who became Professor of Physics at the University of Colorado. He is known for his study of atomic nuclei and application of nuclear physics to astronomical problems, such as stellar evolution, the origin of chemical elements (by the capture of neutrons) and the neutrino theory of supernovae. He developed the theory of radioactive decay based on wave-mechanics, which explained the spontaneous emission of α-rays.

Ganymede Satellite III of **Jupiter**. See also **satellite**.

Gauss, Carl Friedrich (Braunschweig, April 30, 1777 – Göttingen, February 23, 1855) A German mathematician who became Director of the observatory at Göttingen. He calculated the orbits of the minor planets Ceres and Pallas by a new method. He published *Theoria motus corporum coelestium* (Göttingen, 1809) on the determination of cometary and planetary orbits. He was a pioneer of geodesy. His name is used for the unit of magnetic flux density in the c.g.s. system of units. 1 gauss = 10^{-4} tesla in the SI.

G band The band at 4303 Å in the spectrum produced by CH.

Gegenschein (*German:* counterglow) A phenomenon related to the **zodiacal light**. It is visible usually only in the tropics on dark clear nights as a luminous elliptical area directly opposite to the Sun. It is probably caused by the scattering of light by particles in space.

Geiger counter or **Geiger-Müller counter** An instrument developed by J. W. Geiger (1882–1945) and W. Müller for the detection of ionizing radiation (e.g. α-, β-, γ-rays), and by means of auxiliary equipment counting the number of particles producing ionization. The instrument consists of a thin-walled metal tube (the cathode), containing gas under diminished pressure, coaxial with a thin wire (the anode). A potential difference of about 1000 V is maintained between the anode and cathode. The incoming ionizing radiation produces ions in the tube, which cause a voltage pulse in the potential between the two electrodes, which in turn operates the counter. The instrument is often used in satellites and space probes for measuring cosmic rays.

Gemini The name given to a US space research programme as well as to the 12 space vehicles used in it. The Gemini project preceded the Apollo project. Gemini 1 and 2, launched in 1964 and 1965, were unmanned test flights; subsequent ones, launched 1965–66, had a crew of two who carried out various manoeuvres with the vehicles in space.

Geminids An important shower of meteors, rich in fireballs, whose radiant is in the constellation Gemini. The shower appears during the first half of December.

geocentric As viewed from or related to the Earth's centre, as with geocentric co-ordinates, geocentric parallax, geocentric system.

geocentric parallax or **diurnal parallax** The parallax of a body, usually a member of the solar system, determined from simultaneous positional measurements at two places on the Earth's surface a known distance apart. It is the angle subtended at the body between the direction to the centre of the Earth and the direction to the point of observation on the Earth's surface.

geodesy The study of the general figure and dimensions of the Earth.

geographical meridian or **terrestrial meridian** Any great circle passing through the geographical north and south poles.

geoid The terrestrial globe, whose figure cannot be described exactly by a simple geometrical figure such as a sphere or ellipsoid of revolution. It is the theoretical surface which would be formed if the sea, free from temperature, saline, and seasonal variations but subject to the effects of the Earth's rotation and gravitational attraction, were extended over the whole globe. The figure of the geoid is an oblate spheroid, which has been determined by measurements of gravity, and gives the equipotential surface of the Earth's gravitational field. It is used in theoretical calculations.

geomagnetism The branch of physics which is concerned with magnetic phenomena occurring in the Earth and its atmosphere. See also **magnetic storm**.

geophysics The study of the physical properties of the Earth, and related phenomena. It deals with atmospheric physics, geodesy, geomagnetism, oceanography, and seismology; the subject in fact overlaps almost every other scientific discipline.

German mounting see **telescope, mounting of**

Giacobinids A meteor shower associated with the Giacobini-Zinner Comet, and visible in some years about October 9/10. It is also known as the DRACONIDS because the radiant is in the constellation Draco, near ξ Draconis.

giant A star of great size, low density, and high absolute magnitude, whose spectral-luminosity classification places it in the giant sequence of the **Hertzsprung-Russell diagram**. Examination of their spectra enables giants to be distinguished from **supergiants**, **subgiants**, and **main-sequence** stars. Giant stars are in an advanced stage of evolution. See also **red giant**.

giant planets The planets Jupiter, Saturn, Uranus and Neptune, also known as the JOVIAN PLANETS. They are so called on account of their enormous masses as compared with the Earth. See entries under the individual planets for their characteristics.

gibbous A term applied to the Moon or to a planet when its illuminated portion is more than a semicircle and less than a full circle.

Gill, David (Aberdeen, June 12, 1843 – London, January 24, 1914) Astronomer at the Cape and a pioneer of astrometry, with J. C. Kapteyn he made a photographic survey of the southern sky, the results of which were published as the *Cape Photographic Durchmusterung*. He observed the transit of Venus (1874), and determined the solar parallax (1877) and the mass of Jupiter.

G line J. von Fraunhofer's designation for one of the absorption lines in the solar spectrum. It is in the indigo part of the spectrum and corresponds to superimposed lines of iron and calcium at 4308 Å. See also **Fraunhofer lines.**

globular cluster see **cluster**

globule see **nebula**

gnomon A rod, shaft, or pillar mounted perpendicularly to the horizon; it was used in ancient times to find the altitude of the Sun by measuring the length of the shadow cast on the horizontal base. It is the simplest form of sundial, which shows local solar time. In modern sundials the gnomon is the plate mounted perpendicular to the horizon and pointing to the celestial pole; it casts a shadow on the base-plate marked in hours.

Golden Number A number used in fixing the date of Easter. It is the number of a year in the **Metonic Cycle**, a period of 19 solar years almost equal to 235 lunar months. The year 1 B.C. was adopted as the date to commence the succession of 19-year cycles on which the ecclesiastical lunar calendars were based. To find the Golden Number divide the year by 19 and add unity to the remainder: if the remainder is zero the required number is 19; for example, the Golden Number for the year 1980 is 5, being the remainder of 1980/19 plus unity.

Gold, Thomas (Vienna, May 22, 1920–) A physicist who is Professor of Astronomy at Cornell University. He is known for his development of the **steady-state theory** of the expanding universe, and for his studies of dynamical problems connected with the solar system, etc.

Goodricke, John (Groningen, September 17, 1764 – York, England, April 20, 1786) An English astronomer known for his observations on variable stars, and particularly for being the first to give an explanation, as the result of analysing his observations of 1782, of the variation in brightness of **Algol**, namely, that it is an eclipsing binary.

Gould, Benjamin Apthorp (Boston, Massachusetts, September 27, 1824 – Cambridge, Massachusetts, November 26, 1896) An American astronomer who became Director of the Observatorio Nacional Argentina, Córdoba, where his most important work was done in observing and preparing zone catalogues of the stars in the southern sky.

Gould's belt A region of bright stars and gas which is inclined about 20° to the galactic plane and contains the greatest concentration of naked-eye stars of spectral types O and B. It is named after B. A. Gould who studied the region.

granulation The mottled appearance of the Sun's photosphere, caused by gases rising from the Sun's interior. When seeing conditions are very good the granulations can be resolved into small, nearly circular patches, called GRANULES, surrounded by darker areas. Individual granules last only a few minutes. The bright grains observed on the photosphere are **flocculi**.

graticule A system of reference marks or measuring scales consisting of parallel vertical wires, or crosswires, or a reticule (grid squares) which is placed in the focal plane of a telescope and covers the entire field.

grating see **diffraction grating**

gravitation The universal force of attraction existing between all particles of matter. NEWTON'S LAW OF GRAVITATION states that the force of attraction between two masses m_1 and m_2, a distance r apart, is

$$F = \frac{Gm_1m_2}{r^2}$$

G is the CONSTANT OF GRAVITATION, equal to $6{\cdot}6720 \times 10^{-11}\,\mathrm{N\,m^2\,kg^{-2}}$. Newton's Law, formulated in 1687, is the basis of celestial mechanics. The whole concept of gravitation was reinterpreted by Einstein in his Theory of General Relativity in terms of the curvature of space-time.

gravitational acceleration (*symbol: g*) The acceleration acquired by a body falling freely in the gravitational field of a planet towards the planet's centre of mass. Its value varies according to the latitude and altitude; it is affected also by planetary rotation. The standard reference value for the Earth's gravitational acceleration is defined as $9{\cdot}806\,65\ \mathrm{m\,s^{-2}}$.

gravitational collapse The proposed mechanism for the final stage in the evolution of a star, whereby it is compressed under the force of its own gravitational field with the release of an enormous amount of energy.

Stars begin this process of collapse when all their nuclear fuel is exhausted. During collapse the matter of which they are composed is compressed to an extremely high density (see **white dwarf, neutron star**). With the most massive stars the gravitational field becomes so strong that it prevents the escape of photons of radiation, moving at the velocity of light, so the object becomes virtually impossible to detect. When this stage is reached, at which the escape velocity of the star equals the velocity of light, a **black hole**, from which no light, matter, or signal can escape, is formed.

gravitational red shift or **relativistic red shift** A phenomenon which takes place in agreement with Einstein's General Theory of Relativity. Light during emission from a star has to do work to overcome the gravitational field of the star: there is a slight loss of energy which results in spectral lines being shifted towards the red because the decrease in energy results in an increase in wavelength. The phenomenon was first observed by W. S. Adams (1876–1956) at the Mt Wilson Observatory in 1925 for the white dwarf companion of Sirius.

great circle A circular intersection on the surface of a sphere by any plane passing through the centre of the sphere.

Great Nebula in Orion see **Orion Nebula**

green flash An atmospheric phenomenon observable under ideal conditions during the last seconds before sunset. Most of the sunlight has been absorbed in its long passage through the atmosphere and by ozone, so that only two narrow bands of wavelengths in the red and green are left. As the atmospheric refraction of light is greater for short wavelengths, the green light (flash) appears at the last moment of setting. The same phenomenon can occur at sunrise.

Greenwich A district situated on the River Thames about 6·5 km south-east of the City of London, where the Royal Observatory was founded in 1675. In 1884 an international conference held in Washington resolved to adopt 'the meridian passing through the transit instrument of the Observatory of Greenwich as the initial meridian of longitude.' This is the PRIME MERIDIAN (zero degrees) of longitude. The observatory was moved to Herstmonceux Castle, Sussex, between 1948 and 1958; its name was changed in 1948 to The Royal Greenwich Observatory.

Greenwich Mean Astronomical Time (G.M.A.T.) This is Greenwich Mean Time reckoned from zero hour beginning at noon on the same civil date. This avoids, for astronomical purposes, changing the date during the night when observations are being made. Many astronomical records prior to January 1, 1925 employed this system, which was discontinued on that date. To convert G.M.A.T. to U.T. add 12 hours.

Greenwich Mean Time (G.M.T.) Mean solar time for the meridian through Greenwich. From January 1, 1925 it was reckoned from midnight instead of midday and to avoid any risk of confusion with G.M.T. before 1925 the term **Universal Time** was introduced. However, G.M.T. continues in use in navigational publications and elsewhere.

Gregorian calendar see **calendar**

Gregorian telescope A reflecting telescope described by J. Gregory (1638–75) in his *Optica Promota* (London, 1663), five years before Isaac Newton constructed his first reflector. The primary mirror is parabolic and has a hole through the centre. The light received by this mirror is reflected to a concave ellipsoidal secondary mirror mounted on the same axis, and is reflected back through the hole in the primary mirror to an eyepiece. The image is erect. For astronomical purposes this form of telescope has fallen into disuse as it is difficult to mount and handle. It differs from the Cassegrain telescope in having a concave instead of a convex mirror for the secondary. See illus. on page 70.

Grubb, Howard (Dublin, July 28, 1844 – Monkstown, September 16, 1931) A British engineer who was the owner of a scientific instrument business which made seven of the astrographic refractors used in connection with the **Carte du Ciel** programme, besides

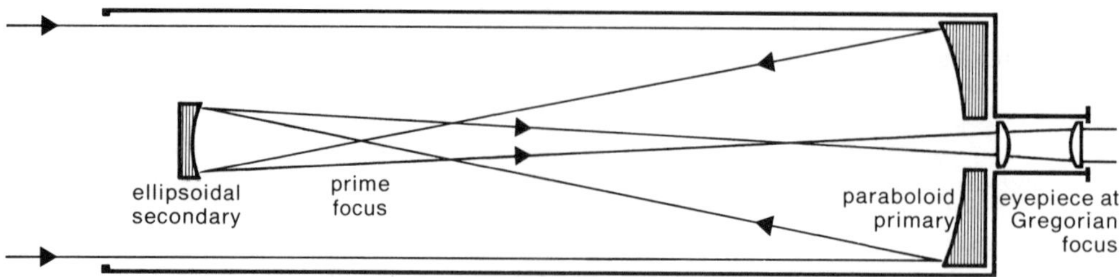

Light path in a Gregorian telescope.

many other medium-sized and large telescopes for leading observatories. On his retirement in 1925 Grubb's business was acquired by C. A. Parsons, who set up a new firm under the name of Sir Howard Grubb, Parsons & Co. Under its present name of Grubb, Parsons it made the 98 inch Isaac Newton telescope (the largest in western Europe), which was inaugurated at the Royal Greenwich Observatory, Herstmonceux Castle, on December 1, 1967.

G star see **stars, spectral classification of**

guider see **finder**

Gum Nebula An immense emission nebula lying in the Milky Way in the southern sky and containing the Vela pulsar and the remnant of the Vela X supernova. It is named after the Australian astronomer Colin Gum (1924–60), known for his survey of radio noise in the southern sky and his discovery of hydrogen emission regions in the southern Milky Way.

HI regions Regions in interstellar space where there are clouds of neutral atomic hydrogen. They are invisible as they do not emit radiation in the visible spectrum, but they can be detected by radio telescopes through their emission of 21-cm radiation.

HII regions Regions in interstellar space where there are clouds of hot ionized hydrogen produced by absorption of ultraviolet radiation from hot stars, or by other mechanisms including shock waves passing through the region.

Hades Name proposed for satellite IX of **Jupiter**. The name Sinope was ultimately adopted.

Hale, George Ellery (Chicago, June 29, 1868 – Pasadena, February 21, 1938) An American astronomer who became Director of the Mount Wilson Observatory. He is noted for his studies of the Sun and development of instruments for solar research. He improved the spectroheliograph and invented the spectrohelioscope. He discovered the dark hydrogen filaments in the Sun, proved that its magnetic and geographic poles do not coincide, established the existence of a magnetic field round the Sun, discovered the **Zeeman effect** in sunspots, and put forward a theory of the nature of sunspots based on their magnetic fields. His vision, foresight, and leadership made possible the realization at Palomar of the 200 inch reflector which bears his name.

Hale Observatories The name by which the Mount Wilson Observatory on Mt Wilson and the Palomar Observatory, about 50 km away on Palomar Mountain, both in South California, have been jointly called since 1970. Las Campanas Observatory, Chile, and Big Bear Solar Observatory, California are also now Hale Observatories.

Hall, Asaph (Goshen, Connecticut, October 15, 1829 – Annapolis, November 22, 1907) An American astronomer who worked on the orbital elements of planetary satellites, the rotation of Saturn, the mass of Mars, and on double stars. On August 11 and 16, 1877 he discovered the satellites Deimos and Phobos of the planet Mars.

Halley, Edmond (London, November 8,

1656 – Greenwich, January 14, 1742) The second Astronomer Royal, he made a survey of the southern sky at St Helena, which was published as *Catalogus stellarum australium* (1679). He observed the transit of Mercury (1677), discovered the proper motions of stars, the periodicity of comets, and the magnetic origin of the aurora borealis. It must also be remembered that it was Halley who persuaded Isaac Newton to publish his *Philosophiae naturalis principia mathematica* through the Royal Society, and that Halley financed the work himself and saw it through the press. He is commemorated by the *Halley Lecture* in astronomy, which is delivered annually at the University of Oxford.

Halley's Comet A comet observed by Edmond Halley in 1682. He deduced that it was a recovery of the comets seen in 1531 and 1607, and predicted its return for 1758. It has a period of 76 years and is due to return next in 1986. There are extensive references to this brilliant naked-eye object, not least in Chinese historical records, as its appearance in the past coincided with a number of notable events. From such records its past appearances have been identified back to the year 240 B.C. The comet is depicted on the Bayeux tapestry which records events connected with the Norman invasion and conquest of England in 1066.

halo 1. A bright whitish ring, complete or partial, around a celestial body, usually the Sun or the Moon. It is caused by refraction of light by ice crystals suspended in the atmosphere. **2.** The **galactic halo.** The spheroidal distribution of stars, globular clusters, and gas clouds in which the disc constituting our Galaxy is embedded. Its radius is about 15 kpc. It emits radio waves.

halo population stars see **Population II stars**

H and K lines Lines of wavelength 3968 Å and 3934 Å respectively in the spectrum of singly ionized calcium.

Harriot, Thomas (Oxford, 1560 – London, 1621) An English mathematician, 'the universal philospher'. So far as is known he was the first person in England to possess telescopes. He seems to have corresponded with Kepler. He made observations of the Moon. He observed the satellites of Jupiter prior to Galileo, who is usually credited with their discovery. He also observed the number and changes in appearance of sunspots before any mention of them by other observers. His findings remained in manuscript.

harvest Moon The Full Moon nearest the autumnal equinox (September 23). The Moon's orbit is then almost parallel with the horizon; consequently it rises only about 15 minutes later each evening instead of the usual half-hour or more, and so provides a succession of moonlit evenings which were formerly of help to harvesters. In the southern hemisphere the harvest Moon is the Full Moon nearest the vernal equinox (March 21).

Hawking, Stephen William (January 8, 1942–) Professor of Gravitational Physics, University of Cambridge. He was awarded the Hughes Medal of the Royal Society in 1976 'in recognition of his distinguished contributions to the application of general relativity to astrophysics, especially to the behaviour of highly condensed matter'. The ultimate fate of matter that has undergone **gravitational collapse** is one of the problems of modern astrophysics. Hawking has collaborated with R. Penrose in work on this subject: they were jointly awarded the Eddington Medal of the Royal Astronomical Society in 1975.

Hay, William Thomson (December 6, 1888 – April 18, 1949) An actor and keen amateur astronomer. He discovered on August 3, 1933 one of the largest and most persistent white spots ever recorded for the planet Saturn. He is author of *Through my Telescope* (1935).

H-D sequence The Harvard-Draper sequence, a classification of stellar spectra in order of decreasing surface temperature of the stars, whose colour ranges from blue through white and yellow to red. It was devised for use in the *Henry Draper Catalogue*. Originally the sequence had six classes designated *B, A,*

F, G, K, and *M,* but has since been extended and is now

$$W-O-B-A-F-G-K\begin{smallmatrix}\nearrow M-S\\ \searrow C\end{smallmatrix}$$

Classes are subdivided decimally by adding a numeral; letters may be prefixed or affixed to designate spectral anomalies. Two other classes have been added to the above, namely *P* for gaseous nebulae and *Q* for novae. See also **stars, spectral classification of**.

Hektor Minor planet 624. It is the largest and brightest of the **Trojans**. It is elongated in shape, and could possibly have been formed by the collision of two minor planets.

heliacal rising In modern usage, the rising of a celestial body simultaneously with the rising of the Sun. Formerly it meant the moment when a bright star, such as Sirius, could just be seen rising before the Sun. Observation of this event was of great importance in ancient times for agricultural purposes, the heliacal rising of Sirius indicating the approach of the annual flooding of the Nile.

heliocentric As viewed from or having relation to the Sun's centre, as with heliocentric co-ordinates (those which a body would have for an observer situated at the Sun's centre) and heliocentric system (of Copernicus). Compare **geocentric**.

heliocentric parallax The parallax of a body determined from observations of its position made six months apart, using the diameter of the Earth's orbit as the baseline.

heliographic latitude see **latitude**

heliographic longitude see **longitude**

heliometer A refractor with its objective lens cut across its diameter to give two semi-lenses, which can be moved relative to each other along the line of separation and rotated as a whole. The instrument was first made by J. Dollond (1754) and produces two images. In the case of two stars with very small separation, their images can be brought to coincidence and the separation measured by a micrometer. Similarly, in the case of the Sun or a planet, the edges of the opposite limbs can be brought to coincidence and the diameter of the object determined. The first measurement of the parallax of a fixed star (61 Cygni) was made by F. Bessel in 1838 using a heliometer. This once important instrument has been superseded by photographic methods.

heliostat see **coelostat**

helium One of the rare gases and the second most abundant element in the universe. On August 18, 1868 during an eclipse of the Sun, J. N. Lockyer (see **Lockyer, Joseph**) observed a bright orange-yellow line in the photospheric spectrum. The line could not be identified as coming from any known terrestrial element, so the existence of an unknown element in the Sun was postulated; it was given the name helium. In 1895 helium was isolated in the laboratory from terrestrial sources by W. Ramsay (1852–1916) and Lord Rayleigh (1842–1919). Helium has the atomic number 2, atomic weight 4·003, boiling point −269°C, and is used in laboratory work for producing very low temperatures. The Earth's atmosphere contains about 0·0005% by volume of the gas; the amount in the Sun and similar stars is about 24% by mass. Most of the helium in the universe is now thought to have formed in the first few minutes of the universe. Helium is also formed in stars during the nuclear reactions involving the transformation of four hydrogen atoms into one helium atom with the release of an enormous amount of energy (see **proton-proton reaction, carbon-nitrogen cycle**). This energy release accounts for the stars' luminosity.

Heraclides of Pontus (*c.* 388 B.C. – *c.* 315 B.C.) A Greek astronomer who originated the geoheliocentric system of the universe which was later developed by Tycho Brahe (see **Brahe, Tycho**).

Hermes A minor planet, about 1 km in diameter, discovered in 1937. On October 28 of that year it approached the Earth to within 800,000 km – the closest approach yet of any asteroid. Although its period is less than three years, it has not been recovered.

Herschel, Caroline Lucretia (Hanover, March 16, 1750 – Hanover, January 9, 1848) The younger sister of William Herschel, she was trained as a concert singer. She joined her brother William at Bath, becoming his housekeeper and colleague in astronomical work, editing and copying his papers, recording his observations, and preparing his catalogues. She became an independent observer, and discovered eight comets (priority of discovery in six of them). She was the first woman to achieve distinction in astronomy.

Herschelian-Cassegrain telescope A variation of the Herschelian telescope. The tilted primary parabolic mirror directs the light to a secondary hyperbolic mirror placed off the main axis of the instrument and located some way up the tube. The secondary directs the light to an eyepiece placed near the primary mirror. This system avoids the making of a hole in the primary mirror, as required in the standard **Cassegrain telescope**. The construction has been used for certain specialized fields of observation.

Herschelian telescope A reflecting telescope in which the primary parabolic mirror is tilted so as to bring the rays of light to the prime focus which is located towards the upper end of the tube where an eyepiece is placed. It was devised by William Herschel to reduce the high loss of light from the secondary mirror of speculum metal used in the Cassegrain and Newtonian reflectors. This type of instrument is not now in use.

Herschel, John Frederick William (Slough, March 7, 1792 – Collingwood, May 11, 1871) An English mathematician and astronomer who was the only child of William Herschel and assisted his father in observational work and in making telescopes. He made measurements on double stars and reexamined nebulae and star clusters in the northern sky. He systematically surveyed the southern sky from the Cape of Good Hope and published the results in 1847. His *Outlines of Astronomy* (1849) was a standard text-book for many decades. He compiled a *General Catalogue of Nebulae and Clusters...*, which in its revision by J. L. E. Dreyer is known as the N.G.C. and remains the standard reference catalogue for those objects. His ambitious *General Catalogue of 10,300 Multiple and Double Stars* was published posthumously.

Herschel, William (Friedrich Wilhelm Herschel) (Hanover, November 15, 1738 – Slough, August 25, 1822) A German-born musician and composer, who came to England in 1757 and later turned astronomer. He became a naturalized Englishman. He made telescopes and mirrors for his own use and for sale. He discovered the planet **Uranus**, which he took for a comet: its true planetary nature was proved by subsequent observation and computation of its orbit. He discovered two satellites of Uranus (1787) and of Saturn (1789) and observed the rotation of Saturn and its rings. He discovered many double stars, nebulae, and clusters, and published catalogues of them. He formed a true conception of the Milky Way as the plane of the disc-shaped stellar universe, and also noted the motion of the Sun towards a point in the constellation Hercules. He discovered and investigated the properties of infrared radiation (1800).

A Herschelian-Cassegrain telescope.

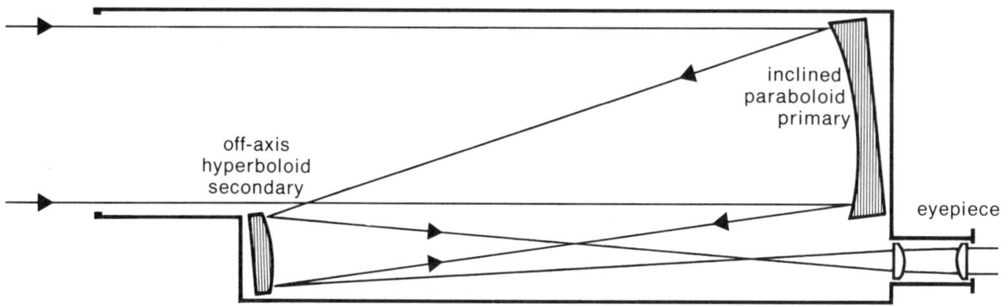

Sir William Hershel, the discoverer of Uranus. Painting by L. F. Abbott, National Portrait Gallery, London.

hertz (*symbol:* Hz) The derived SI unit of frequency. It is the number of repetitions of a regular occurrence in one second. British practice originally used cycles per second (1 Hz = 1 c.p.s.). It is named after Heinrich Hertz (1857–94), experimental demonstrator of electromagnetic waves.

Hertzsprung, Ejnar (Frederiksberg, October 8, 1873 – Roskilde, October 21, 1967) A Danish astronomer who became Director of the Leiden Observatory. He is best remembered for his study of the relationship between luminosity and spectral type of stars. By plotting one against the other he discovered the existence of giant, supergiant and main-sequence stars. This graph, now known as the **Hertzsprung-Russell diagram**, was developed independently and in a slightly different form by H. N. Russell (see **Russell, Henry**). Hertzsprung is also known for his work on double and variable stars.

Hertzsprung gap A region in the Hertzsprung-Russell diagram between the main sequence and the giant branch where there are noticeably few stars. Those that do appear there are RR Lyrae and other variable stars with spectral types from about A0 to G0.

Hertzsprung-Russell diagram (*abbrev.:* H-R diagram) A diagram, developed independently by E. Hertzsprung and H. N. Russell, on which the luminosity of stars is plotted against their spectral type; this is equivalent to plotting surface temperature or colour index against absolute magnitude. The diagram reveals a pattern in which most stars lie on a diagonal band, now called the **main sequence**, which includes our Sun. There is another region of **giant stars**, mainly M- and K-type stars, which is separated from the main sequence by the **Hertzsprung gap**. There are further groupings of **supergiants**, **subgiants**, **subdwarfs**, and a special class of **white dwarfs**. The H-R diagram has become an important aid to astrophysics for interpreting astronomical data; it has prompted various theories of stellar evolution.

Hesperus An old name for the planet Venus when visible in the western sky as the Evening Star.

Hestia Name proposed for satellite VI of **Jupiter**. The name Himalia was ultimately adopted.

Hevelius (or **Hevel**), **Johann** (Danzig, January 28, 1611 – Danzig, January 28, 1687) A German astronomer who was the last great astronomer to make observations with naked-eye sighting instruments for positional measurements.

Hewish, Antony (May 11, 1924–) A British astronomer who is Professor of Radio Astronomy in the University of Cambridge. With his team of assistants at the Mullard Radio Astronomy Laboratory, near Cambridge, he discovered (1967) the first **pulsar**, now known as CP 1919.

Hidalgo Minor planet 944, discovered by W. Baade in 1920. It has the second largest orbit, and the third highest eccentricity of any known minor planet.

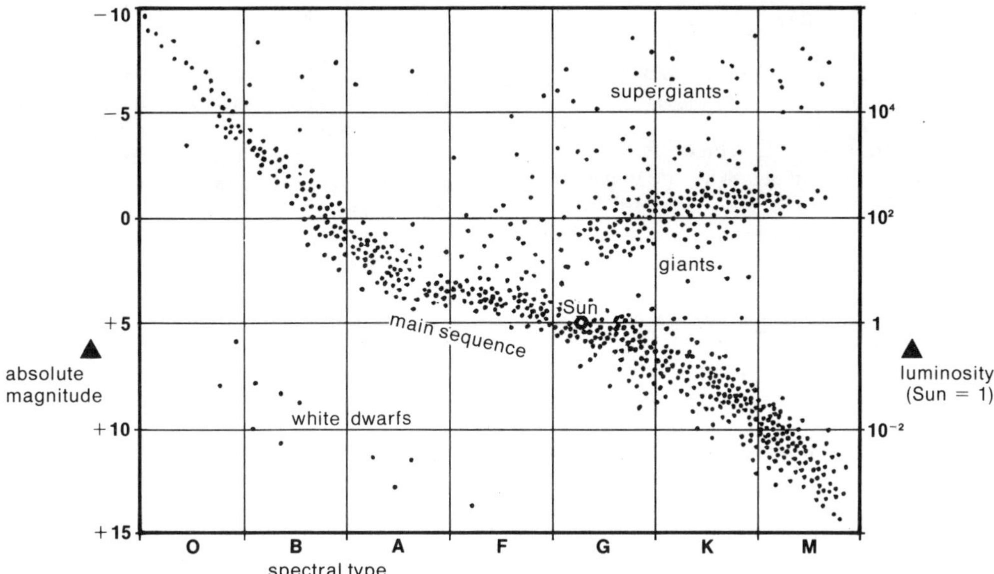

Hertzsprung-Russell diagram of the nearest and brightest stars.

hierarchical universe A cosmological system put forward by C. V. L. Charlier (1862–1934) consisting of clusters of galaxies, which are grouped into larger-scale clusters, which in their turn are within even larger clusters, and so on. The radius of a given cluster is greater than the square of the preceding cluster. Accordingly, this concept implies that the universe is infinite in respect of distance and mass.

high-velocity stars Mainly **Population II stars** of the galactic halo which do not share the rotation of the majority of stars of the Galaxy, including the Sun, around the galactic centre. Consequently, they seem to have large proper motions and high velocities relative to the Sun. Their orbits are usually eccentric and have a large inclination to the galactic plane. Arcturus, the brightest star in the constellation Boötes, is a typical example.

Hill, George William (New York, March 3, 1838 – April 16, 1914) An American astronomer and mathematician, known for his work on celestial mechanics. He developed theories relating to the motions of the Moon, Jupiter, and Saturn.

Himalia Satellite VI of Jupiter. See also **satellite**.

Hipparchos of Nicaea (2nd cent. B.C.) A Greek astronomer who made many accurate astronomical observations. He compiled a star catalogue (now lost but incorporated in Ptolemy's **Almagest**) giving co-ordinates and magnitudes. He discovered the precession of the equinoxes and also an irregularity in the Moon's motion.

H line J. von Fraunhofer's designation for one of the absorption lines in the solar spectrum; it is produced by singly ionized calcium and occurs at 3968 Å. See also **Fraunhofer lines**.

horizontal system see **co-ordinates**

Horrocks, Jeremiah (Toxteth, near Liverpool, *c.* 1619 – January, 1641) An English astronomer who made the first observation of a transit of Venus (November 24, 1639) and wrote an account, *Venus in Sole visa*, published by J. Hevelius in 1662 with *Mercurius in Sole visum*, written by the latter. Horrocks developed the theory of lunar motion.

horseshoe mounting see **telescope, mounting of**

Horsehead Nebula (NGC 2024) A dark nebula near ζ Orionis, so-called because of its characteristic shape as seen against a bright gaseous nebula.

hour angle The angle which the declination circle of an object makes with the meridian. It is usually measured westwards in the plane of the equator from the meridian and reckoned from 0° to 360°.

hour circle In the equatorial system of **coordinates**, a great circle perpendicular to the celestial equator and passing through the celestial pole as well as a specific point on the celestial sphere. It serves for the measurement of **Declination**. It is also the graduated circle fitted to the polar axis of an equatorially mounted telescope for the purpose of showing the **hour angle** of an object.

Hoyle, Fred (June 24, 1915–) An English astrophysicist distinguished for his work in cosmology, his research on the evolution of stars, and his development of the steady-state theory of the universe and a new theory of gravitation.

H-R diagram see **Hertzsprung-Russell diagram**

Hubble classification A classification of galaxies by shape or structure, introduced by Edwin Hubble in 1925. The three major categories are elliptical (E), spiral (S), and barred spiral (SB); each category has several subdivisions depending on the observable shape or structure of the galaxy (see diagram). Irregular galaxies were not included in Hubble's original classification.

Hubble constant (*symbol:* H_0) The linear rate at which the velocity of recession of galaxies changes with respect to distance from us, according to the Hubble law. The value of the constant as given by Hubble was about $500\,\mathrm{km\,s^{-1}\,Mpc^{-1}}$; this value has been reduced several times and in 1974 was calculated as $56{\cdot}9\,\mathrm{km\,s^{-1}\,Mpc^{-1}}$. This value leads to an age of $1{\cdot}7 \times 10^{10}$ years for the universe, assuming that all the galaxies started to recede from a common point in the past.

Hubble diagram A diagram in which the apparent magnitude of galaxies is plotted against the red shift of their spectral lines.

Hubble, Edwin Powell (Marshfield, Maryland, November 20, 1889 – San Marino, California, September 28, 1953) An American astronomer who is known for his discovery that many nebulae are extragalactic objects: he found that although some nebulae were clouds of gas lying within our Galaxy, many others were separate and distant star systems. He also established that these star systems, i.e. galaxies, are uniformly distributed. He instituted a classification of galaxies. He attributed the **red shift** of the spectral lines of galaxies (the HUBBLE EFFECT) to the recession of galaxies, and hence to the expansion of the universe, and established the velocity-distance relationship of recession. See **Hubble Law**.

Hubble Law The law proposed by Edwin Hubble in 1929 claiming a linear relationship between the distance (D) of galaxies from us

Hubble classification of galaxies.

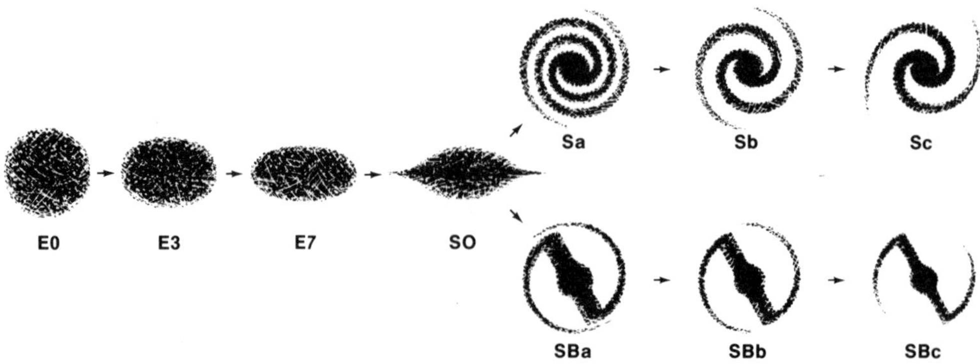

and their velocity of recession (V), deduced from the red shift exhibited by their spectra. The constant of proportionality is the **Hubble constant**. Thus

$$V = H_0 D.$$

Huggins, William (London, February 7, 1824 – London, May 12, 1910) An English astronomer who was a pioneer of stellar spectroscopic photography. He demonstrated the virtual absence of an atmosphere on the Moon. By spectral analysis he demonstrated that terrestrial elements are present also in stars, and confirmed the gaseous nature of a diffuse nebula. He investigated the motion of Sirius and other stars in the line-of-sight and discovered the first moving cluster (the Ursa Major cluster). He showed that the spectra of comets and meteors are similar to the spectrum of a hydrocarbon flame.

Hulst, Hendrik Christoffel van de (Utrecht, November 19, 1918–) A Dutch astronomer, distinguished for his research on interstellar matter and the solar corona. He predicted (1944) and discovered (1951) the emission of 21 cm radio waves by hydrogen.

Humason, Milton Lasell (August 19, 1891 – Mendocino, California, June 18, 1972) An American astronomer known for his work on galaxies. He discovered comet 1962 VIII.

hunter's Moon The full Moon which follows the **harvest Moon** and provides a succession of moonlit evenings in early October. In the southern hemisphere the hunter's Moon occurs in April/May.

Huygens, Christiaan (The Hague, April 14, 1629 – The Hague, June 8, 1695) A Dutch natural philosopher, mathematician, and astronomer, who made many important contributions to dynamics, optics, mathematics, and astronomy. He established the wave theory of light. He was concerned with the development of the telescope and introduced the convergent eyepiece. He made important astronomical observations on the rotation and oblateness of Mars, on Saturn's rings (1656), and on double stars. He discovered Titan, the largest of Saturn's satellites (1655). He also invented the pendulum clock (1656).

Hyades An open cluster of about 200 stars visible to the naked eye in the constellation Taurus. It is a moving cluster with a radial velocity of about $+36\,\mathrm{km\,s^{-1}}$, distant about 140 light-years. The star Aldebaran appears to belong, but is in fact much closer to us and does not form a constituent of the cluster.

hydrogen The lightest element and the one with the simplest structure. It has an atomic number 1 and an atomic weight 1·008. It is the most abundant element in the universe, being relatively uniformly distributed over the whole galactic system. It occurs in interstellar space as neutral hydrogen in **HI regions**, as ionized hydrogen in gaseous nebulae and **HII regions**, and as molecular hydrogen in dense clouds.

hydrogen-alpha line (*symbol:* Hα) The absorption or emission line of hydrogen occurring in the red portion of the spectrum with a wavelength of 6563 Å. It is Fraunhofer's line C. It is the third most conspicuous absorption line in the solar spectrum and is responsible for the pink colour of the solar chromosphere when viewed during a total eclipse. Light of this wavelength is often used for monochromatic study of the Sun with the spectrograph.

hydrogen-beta line (*symbol:* Hβ) The emission or absorption line of hydrogen occurring at 4861 Å in the spectrum of the Sun and other celestial bodies. It was designated the F line by J. von Fraunhofer (see **Fraunhofer lines**).

hydrogen ion, negative (H^-) A hydrogen atom which has acquired a second negative electron. These negative hydrogen ions are considered to contribute considerably to the continuous solar spectrum.

hydrogen radiation, 21 cm line Radiation emitted by neutral hydrogen at a wavelength of about 21 cm. Its existence at radio wavelengths was predicted in 1944 by H. C. van de Hulst. It was detected by three independent sets of observers in 1951. This emission line, at a frequency of 1420 megahertz, is caused by the electron in a hydrogen atom spontaneously changing its direction of spin so that the atom assumes a lower energy level. Although the change occurs extremely rarely, the number of neutral hydrogen atoms in the universe is so enormous that the radiation can be detected by radio telescopes. Regions of neutral hydrogen both in our Galaxy and beyond have thus been mapped. The corresponding line for deuterium has a wavelength of 91·6 cm.

hydrogen spectrum The spectrum of hydrogen, whose atom has the simplest structure of any (one proton as nucleus with one single orbiting electron), has been studied in great detail. The spectrum contains five principal sets of lines associated with different levels of energy of the hydrogen atom. They are the **Lyman series, Balmer lines, Paschen series, Brackett series,** and **Pfund series**. Each contains very many lines. Since hydrogen is the most abundant element in the universe, the spectra of celestial objects show many lines of that element, from which it is possible to draw conclusions concerning the nature of the objects.

hydrogen stars Alternative name for stars of spectral type A. They are so called because of the prominence of hydrogen absorption lines in their spectra.

hyperbola One of the conic sections formed by cutting a right circular cone at an angle steeper than its slope: its eccentricity is greater than unity. It is an open curve and is the path followed by a celestial body which passes but is not captured into an orbit around another body.

hyperbolic comet A comet whose orbital eccentricity seems to be greater than unity, so the orbit is hyperbolic. As the orbits in question have been observed only over a very short portion, it is likely that more accurate calculations would in fact show such orbits to be elliptical and of very long period.

Hyperion Satellite VII of **Saturn**. See also **satellite**.

Iapetus or **Japetus** Satellite VIII of **Saturn**. See also **satellite**.

Icarus Minor planet 1566, discovered by W. Baade in 1949. It has a small orbit and large eccentricity and is the only minor planet known whose perihelion is closer to the Sun than the orbit of Mercury. In June 1968 at opposition it was about six million km away from the Earth.

image tube see **camera**

immersion The disappearance through an eclipse or an occultation of a celestial body by passing behind or into the shadow of another. Two kinds of occultation are distinguished, firstly when the Moon temporarily obscures a star or occasionally a planet, and secondly when planetary satellites are temporarily obscured by the planet itself, as can happen with the satellites of Jupiter and Saturn. The reappearance of the celestial body after being eclipsed or occulted is termed EMERSION.

inequality The modification or calculation made necessary by the unequal motion of a planet or satellite.

inferior conjunction One of the planetary **aspects**. It is the position in which the Sun, the planet (Mercury or Venus) and the Earth are in a straight line, the planet lying between the Sun and Earth. An aspect of this kind is possible only for a celestial object lying nearer to the Sun than the Earth.

inferior culmination see **culmination**

inferior planet see **planet**

infrared radiation Electromagnetic radiation beyond the visible red portion of the spectrum. Infrared radiation is not visible; its wavelengths are shorter than about 8000 Å and extend to radio wavelengths. Studies of infrared radiation from space are made by means of ground-based telescopes and instruments carried in satellites, rockets, etc. These studies have revealed thousands of sources emitting infrared radiation, most of which lie within the Galaxy.

Innes, Robert Thorburn Ayton (Edinburgh, November 10, 1862 – March 13, 1933) A British astronomer who became Director of the Transvaal Observatory, Johannesburg, and subsequently Union Astronomer. He specialized in the study of double stars and compiled the *Southern Double Star Catalogue* (Johannesburg, 1927).

intensity interferometer The apparatus at Narrabri, New South Wales, Australia, which uses a photometric method for examining interference fringes. This phenomenon is used to increase the resolving power of telescopes in connection with measuring the diameters of stars or the separation of double stars. The linear diameter of a star can then be calculated if its distance is known from measurements of trigonometrical parallax. The apparatus was developed by R. Hanbury-Brown in collaboration with R. Q. Twiss.

interference A phenomenon observed when trains of light waves, radio waves, or other electromagnetic radiation from the same source, and hence of the same wavelength and amplitude, are superimposed. The resulting effect of combining two such beams will depend upon the phase relationship of the two beams when they combine. If they are fully in phase at some point, then the amplitude of the resultant beam will be equal to the sum of the amplitudes of the individual beams; if they are completely out of phase, then the amplitude of the resultant beam will be equal to the difference of the individual amplitudes. In the former case there will be reinforcement or enhancement of the image produced by the combined beams; in the latter case there will be cancellation or extinction of the image. The full image produced by the combined beams will be a regular pattern of alternating high and low intensities.

The UK Infrared Telescope which is positioned at an altitude of 4200 m above sea level near the summit of Mauna Kea in Hawaii.

interference filter An optical device utilizing the principle of **interference** to isolate a narrow spectral band. It is used for monochromatic photography of the Sun, for example, in the light of the Hα line.

interferometer, radio see **radio interferometer**

interferometer, stellar see **stellar interferometer**

intergalactic medium Matter occurring in the regions between galaxies. It has been detected, in the form of ionized gas, in clusters of galaxies.

International Geophysical Year The period July 1, 1957 to December 31, 1958, during which there was a co-operative enterprise by 60 nations to make an intensive study of the whole Earth by simultaneous observations. Studies were made of the Earth's fluid envelope in respect of its globe, glaciers, ionosphere, aurorae, magnetism, cosmic rays, radioactivity, etc. The first artificial satellites were launched during this year.

International Polar Sequence A group of about 150 stars visible from all observatories in the northern hemisphere. The stars have magnitudes from 2·5 to 20 and serve as standards of reference for brightness. They have the advantage that at any given place they are observable with little change in elevation.

International Quiet Sun Years The years 1964 and 1965, when the Sun was at its minimum activity, during which time solar and geophysical conditions were intensively studied by co-operation between observatories on an international basis.

interplanetary matter That which occurs in space between the planets. It comprises very tenuous gas (mainly neutral and ionized hydrogen), atomic particles ejected from the Sun via the solar wind (mainly protons and electrons), dust, and meteoroids.

interstellar extinction see **interstellar matter**

interstellar dust grains see **interstellar matter**

interstellar gas see **interstellar matter**

interstellar matter The matter contained in the regions between celestial objects in the Galaxy. The space is not void as was originally thought. It contains clouds of INTERSTELLAR GAS, mainly hydrogen with some helium, and a small percentage of dust particles, now called INTERSTELLAR DUST GRAINS, whose composition and structure still await elucidation. The hydrogen is mainly in a relatively cool neutral form at a temperature of 10–100 K. Some very much hotter and more diffuse regions of ionized gas also exist together with dense clouds of molecular hydrogen and other molecules. The clouds of gas and dust are largely confined to the plane of the Galaxy and tend to be concentrated in the spiral arms. The density of these clouds is very low, being 10^7–10^9 atoms per m^3, but nevertheless they have an obscuring effect on the light of stars shining from behind: some clouds are thick enough to be almost opaque. These clouds are responsible for the reddening of starlight which passes through them; the phenomonon is known as INTERSTELLAR EXTINCTION. The spectra of remote stars show absorption lines superimposed on the stellar spectra. These lines are produced by interstellar dust and gas. Silicon, magnesium, iron, calcium, titanium, and other elements have been detected from these absorption lines.

interstellar molecules Molecules of compounds, mainly organic, that have been identified by spectroscopy primarily at microwave frequencies. The first molecules, including the cyanogen radical (CN) and the methylidyne radical and ion (CH, CH^+), were detected in the 1940s at optical frequencies. The hydroxyl radical (OH) was the first (in 1963) to be identified by radio astronomy. The list of identified radicals has been greatly extended since then, and includes hydrogen cyanide (HCN), methyl alcohol (CH_3OH), formaldehyde (H.CHO), ammonia (NH_3), carbonyl sulphide (COS), and many others.

intrinsic variable A star whose radiation of light is in fact variable in intensity, as opposed to an EXTRINSIC VARIABLE whose change in radiation is the result of something external to the star, as in the case of eclipsing binaries.

Io Satellite I of **Jupiter**. See also **satellite**.

ionization The formation of ions. Ions are atoms or molecules that have lost one or more electrons from their outer orbital and become positively charged. 'Doubly ionized' and 'triply ionized' mean that two or three electrons respectively have been lost by the atom or molecule.

ionization temperature see **celestial bodies, temperature of**

ionosphere A region in the Earth's **atmosphere** extending from about 60 km to 500 km above the surface, in which atoms and molecules are ionized by ultraviolet radiation and X-rays from the Sun. The degree of ionization is greatly affected by solar activity. The presence of charged particles causes the ionosphere to reflect radio waves of long wavelengths although shorter wavelengths can pass through undisturbed. The ionosphere is usually divided into three layers: the D-LAYER, 60 to 110 km above the Earth, the E-LAYER (or HEAVISIDE LAYER), 110 to 160 km high; the F-LAYER (or APPLETON LAYER), above 160 km. The reflecting power of these layers makes long-range broadcasting and telephony possible up to frequencies of about 30 megahertz. The radio waves and microwaves that pass through the ionosphere are used in radio astronomy and satellite communications; they make up the **radio**

window. Their frequency ranges from about 20 megahertz to about 100 gigahertz. The free electrons formed as a result of ionization do at times have serious adverse effects on the reception of radio signals.

iris diaphragm An adjustable obstruction placed in an optical system to cut off marginal rays of light. It consists of a number of overlapping leaves of thin metal. One end of each leaf is located on the periphery of the mount. The other is located on a circular ring, which on rotation moves the centre of each leaf towards or away from the axis of the optical system thereby reducing or enlarging the opening for the passage of light.

irradiation A physiological phenomenon by virtue of which any bright object viewed against a dark background appears larger than it really is. The effect can introduce errors into telescopic measurements. The errors can be reduced by using an instrument of large aperture, and by illuminating the field of the eyepiece so as to reduce the contrast. A black object viewed against a bright background appears smaller in size.

irregular variable A variable star whose variation in emission of light does not conform to any regular or predictable pattern. U Geminorum is an example.

Isaac Newton telescope The versatile 98-inch reflector which was originally sited at the Royal Greenwich Observatory, Herstmonceux, but is in the process of being removed to the Northern Hemisphere Observatory in the Canary Islands, a new observatory for British and other astronomers. The Pyrex glass blank for the primary mirror was cast by the Corning Glass Company in preliminary preparation for casting the 200-inch mirror for the Hale telescope at Palomar. The telescope was built by the firm of Grubb Parsons, and was inaugurated by Queen Elizabeth II on December 1, 1967

island universe An obsolete term for the Milky Way System and other galaxies.

isostasy The plasticity of the surface crust of a planet.

jansky (*symbol:* Jy) The unit of flux density of radio emission, adopted by the IAU in 1973. It is named in honour of Karl Jansky.

$$1\ \text{Jy} = 10^{-26}\ \text{W m}^{-2}\ \text{Hz}^{-1}.$$

Jansky, Karl Guthe (Oklahoma, October 22, 1905 – Red Bank, New Jersey, February 14, 1950) An American radio engineer whose discovery published in 1932 that a source of radio static was located in the constellation of Sagittarius led directly to the development of radio astronomy.

Janssen, Pierre Jules César (Paris, February 22, 1824 – Meudon, December 23, 1907) A French astronomer who became Director of the observatory at Meudon. He was noted for his spectroscopic observations of the Sun and planets. He also made a photographic study of the Sun and invented a method of observing solar prominences in daylight.

Janus Satellite X of Saturn. See also **satellite**.

Jeans, James Hopwood (Ormskirk, September 11, 1877 – Dorking, September 16, 1946) A British mathematician and physicist who is noted for his work on stellar dynamics, the evolution of stars, double-star systems, and spiral galaxies, and the origin of the solar system (from matter drawn out of the Sun by a passing star). He is author of *Problems of Cosmogony and Stellar Dynamics* (1919).

Jeffreys, Harold (Fatfield, Co. Durham, April 22, 1891–) An English mathematician, astronomer, and geophysicist. He is distinguished for his investigations on the origin of the solar system, developing J. H. Jeans's theory thereon, and also on geophysical problems and the travel times of seismic waves. His 70th birthday was commemorated by establishing *The Harold Jeffreys Lecture*, which is devoted to subjects of geophysical interest and is an annual lecture to the Royal Astronomical Society. It is endowed out of the proceeds from sales of *The Earth Today* (1961), published as a tribute to Jeffreys.

Jewel Box The open cluster κ Crucis (NGC 4755) in the southern constellation

Crux. The name derives from the varied colours exhibited by the stars in this cluster.

Jones, Harold Spencer (London, March 29, 1890 – London, November 3, 1960) An English astronomer who was appointed Astronomer Royal. He is distinguished for his determination of the solar parallax and hence the **astronomical unit**, and of the mass of the Moon, the oblateness of the Earth, and the motions of the Sun, Moon, and planets.

joule (*abbrev.:* J) A derived unit in the SI. One joule is the energy which is equal to the work done when a force of one **newton** is displaced through a distance of one metre in the direction of the force.

Jovian planets The four giant planets, namely, Jupiter, Saturn, Uranus, and Neptune.

Julian calendar see **calendar**

Julian Day A means of providing a continuous measure of time, so obviating the need for division into years and months. The Julian Day and JULIAN PERIOD were conceived in 1582 by the French chronologist Joseph Justus Scaliger (1540–1609). They were so-called by Scaliger in memory of his father, Julius Caesar Scaliger, and are not to be confused with the Julian calendar devised by Julius Caesar. The beginning of the Julian period was taken as that day on which the solar cycle of 28 years, the lunar cycle of 19 years, and the cycle of indiction comprising 15 years all began together: this beginning was found to be in 4713 B.C. The Julian period contains $28 \times 19 \times 15 = 7980$ Julian years. The Julian Day Numbers count the number of days that have elapsed at Greenwich Mean Noon by a given day since the beginning of the epoch on 4713 B.C. January 1 (Julian calendar), which is defined as day zero. The Julian Date used in astronomy, for example in records concerning the luminosity of variable stars, is given by the Julian Day Number for the preceding noon at Greenwich followed by the fraction of the mean solar day that has elapsed. Thus, the Julian Date of an observation made at 6 p.m. on 1978 September 8 is 2,443,760·25, the Julian Day Number being 2,433,760.

It is to be noted that in historical usage the year 1 B.C. is followed by the year A.D. 1: in astronomical usage A.D. 1 is preceded by year zero, which in turn is preceded by year −1 (corresponding to the year 2 B.C. in historical usage). Thus

$$1 \text{ A.D.} \equiv \text{year} + 1$$
$$1 \text{ B.C.} \equiv \text{year } 0$$
$$2 \text{ B.C.} \equiv \text{year} - 1$$

To convert years B.C. to astronomical years, subtract unity and prefix the minus sign.

Juno Minor planet 3, discovered by K. L. Harding (1765–1834) in 1804.

Jupiter The largest and nearest of the giant planets. Its mass is greater than that of all the other planets and satellites in the solar system put together. It has a rapid axial rotation which has caused it to become flattened, its equatorial diameter greatly exceeding its polar diameter. Its low density ($1{\cdot}33\,\text{g cm}^{-3}$) indicates that it is composed primarily of hydrogen and helium. The main data concerning Jupiter are given in the table. It is one of the brightest objects in the sky, and is therefore a conspicuous naked-eye object. When viewed through a telescope, the yellowish elliptical disc is seen to be crossed by a number of continually changing bands of clouds, alternately brownish-red and bright yellow. The darker bands are called BELTS while the lighter ones are known as ZONES. Spots and streaks occur in the bands. The most distinctive feature is the RED SPOT, which was observed by R. Hooke in 1664, since when it has disappeared and reappeared several times. It has been observed regularly since 1831, during which time it has changed shape and colour. It is a vortex of cold anticlockwise-rotating clouds that does not stay fixed with respect to the surroundings, but rather floats, with variations in height, above the surrounding cloud. The Red Spot is surrounded by a lighter region called the HOLLOW. Another prominent feature in the same latitude was the SOUTH TROPICAL DISTURBANCE, which was first observed in 1901, and subsequently disappeared in 1940 after having exhibited various changes in

A view of Jupiter from a Voyager space probe taken in 1979.

length, colour, and structure. Considerable disturbance has been observed at times in the equatorial belts.

Spectral analysis of the light from the planetary disc shows it to be reflected sunlight with absorption lines that indicate the presence of hydrogen, helium, methane, and ammonia in the Jovian atmosphere; the cloud zones are thought to be composed of crystals of frozen ammonia, while sulphur compounds and other complex molecules occur in the cloud belts. There is no solid planetary surface beneath this atmosphere. Jupiter is a source of continuous and sporadic radio emissions. The planet also has a strong magnetic field and an extensive magnetosphere within which lie the orbits of the five innermost satellites. Radiation belts, thousands of times more intense than the Earth's **Van Allen belts**, occur in the magnetosphere. Jupiter radiates about 2·5 times more heat than it receives from the Sun; it must therefore possess a high internal temperature. This is estimated at about 50,000 K, and is presumably the result of gravitational contraction. The nature of the planetary core is debatable: it may consist of metallic hydrogen surrounding a central mass of iron and silicates, this whole core being in turn surrounded by liquid hydrogen.

A great deal of information has been obtained about the physical nature of the planet from the U.S. **Pioneer** and **Voyager** space probes.

Globe	
Diameter (equatorial)	142,800 km
Diameter (polar)	133,500 km
Density (water = 1)	1·33 g cm^{-3}
Mass	1·90 × 10^{30} g
Volume	1·43 × 10^{18} km^3
Sidereal period of axial rotation (equatorial)	9^h 50^m 30^s
Escape velocity	60·22 km s^{-1}
Albedo	0·73
Inclination of equator to orbit	3° 04′
Surface temperature	173 K (maximum)
Surface gravity (Earth = 1)	2·643

Orbit	
Semi-axis major	5·203 A.U. = 777·8 × 10^6 km
Eccentricity	0·0484
Inclination to ecliptic	1°·3
Sidereal period of revolution	4332·59 d
Mean orbital velocity	13·05 km s^{-1}
Mean synodic period	399 d

Jupiter has 14 satellites – more than any other planet. The four bright ones, Io (I), Europa (II), Ganymede (III), and Callisto (IV), are often referred to as the GALILEAN SATELLITES in honour of Galileo, who discovered them. The remaining satellites are faint objects. Satellites I–V form a compact group in orbit close to Jupiter and in the planet's equatorial plane; satellites VI, VII, X, and XIII form a second group further away; the remaining satellites form a third group at an even greater distance from the planet. It has been suggested that the third group consists of minor planets that have been captured by Jupiter. Alternatively it has been proposed that the second and third groups are fragments of two larger satellites. For data on the satellites see the table at **satellite**.

Much has been learnt of the Galilean satellites and the satellite Amalthea (V) from the Voyager probes which flew past Jupiter in March and July 1979. The four Galileans are markedly different in appearance. Io, with a mottled red-orange-brown surface devoid of craters, was found to be volcanically active. Europa has a smooth featureless surface covered in a network of cracks; this surface is possibly a thick layer of ice concealing any topographical features. Ganymede is the largest Jovian satellite. Its surface exhibits dark heavily cratered regions crossed with bands of brighter, younger terrain. Callisto, which is the least dense of the Galileans, has been found to be the most cratered body known.

Inside the orbit of the greatly elongated Amalthea, Voyager 1 discovered a ring of particles about 55,000 km above the cloud level and a few tens of km thick. Jupiter is therefore the third planet found to have a ring system.

Kant, Immanuel (Königsberg, April 22, 1724–Königsberg, February 12, 1804) A German philosopher who propounded in his work *Allgemeine Naturgeschichte und Theorie des Himmels* (Königsberg, 1755) the earliest **nebular hypothesis** for the origin of the solar system. According to this hypothesis a rotating primeval nebula composed of gas and cosmic dust gradually condensed under the action of gravitation to become a disc in which the planetary bodies in orbit around the central condensation, the Sun, were evolved.

Kapteyn, Jacobus Cornelius (Barneveld, January 19, 1851 – Amsterdam, June 18, 1922) A Dutch astronomer who did pioneering work on the motions and distributions of stars. He collaborated with D. Gill in the preparation of the *Cape Photographic Durchmusterung*. He discovered (1904) two streams of stars moving in opposite directions in the plane of the Milky Way, and discovered the effect known as star-streaming, which is the observed effect of the rotation of the Galaxy. Through his method of selected areas he originated statistical astronomy (1906). He also discovered the star named after him.

Kapteyn selected areas The 206 random areas chosen by J. C. Kapteyn for making statistical counts of stars for the purpose of ascertaining the shape of the Galaxy.

Kapteyn's star (HD 33793) A nearby red subdwarf star with the extremely high radial velocity of +242 km s^{-1}. It was discovered by J. C. Kapteyn.

K-component (K is from the German *Kontinuum*) The inner part of the solar corona, which gives a continuous spectrum.

κ-Cygnids A meteor shower that occurs between August 19 and 22.

Keeler, James Edward (La Salle, Illinois, September 10, 1857 – San Francisco, August 12, 1900) An American astronomer who became Director of Lick Observatory. He is known for his work on the rotation of

Saturn's rings, and confirmed J. C. Maxwell's theory that the rings consist of meteoritic particles. He carried out much observation on nebulae, and showed their close relationship to stars.

K-effect or **K-term** An unexplained effect discovered by J. C. Kapteyn and E. B. Frost in 1910. When the radial velocities of O and B stars are averaged over the whole sky and are corrected for the Sun's motion, the value is not reduced to zero but is positive.

kelvin One of the base units of the SI. It is the unit of thermodynamic temperature, and is the fraction 1/273.16 of the thermodynamic temperature of the triple point of water (the temperature at which ice, water, and water vapour are in equilibrium). A temperature interval of one kelvin has the same magnitude as one degree Celsius, and is now designated K and not °K. The triple point of water is at 273·16 K. See also **temperature scale**.

Kelvin (Baron Kelvin of Largs) see **Thomson, William**

Kelvin contraction or **Kelvin-Helmholtz contraction** The decrease in size of a star through radiation of thermal energy at the expense of potential energy. It is now believed that this is an evolutionary stage prior to a star moving into the main sequence of the **Hertzsprung-Russell diagram**.

Kepler, Johannes (Weil der Stadt, December 27, 1571 – Regensburg, November 15, 1630) A German mathematician and astronomer renowned for his discovery of the fundamental laws of planetary motion named after him. He wrote the *Mysterium Cosmographicum* (Tübingen, 1596) to show the simplicity of the Copernican theory of the solar system. In 1600 he left Graz, where he taught mathematics, on account of religious persecution. He went to Prague to work with Tycho Brahe whom he succeeded in 1601 as Imperial Mathematician with the task of completing the **Rudolphine Tables**. He wrote notable works on optics – *Ad Vitellionem Paralipomena* (Frankfurt, 1604) and *Dioptrice* (Augsburg, 1611), which deals with the theory of the telescope. *The Astronomia Nova* (Prague, 1609) contains the first two of his laws of planetary motion; the third law appeared in his *Harmonices Mundi Libri V* (Linz, 1619).

Kepler's Laws The three fundamental laws governing the motions of the planets around the Sun, enunciated by J. Kepler in 1609 and 1619 and based on observations made by Tycho Brahe.

Law 1 deals with the shape of a planetary orbit: the orbit of each planet is an ellipse with the Sun at one of the foci.

Law 2 explains the varying speed of planetary motion such that the planet moves faster the nearer it is to the Sun: the line joining the planet to the Sun (i.e. the radius vector) sweeps out equal areas in equal times. This is known as the law of areas.

Law 3 relates the size of the orbit to the period of revolution: the square of the period of revolution in orbit is directly proportional to the cube of the mean distance of the planet from the Sun.

Kepler's star (3C 358) A supernova that appeared in October 1604 in the constellation Ophiuchus and remained visible to the naked eye until March 1606. It was studied by J. Kepler, who gave an account of it in his *De stella nova in pede Serpentarii* (Prague, 1606). Evidence of the explosion is still detectable.

kilogram (*symbol:* kg) The unit of mass in the SI. It is equal to the mass of the international prototype of the kilogram, which is a cylinder of platinum-iridium preserved at the Bureau International des Poids et Mesures, Paris.

kiloparsec (*symbol:* kpc) One thousand parsecs, equal to 3261·6 light-years.

kinetheodolite An instrument of greater accuracy than a **theodolite**, designed primarily for use in trials of high-speed missiles. Two observers are required, one to control the azimuth and the other the altitude. Photographs are taken automatically on film, which also records the azimuth and altitude readings, the time, and an image of the object under observation. The instrument has been used to track artificial satellites. When used for this purpose it takes about five photographs per second; the direction of the object as measured is correct to 0·01° and the time to 0·01 second.

Kirchhoff's Laws Two laws of spectroscopy put forward in the 1850s by G. R. Kirchhoff (1824–87) and R. W. Bunsen (1811–99). The first states that each chemical element gives a characteristic series of spectral lines. The second states that each chemical element absorbs the radiation which it is capable of emitting. This gives rise to the phenomenon of dark spectral lines (known as reversal of lines) observed by J. von Fraunhofer. KIRCHHOFF'S RADIATION LAW, which is of importance for astrophysics, states that the ratio of emissivity to absorptivity at the same frequency of radiation is independent of the properties of the body, and is a function only of frequency and temperature.

Kirkwood, Daniel (Harford County, Maryland, September 27, 1814 – Riverside, California, June 11, 1895) An American astronomer known for his work on commensurabilities and resonance effects in the solar system. The KIRKWOOD GAPS, named after him, are regions in the minor-planet zone where minor planets are absent. The period of a planet in orbit at these distances would be a simple fraction of that of Jupiter, with the result that the cumulative effect of Jupiter's perturbations would force the planet into another orbit. He suggested that the non-uniform distribution of particles in the ring system of Saturn was caused by the perturbing effect of its satellites, in particular Mimas whose period is twice that of a body at the distance of the Cassini division.

K line see **calcium K line**

KL nebula An infrared nebula in the constellation Orion discovered by D. E. Kleinmann and F. J. Low in 1967.

Kramers' opacity The opacity of stellar material as a function of chemical composition. It is named after H. A. Kramers (1894–1952), a Dutch physicist, who worked on atomic spectra.

Krzemiński's star The optical component of **Centaurus X-3**, identified in 1974 by W. Krzemiński.

K star see **stars, spectral classification of**

K term see **K-effect**

Kuiper, Gerard Peter (Harencarspel, Netherlands, December 7, 1905–) An astronomer of Dutch birth who became a naturalized American. He is known for his studies of lunar and planetary surface features. He discovered satellite V (Miranda) of Uranus in 1948, and satellite II (Nereid) of Neptune in 1949. The KUIPER BANDS are features in the spectra of Uranus and Neptune at about 7500 Å which have been identified as belonging to methane.

Lacaille, Nicolas Louis de (Rumigny, May 15, 1713 – Paris, March 21, 1762) A French astronomer known for his work in positional astronomy. The results of his surveys of the sky in the southern hemisphere were published in his *Coelum australe stelliferum* (Paris, 1763). He added 14 constellations to the southern sky and made the first measurement of a South African arc of meridian.

Lagrange, Joseph Louis (Turin, January 25, 1736 – Paris, April 10, 1813) A French-Italian mathematician noted for his contributions to theoretical mechanics. In astronomy he studied the three-body problem, the Moon's libration, the motion of Jupiter's satellites, and cometary problems.

Lagrangian points One of the special cases in which a strict algebraic solution of the **three-body problem** is possible, as was

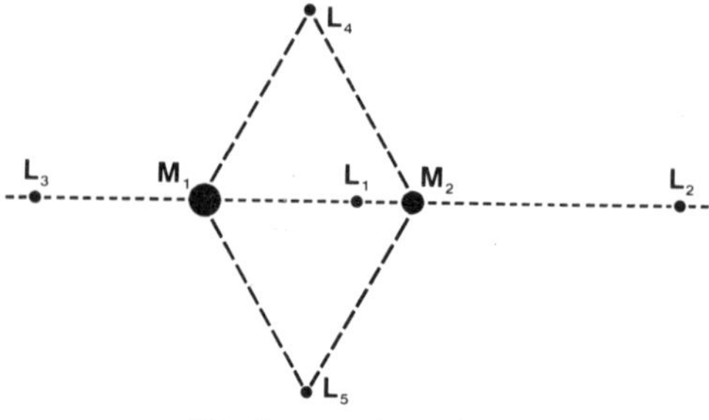

The Lagrangian points.

demonstrated by Joseph Lagrange in 1772. In this case one of the bodies is in one of the five points of libration with respect to the other two bodies. The five points lie in the orbital plane of two bodies, each of large mass, in circular orbits round a common centre of mass, with a third body of negligible mass in equilibrium. Three of the five points lie on the straight line joining the centres of mass of the two main bodies; the other two points form equilateral triangles with the main masses. M_1 is more massive than M_2. Points L_1, L_2, L_3 are unstable, and their positions are determined by the ratio of M_1 to M_2. Points L_4 and L_5 are stable and form equilateral triangles with M_1 and M_2.

The configuration is exemplified by the Sun (larger mass), Jupiter (smaller mass) and a minor planet (negligible mass) situated at the vertex of an equilateral triangle. In this instance one of the Lagrangian points leads Jupiter by 60°, the other follows by the same amount (see **Trojans**).

Lalande, Joseph Jérome le François de (Bourg, July 11, 1732 – Paris, April 4, 1807) A French astronomer noted for his planetary tables, his account of the transit of Venus, his editorship of the *Connaissance des Temps*, and numerous writings on astronomy. He observed Neptune in 1795, 51 years before its optical discovery by J. G. Galle, without realizing it was a planet.

Laplace, Pierre Simon (Beaumont-en-Auge, March 23, 1749 – Paris, March 5, 1827) A French mathematician and astronomer noted for his contributions to celestial mechanics, in particular his great work *Traité de Mécanique Céleste* (Paris, 1799–1825). He studied the perturbations of planets and satellites, the shape and rotation of Saturn's rings, and the stability of the solar system. In 1796 he propounded his historically important nebular hypothesis on the origin of the solar system, which has now been shown on theoretical grounds not to be possible. Modified forms, however, form the basis of most modern theories on the subject.

Large Magellanic Cloud see **Magellanic Clouds**

laser A device which operates in a similar manner to a maser and emits a narrow beam of powerful monochromatic coherent radiation at optical, infrared and ultraviolet wavelengths. Lasers have been used to give a precise measurement of the Moon's distance from Earth. The word is an acronym from *l*ight *a*mplification by *s*timulated *e*mission of *r*adiation.

Lassell, William (Bolton, June 18, 1799 – Maidenhead, October 5, 1880) An English brewer and keen amateur astronomer. He discovered Triton, a satellite of Neptune, on October 10, 1846, only 17 days after the discovery of the planet itself on September 23. He discovered also Hyperion, which is satellite VIII of Saturn (independently of G. P. Bond in America), and Ariel and Umbriel, the innermost two satellites of Uranus. He was the first to use the equatorial system of mounting for reflecting telescopes.

Last Quarter One of the Moon's phases when it has almost reached western quadrature and is half illuminated as seen from the Earth.

late-type stars Stars of spectral types K, M, S, and C, which are near the end of the Harvard-Draper sequence.

latitude A concept in systems of co-ordinates. It is the angular distance of a body measured from some reference point or plane. CELESTIAL (or ECLIPTIC) LATITUDE is the angular distance of a celestial body from the ecliptic measured along a secondary to the ecliptic. It is measured from 0° to 90°, positive northwards, negative southwards. GALACTIC LATITUDE is the angular distance of a body from the galactic plane. It is measured from 0° to 90° from the galactic equator, positive to the north galactic pole, negative to the south galactic pole. GEOCENTRIC LATITUDE is the angle subtended at the Earth's centre by the arc of meridian contained between the place in question and the equator. HELIOCENTRIC LATITUDE is the latitude of a body measured with reference to the centre of the Sun. It is the latitude a body would have as seen by an observer at the centre of the Sun. It is reckoned from the solar equator, positive towards the north. HELIOGRAPHIC LATITUDE is the angular distance of a point on the Sun measured from the Sun's equator. TERRESTRIAL (or GEOGRAPHICAL) LATITUDE is the angular distance of a place from the Earth's equator measured along the meridian of that place. It is equal to the inclination of the Earth's axis to the horizon at the place in question. It is measured from 0° to 90° from the equator, positive northwards, negative southwards.

latitude, variation of A small change in geographical latitude resulting from a change in the point of intersection of the Earth's axis of rotation with the surface of the Earth. The axis of the figure of the Earth defines the Earth's geographical poles and normally coincides with the axis of the Earth's rotation, which defines the celestial poles. If this symmetry is disturbed, as by earthquakes, then one of the poles will rotate about the other until the effect of the disturbance is damped out by friction. The effect was predicted by L. Euler (see **Euler, Leonhard**) in 1765, pointed out by M. Nyrén (1837–1921) in 1873, and detected in 1888 by F. Küstner (1856–1936), who found variations in the geographical latitude of Berlin. The geographical latitude of a spot being equal to the altitude of the celestial pole, i.e. the angular distance of the pole from the horizon of the place of observation, it follows that if the altitude of the pole varies, then the latitude of the spot will vary by the same amount. The effect is the result of at least two motions: one has a period of about 14 months and variable amplitude, the other a period of a year and nearly constant amplitude. The effect of this latter component is ascribed to disturbance of terrestrial masses by changes in meteorological conditions. See also **Chandler period**.

L component That part of the solar corona which gives a spectrum consisting of emission lines.

leap day The extra day added to the leap year. See **calendar**.

leap second The adjustment of one second in the radio time signal at midnight on June 30 or December 31. By international agreement the rate of radio time signals was maintained since January 1972 in agreement with the international atomic time scale. As the Earth's rotational period is slowing down by about 0·003 of a second per day, the radio time signals gradually diverge from G.M.T., which is based on the Earth's axial diurnal rotation. In order to correct this divergence the radio time seconds are retarded by one leap second when necessary.

leap year A year of 366 days. See **calendar**.

Leavitt, Henrietta Swan (Lancaster, Massachusetts, July 4, 1868 – Cambridge, Massachusetts, December 12, 1921) An American astronomer, distinguished for her work at Harvard College Observatory on the photographic magnitude of stars and her derivation of the **period-luminosity law** of Cepheid variable stars.

Lemaître, Georges (Charleroi, Belgium, July 17, 1894 – Louvain, June 20, 1966) A Belgian mathematician known for his work on celestial and stellar dynamics, cosmic rays, the Theory of General Relativity, and in particular for his famous memoir on a homogeneous universe of constant mass and increasing radius. He put forward the notion of the Primeval Atom, a highly condensed state from which the universe started. See **big bang theory**.

lens An optical element which utilizes the refracting property of a transparent material, usually glass. Single lenses have either two curved surfaces, or one curved and one plane surface. Various combinations of lenses are used for the purpose of satisfying the requirements of scientific instruments. Lenses fall into two main groups: converging and diverging.

lens stop A fixed diaphragm consisting of a metal plate with a circular hole centrally disposed, placed at a suitable position in a system of lenses for altering the working aperture of the lens. It is painted a dull black to prevent unwanted reflections. By restricting the rays transmitted by the lens to those close to the optical axis, definition is improved but at the expense of illumination of the image.

Leonardo da Vinci (Vinci, near Florence, April 15, 1452 – Amboise, France, May 2, 1519) Artist, inventor, anatomist, and military engineer. His notebooks record his scientific investigations, and show that he had a reasonably correct knowledge of astronomy. He correctly explained **earthshine** and adduced evidence for the Earth's rotation.

Leonids A meteor shower appearing between November 15 and 19 whose radiant is in the constellation Leo. The shower is produced by debris from the comet P/Tempel–Tuttle (see **meteor stream**), and has provided magnificent displays in the past at intervals of 33 years. That seen on the night of November 12–13, 1833 was the heaviest shower ever recorded, with an hourly rate of about 35,000 meteors.

level error The difference between the apparent elevation and the true elevation of a celestial body, the former being measured above the apparent horizon (the plane which intersects the vertical at the point of observation) and the latter being measured above a plane passing through the centre of the Earth. The magnitude of the error depends on the altitude of the body being observed. It is of significance in respect of bodies in the solar system, and the difference between the altitudes measured from these two horizons is known as the PARALLAX OF THE BODY. In the case of stars the difference between the two horizons can be neglected because of the remoteness of stars.

Lenses. Top: converging. Bottom: diverging.

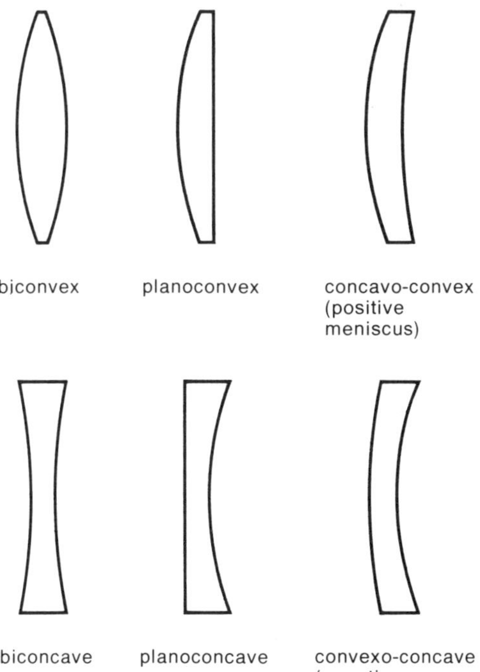

Le Verrier, Urbain Jean Joseph (Saint-Lô, March 11, 1811 – Paris, September 23, 1877) A French mathematical astronomer who became Director of the Observatoire de Paris. He is distinguished for his theoretical work on planetary and cometary motions. He is remembered in particular for his accurate prediction of the position of an unknown planet (Neptune) responsible for perturbation of the orbital motion of Uranus, which enabled the planet to be optically discovered by J. G. Galle and H. L. D'Arrest.

Libra, First Point of One of the equinoctial points, the other being the First Point of Aries (see **Aries, First Point of**). It is the point of intersection of the ecliptic and the celestial equator through which the Sun passes when crossing from north to south of the equator. See also **equinox**.

libration The small oscillation of a celestial body about its mean position. The term is most frequently applied to the Moon. As a result of libration it is possible to see 59% of the Moon's surface. There are three quite different components. LIBRATION IN LATITUDE arises because the axis of the Moon's rotation is not exactly perpendicular to the orbital plane. The two poles therefore tilt alternately to and from the Earth so that a little more of the lunar surface can be observed beyond the north and south poles. LIBRATION IN LONGITUDE arises from the fact, discovered by J. Hevelius (see **Hevelius, Johann**) in 1647, that the angular velocity of the Moon's axial rotation and the angular velocity of the Moon's orbital motion are not always equal. As a result, at times a little more of the lunar surface is observed at the eastern and western boundaries than corresponds to the mean position. PARALLACTIC or DIURNAL LIBRATION is caused by the different aspect of the Moon as seen by observers at different places on the Earth's surface.

light Visible **electromagnetic radiation** covering approximately the wavelength region 4000 Å to 8000 Å. Shorter (ultraviolet) and longer (infrared) invisible radiations are sometimes included in a wider definition of light; the extended range covers wavelengths from about 1·4 mm to 30 Å. The **velocity of light** *in vacuo* is 299,792·5 km s^{-1} or 186,281 miles per second. This is the speed at which all electromagnetic radiation travels in space. The propagation of light and other electromagnetic radiation can be considered to be by means of transverse vibrations, i.e. perpendicular to the direction of propagation. The vibrations are normally in all planes perpendicular to the direction of propagation. When, however, the vibrations are all in one particular plane the radiation is said to be POLARIZED. Because of the particle-wave duality of electromagnetic radiation, light can also be considered as a stream of **photons** travelling at the velocity of light.

White light, the physiological experience of the human eye to sunlight, can be decomposed into its component colours by a prism, thus producing a spectrum in which the colours in order of decreasing wavelength are red, orange, yellow, green, blue, indigo, and violet. The LUMINOUS INTENSITY of light is the amount of radiation falling on unit area in unit time; its unit of measurement is the **candela**.

Light falling on a substance may be partly reflected, absorbed, and/or transmitted. **Scattering of light** occurs when it encounters small particles. Light normally travels in a straight line, but on passing by the edge of an object it can be deviated, producing the phenomenon of **diffraction**. If two trains of light waves from a coherent source are superimposed, the phenomenon of **interference** can be caused under certain conditions. Light falling on a body exerts a very small pressure (see **radiation pressure**). One of the predictions from Einstein's General Theory of Relativity is that light can be deflected by a gravitational field: the effect has been observed during a solar eclipse.

light curve A presentation in graphical form of the change in brightness with time of a **variable star**.

light-time The time required for light to travel from any celestial object to the Earth. It takes light 8·3 minutes to travel the distance between the Sun and Earth. If an object is observed at the time T and the light-time is t, then the observation relates to the state of the object at time $(T-t)$.

light, velocity of (*symbol: c*) The constant speed at which light and other **electromagnetic radiation** travels in a vacuum. It was first determined by astronomical methods by O. Rømer (1676) and J. Bradley (1728). Its value is $2{\cdot}997\,924\,580 \times 10^{8}$ m s^{-1}.

light-year (*abbrev.:* ly) A measure of distance used in astronomy. It is the distance travelled by light *in vacuo* in one tropical year.

$$1 \text{ ly} = 9{\cdot}4607 \times 10^{12} \text{ km} = 0{\cdot}3066 \text{ pc} = 63{,}240 \text{ A.U.}$$

limb The apparent edge of the disc of a celestial body such as the Sun, Moon, or a planet. When a telescope is trained on a body of this kind it will move across the field of view on account of the Earth's diurnal motion, unless the telescope is equatorially mounted and driven so as to nullify the Earth's axial rotation. The leading limb in the uncompensated motion is called the PRECEDING LIMB; the trailing limb is called the FOLLOWING LIMB.

limb brightening The increase in the intensity of radiation of radio and X-ray frequencies towards the limb of the Sun.

limb darkening The decrease in the brightness of the Sun's disc towards the limb, giving the appearance of an annular shadow.

line of apsides see **apsis**

line of nodes see **nodal line**

line-of-sight velocity see **radial velocity**

line profile see **spectrophotometry**

line spectrum see **spectrum**

Littrow, Joseph Johann von (Bischof-Teinitz, March 13, 1781 – Vienna, November 30, 1840) An Austrian astronomer who became director of the observatory at Vienna. He is known for his work on the construction of telescopes and his studies on refraction of light.

local arm see **Orion arm**

Local Group The group of at least 25 galaxies, which includes the Galaxy, the two Magellanic Clouds, the Andromeda Galaxy, and the Triangulum Spiral Galaxy. They are distributed over a nearly ellipsoidal space having a maximum diameter of about 1500 kpc. The member galaxies are gravitationally bound, so that, unlike more distant galaxies, they are not receding from us or from each other.

local star system The stars of the Galactic system in the immediate vicinity of the Sun.

local time see **time**

Lockyer, Joseph Norman (Rugby, May 17, 1836 – Salcombe Regis, August 16, 1920) An English astronomer and astrophysicist who was the founder and first editor of *Nature* (1869). He was the Director of the Solar Physics Observatory, South Kensington. A pioneer of solar spectroscopy, he devised a method of observing the spectra of solar prominences in the absence of an eclipse. With E. Frankland (1825–99) he discovered helium in the Sun's atmosphere (1868). He founded the Norman Lockyer Observatory in Sidmouth in 1913; it is now the observatory of the University of Exeter.

longitude A concept in systems of co-ordinates. CELESTIAL (or ECLIPTIC) LONGITUDE is measured from the First Point of Aries (the vernal equinox) to the great circle passing through the pole of the ecliptic and the object to be measured. It is reckoned in degrees eastwards from 0° to 360°. GALACTIC LONGITUDE is the angle measured along the galactic equator between the direction of the galactic centre and the point of intersection of the galactic equator with the galactic circle of longitude of the object to be measured. TERRESTRIAL (or GEOGRAPHICAL) LONGITUDE is the angle subtended at the centre of the Earth by the arc measured along the Earth's equator between the meridian of Greenwich and the meridian of the observer. It is measured from 0° to 360° eastwards, or from 0° to 180° eastwards or westwards. HELIOCENTRIC LONGITUDE is the longitude of a body measured with reference to the centre of the Sun. It is the longitude that a body would have as seen by an observer at the centre of the Sun. HELIOGRAPHIC LONGITUDE is the angle subtended at the centre of the Sun, and is measured from the solar meridian that passed through the ascending node of the solar equator on the ecliptic on January 1, 1854 at Greenwich mean noon; it is reckoned from 0° to 360°, in the direction of rotation.

long-period variables Also known as MIRA STARS after Mira Ceti, which was the first variable to be discovered (by D. Fabricius in 1596). They are variable, probably pulsating, stars with periods in excess of 100 days, usually of spectral type M or C.

Lorentz-Fitzgerald contraction or **Lorentz contraction** A hypothesis put forward by G. F. Fitzgerald to explain the results of the Michelson-Morley experiment on aether drift, and extended to electromagnetic phenomena by H. A. Lorentz. According to the hypothesis all matter might undergo distortion as a result of motion at a very high velocity through space: all bodies will contract along the direction of travel when moving at velocities comparable with that of light. Einstein subsequently showed that such contraction was a necessary consequence of the Special Theory of Relativity.

Lorentz, Hendrik Antoon (Arnhem, July 18, 1853 – Haarlem, February 4, 1928) A Dutch physicist who discovered independently of G. F. Fitzgerald the Lorentz-Fitzgerald contraction. He worked on electromagnetic theory, kinetic theory, gravitation, and thermodynamics. He gave the explanation of the **Zeeman effect** and derived the equations known as the **Lorentz transformations**.

Lorentz transformations Equations derived by H. A. Lorentz which make possible the linear transformation of Cartesian co-ordinates of time and space into another set of co-ordinates appropriate to an observer moving with a uniform high velocity in that frame of reference.

Lovell, Alfred Charles Bernard (Oldland Common, Gloucestershire, August 31, 1913 –) A British radio astronomer who is the founder and Director of the Nuffield Radio Astronomy Laboratories, Jodrell Bank, Macclesfield, Cheshire. He is distinguished for his pioneering work in exploring radio emission from space.

Lowell, Percival (Boston, March 13, 1855 – Flagstaff, November 13, 1916) An American astronomer who established the Lowell Observatory, Flagstaff, Arizona (1894). He is distinguished for his observations and discoveries concerning planets. His mathematical investigations led to the discovery, 14 years after his death, of the planet **Pluto**. He was a firm believer in the existence of 'canals' on Mars, which he ascribed to the activities of intelligent beings.

lower culmination see **culmination**

low-velocity stars Population I stars of abnormally low space velocities.

Lucifer Old name for the planet Venus when visible in the eastern sky as the Morning Star.

luminosity The absolute brightness of a star given by the total amount of energy radiated from its entire surface per second. It is expressed in watts (joules per second), ergs per second, or in terms of **absolute magnitude**.

luminosity class A classification of stellar spectra according to luminosity (absolute magnitude) and colour (spectral type). The luminosity classes of the MKK SYSTEM are listed in the table. See also **spectral-luminosity classification**.

Ia	supergiants of high luminosity
Ib	supergiants of low luminosity
II	luminous giants
III	normal giants
IV	subgiants
V	main-sequence stars
VI	subdwarfs

luminosity function A measure of the distribution of stars in space according to their absolute magnitude. It is usually expressed as the number of stars per cubic parsec for each unit of absolute magnitude.

Luna or **Lunik** The name of lunar space vehicles launched by the USSR. 'Lunik' is derived from two Russian words: *luna* and *sputnik*. Luna 1 (Lunik I) was launched on January 2, 1959. Luna 3, launched on October 4, 1959, obtained the first photographs of the averted side of the Moon. Luna 9, launched on January 31, 1966, achieved the first soft landing on the Moon and transmitted some historic photographs of the lunar terrain. Luna 10, launched on March 31, 1966, was the first craft to enter lunar orbit. Luna 16, launched on September 12, 1970, automatically returned the first soil sample to Earth. Luna 17, launched on November 10, 1970, landed the first unmanned lunar roving vehicle, **Lunokhod** 1.

lunabase The dark-coloured basic rocks of the Moon's maria, heavy and compact in texture.

lunar eclipse An example of the obscuration of a non-luminous reflecting body. It occurs when the Earth lies between the Sun and the Moon, thereby preventing illumination of the latter by the former. As the Moon's orbital plane is inclined to the plane of the ecliptic, an eclipse can occur only when the Moon is at opposition (i.e. at Full Moon) and at the same time at or near one of its nodes. The Sun, Earth, and Moon are then very nearly in a straight line. Lunar eclipses are visible from any part of the Earth provided that the Moon is above the horizon. The Moon does not become completely invisible during an eclipse; it appears as a coppery coloured disc because it receives some sunlight through refraction by the Earth's atmosphere. A

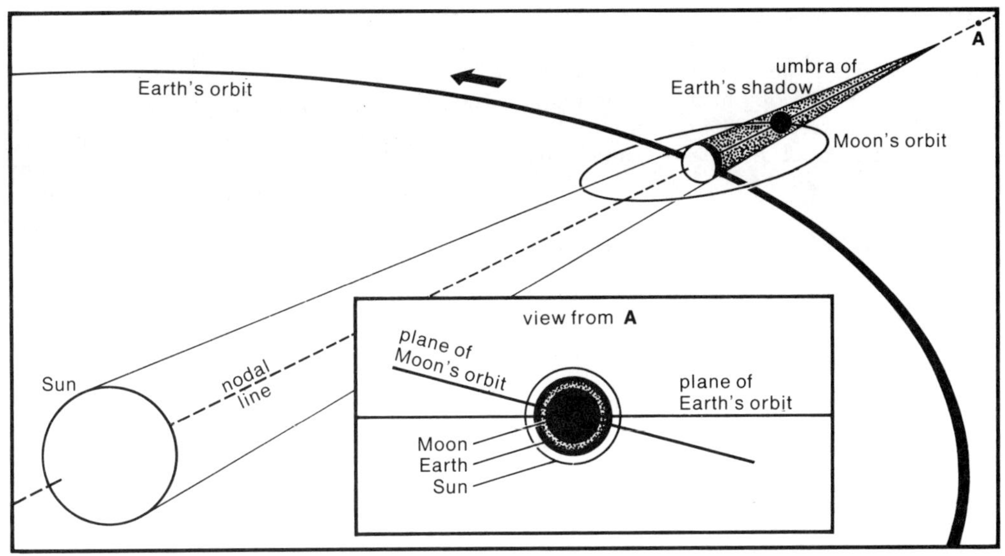

A total lunar eclipse.

TOTAL LUNAR ECLIPSE occurs when the Moon is entirely within the Earth's **umbra**. It is PARTIAL when the Moon is partly within the umbra. It is PENUMBRAL when the Moon is entirely within the **penumbra**. The duration of a total lunar eclipse from first to last contact with the Earth's shadow lasts up to 3·5 hours; totality lasts up to 1·75 hours.

lunarite Light-coloured acidic rocks of the Moon's 'continental' areas. It is of lower density than lunabase, which contains heavier metals.

Lunar Orbiter The name given to a series of American lunar probes put in orbit around the Moon for the purpose of photographing the surface features, and carrying out instrumental determinations relating to radiation from the Sun, the magnetic field of the Moon, and micrometeorites. The photographs taken by the probes were processed during flight, scanned, and telemetered back to Earth. Lunar Orbiter 1 obtained fine photographs during August 1966. Subsequent Orbiters continued the photographic mapping and obtained results superior to anything obtainable from Earth. Lunar Orbiter 1 was launched on August 10, 1966, Lunar Orbiter 5 on August 2, 1967.

lunation Also known as the Moon's SYNODIC PERIOD or the SYNODIC MONTH. The time interval between two identical phases of the Moon, e.g. from New Moon to New Moon. Its value is 29·530 59 days.

luni-solar precession The main component of the **precession** of the equinoxes, resulting from the gravitational effects between the Moon, the Sun, and Earth.

Lunokhod A Russian word meaning 'a lunar research vehicle'. The first was placed on the Moon by Luna 17 on November 17, 1970. All operations of the vehicle, such as travel over the lunar surface, surveying the area, testing the mechanical properties of the ground and making chemical analysis of it were controlled from a station on Earth.

Lyman series The series of lines in the spectrum of hydrogen analogous to the series in the visible part of the spectrum discovered by J. J. Balmer. The lines are associated with transitions of electrons to or from the ground state of the atom (producing emission or absorption lines, respectively) and lie entirely in the ultraviolet region. The first Lyman line, Lα, is at 1216 Å, and was first detected in the solar spectrum in 1959 by photography from a rocket. Lβ is at 1026 Å; Lγ is at 972 Å; the limit is at 912 Å.

Lyman, Theodore (Boston, Massachusetts, November 23, 1874 – Boston, October 11, 1954) An American physicist who was a pioneer of spectroscopy in the far ultraviolet range. He discovered (1906) the series of lines in the spectrum of hydrogen named after him.

Lyot, Bernard Ferdinand (Paris, February 27, 1897 – Cairo, April 2, 1952) A French astronomer known for his work in solar physics. He invented the **coronograph** for the study of the Sun's corona and the filter named after him.

Lyot filter A most important instrument for solar research developed by B. F. Lyot, who added it to the **coronograph** which was also of his invention. It is a polarizing monochromatic filter, able to provide a passband less than 1Å wide. It consists of alternate quartz plates and polaroid sheets, each successive quartz plate being twice the thickness of the preceding one.

Lyrids A meteor shower that normally appears between April 19 and 24 and has its radiant on the border of the constellations Lyra and Hercules. Observations of the Lyrids have been made over a very long time. They are associated with Comet Thatcher (1861 I).

Lysithea Satellite X of Jupiter. See also **satellite**.

Lyttleton, Raymond Arthur A contemporary British astronomer known for his researches on comets, cosmogony, geophysics, and theoretical astrophysics. He was a pioneer of the steady-state theory of the universe.

Mädler, Johann Heinrich von (Berlin, May 29, 1794 – Hanover, March 14, 1874) A German astronomer who became Director of the Dorpat Observatory. He is known for his collaboration with W. Beer (see **Beer, Wilhelm**) at the latter's observatory at Berlin in producing (1834–36) the then most accurate map of the Moon's surface, which was followed by the descriptive work *Der Mond* (1837). They produced also the first map of Mars (1840).

Maffei 1 A distant celestial object in Cassiopeia, in the region of I.C. 1805 and near the radio source 3C 69. It is characterized by pronounced brightness in the infrared. MAFFEI 2 is a similar object to Maffei 1 but Bsmaller and fainter. Both were discovered by P. Maffei in 1968. Maffei 1 seems to be a member of the Local Group, but Maffei 2 is far more distant.

Magellanic Clouds Two of the three galaxies visible to the naked eye; the third one is the Andromeda galaxy. The two Clouds are in the southern hemisphere. They are named after the Portuguese explorer, F. Magellan (1480–1521) who observed them. They are important for the study of Population I and II stars. The period-luminosity relationship for Cepheid variables was discovered by observation of such stars in the Clouds. Radio-frequency radiation shows that both Clouds rotate. The LARGE MAGELLANIC CLOUD lies in the constellations Dorado and Mensa and consists mainly of Population I stars; it is about 50–60 kpc distant and has a mass of about 10^{10} solar masses. The SMALL MAGELLANIC CLOUD lies in the constellation Tucana and contains relatively more Population II stars than the Large Magellanic Cloud; it is about 60–70 kpc distant and has a mass of about 2×10^9 solar masses.

magnetic storm A world-wide irregular disturbance and fluctuation of the Earth's magnetic field lasting from several hours to days, the intensity of the disturbance being greater in high latitudes than in low latitudes. The abrupt start of some storms is known as a 'sudden commencement'. There is a marked tendency for less severe storms to recur after an interval of 27 days (the Sun's period of synodic rotation), and it has been shown that storms are related to solar activity. Intense storms occur after the appearance of intense **solar flares** in the central area of the Sun's disc; milder storms are thought to be caused by activity in other areas of the Sun's surface. The time lag of a day or more between a flare and the incidence of a magnetic storm shows that electromagnetic radiation is not the cause. This would require only about eight minutes to reach the Earth. The disturbance is caused by a stream of ionized particles ejected from the Sun.

magnetic variation or **magnetic declination** The angle between the planes of the geographic and magnetic meridians, i.e. the angle between the direction of magnetic north as shown by a magnetic compass and the true geographic north. Its value is slowly varying.

magnetograph An instrument used to obtain continuous records of the Earth's magnetic field. It generally comprises three magnets suspended so that their axes are horizontal. One of the magnets has its axis in the magnetic meridian and records changes in magnetic declination. A second magnet has its axis perpendicular to the meridian and records changes in the horizontal magnetic force. The third magnet can move in the vertical plane and records changes in the vertical magnetic force.

magnetohydrodynamics (*abbrev.:* MHD) The study of the motion of charged particles in a magnetic field, which results in a coupling between mechanical and electrical forces. Such action was first demonstrated by H. Alfvén (see **Alfvén, Hannes**) who put forward an MHD theory to explain sunspots on the basis that magnetohydrodynamic waves (Alfvén waves) are propagated from the Sun's core outwards to the surface, where vortex rings are formed.

magnetopause That region in the ionosphere where the Earth's magnetic field ceases and interplanetary space begins. It is the boundary of the **magnetosphere**.

magnetosphere The region of space surrounding a celestial body in which there is a magnetic field associated with that body. In the case of the Earth, Jupiter and Saturn, it contains also charged particles ejected by the Sun and trapped in the planet's magnetic field.

magnitude An arbitrary measure of the brightness of stars and other celestial objects. **Hipparchos of Nicaea** in antiquity classified the stars in six groups, the first being the brightest and the sixth the faintest discernible by the naked eye. The concept has since been refined and extended to include the radiation from all celestial bodies. **Apparent magnitude** is the brightness of an object as seen from Earth, and no matter how determined gives no indication of the intrinsic brightness. **Absolute magnitude** is the true brightness of an object, taking into account its distance from Earth. See also **stellar magnitude, determination of**.

main sequence That part of the **Hertzsprung-Russell diagram** containing the majority of stars. It starts at the upper left-hand side of the diagram with high-temperature, high-luminosity stars, and continues to the bottom right-hand side with low-temperature, low-luminosity stars. A star enters the main sequence when it has started the nuclear transformation of hydrogen into helium, and remains there until about 12% of its available hydrogen has been consumed. The position of a star on the main sequence depends primarily on its mass. The upper limit is at stars of about 60 solar masses and of spectral types O and B. The lower limit is at 0·085 solar mass with stars of spectral type M.

major planets The nine largest planets of the solar system, namely Mercury, Venus, Earth, Mars, Jupiter, Saturn, Uranus, Neptune, and Pluto, thus distinguishing them from the **minor planets** or asteroids. The name is sometimes used with reference to the giant planets alone, namely Jupiter, Saturn, Uranus, and Neptune.

Maksutov, Dmitrii Dmitrievich (Odessa, April 23, 1896 – Leningrad, August 12, 1964) A Soviet optician known for his work on astronomical optics.

Maksutov telescope A modification of the **Schmidt camera** devised independently by A. Bouwers (1940) and D. D. Maksutov (1944). The aspherical correcting plate of the standard Schmidt camera is replaced by a concave **meniscus lens** of nearly constant thickness, figured to be nearly free from chromatic

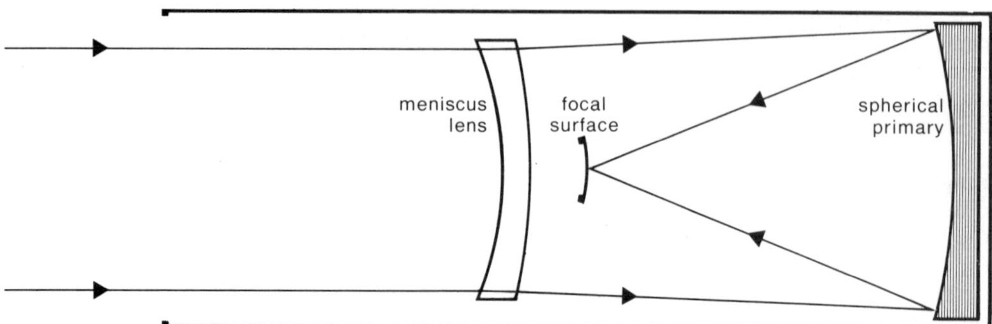

Light path in a Maksutov telescope.

aberration and having spherical aberration in the opposite sense to that of the primary concave spherical mirror; the combination of meniscus lens and primary mirror is thus free from both spherical and chromatic aberrations. The optical parts are more easily manufactured than those in the Schmidt camera and the length of the telescope tube is shorter. Numerous modifications to the basic design are possible to permit the optical system to be used as a Gregorian, Newtonian, Cassegrain, or Herschelian telescope. They are described in the article on 'New Catadioptric Meniscus Systems' by D. D. Maksutov in the *J. Optical Soc. America* (1944) *34,* 270–284.

many-body problem The determination of the motion of *n* bodies in space under the influence only of their mutual gravitational attraction. It is a fundamental problem in celestial mechanics. The **two-body problem** and the **three-body problem** are classical.

mare (*Latin:* sea; *plural* maria) A dark feature on the Moon and on Mars. These features are not seas. Lunar maria are in fact regions of iron-rich basaltic lava of low albedo that erupted on to the Moon's surface some 3000 to 4000 million years ago. Martian maria are areas of darker bedrock.

Mariner The name given to a series of spaceprobes launched by the United States. Mariner 2, launched from Cape Canaveral on August 27, 1962, passed by Venus on December 14, 1962 and transmitted data on the physical conditions on the planet. Mariner 4, launched on November 28, 1964, passed by Mars on July 15, 1965. It successfully transmitted some excellent photographs of the surface of that planet and measured the atmospheric density. Mariner 5, launched on June 14, 1967, passed Venus on October 19, 1967 and made measurements of the planet's mass, diameter, magnetic field, and atmospheric properties. Mariners 6 and 7 were launched on February 25 and March 27, 1969 to Mars and obtained further excellent photographs of the planet.

Mariner 9, launched on May 30, 1971 went into orbit around Mars on November 13, 1971. It was the first probe to enter a planetary orbit. In addition to atmospheric and surface measurements, a year-long photographic reconnaissance of the entire surface was made together with close views of the two moons, Phobos and Deimos. Mariner 10, launched on November 3, 1973, achieved the first two-planet mission by passing Venus on February 5, 1974 and then encountering Mercury on March 5, 1974, September 21, 1974 and March 19, 1975. Considerable information and thousands of photographs were transmitted to Earth. The proposed Mariners 11 and 12 were renamed Voyagers 1 and 2.

marine sextant An instrument constructed to have a graduated scale on a 60° or 70° arc. The altitude as measured by its use at sea must be corrected for the dip of the horizon, i.e. the curvature of the Earth's surface, and for atmospheric refraction. Altitudes of stars can be obtained not only while the horizon is visible, i.e. during twilight, but also during darkness; measurements are then made with the aid of an artificial horizon, which is a shallow pool of mercury, when the angle measured is that between the star and its image seen in the mercury.

Mars The fourth planet from the Sun. When viewed through a telescope it appears as a small orange-red disc with blue-green markings, and frequently with white patches at the poles. Mars is sufficiently close to the Earth for surface features to be distinguished in considerable detail by telescopic and photographic methods. The yellow clouds seen on Mars are apparently caused by violent dust storms. The blue/violet clouds revealed on viewing the planet through suitable colour filters are probably caused by small crystals of condensed matter in the atmosphere; white clouds over the polar caps may contain larger crystals. There are seasonal variations in the appearance of the blue-green surface markings and of the polar caps; the latter change probably indicates the melting and re-forming of snow.

The earlier **Mariner** space-probes revealed the great similarity between Mars and the Moon in respect of cratered areas. They also provided a wealth of information on the areas classed as maria, desert, or oasis, as well as on the atmoshere, ionosphere, and physical properties such as temperature, size, etc.

Very much more detailed information was obtained from the orbiting Mariner 9 and the two Orbiters and Landers of **Vikings** 1 and 2. It is now known that although the southern hemisphere of Mars is heavily cratered, the northern hemisphere exhibits a more volcanic appearance: there are extensive lava plains, only sparsely cratered, and several regions of former volcanic activity; four huge shield volcanoes lie close to the equator, the largest being the 25-km high Olympus Mons.

Surface of Mars taken from Viking Lander 1.

Immense canyons together with smaller meandering channels have been discovered in equatorial regions. The canyons have apparently resulted from faulting and collapse of the Martian surface. The channels are thought to have been formed by running water which was present on the planet in a previous era, indicating a very different climate from the present one. The variable polar ice caps appear to be composed of solid carbon dioxide with underlying caps of water ice that are revealed in the summer months on the Martian hemispheres. Analysis of soil samples by the two Viking Landers showed a high proportion of silicon and iron with smaller amounts of magnesium, aluminium, sulphur, cerium, calcium, and titanium. No form of life was, however, detected. The atmosphere consists mainly of carbon dioxide (about 95%) with small quantitites of nitrogen, argon, oxygen, carbon monoxide, and water vapour. The atmospheric pressure at the surface of the planet is less than 1·% of that at the surface of the Earth. A very weak magnetic field has been detected.

Mars has two satellites very close to it. They are Phobos (Fear) and Deimos (Terror), whose existence was 'discovered' by the Laputan astronomers, as recorded by Jonathan Swift in *Gulliver's Travels* (1727). This satirical fictional account is remarkable for being so close to fact. They were positively discovered by A. Hall (see **Hall, Asaph**) in 1877 by a deliberate search. The main data relating to Mars are given in the table.

Globe	
Diameter (equatorial)	6790 km
Diameter (polar)	6750 km
Density (water = 1)	3·94 g cm^{-3}
Mass	6·45 × 10^{23} kg
Volume	1·637 × 10^{11} km^{3}
Sidereal period of axial rotation	24^{h} 37^{m} 23^{s}
Escape velocity	5·03 km s^{-1}
Albedo	0·16
Inclination of equator to orbit	24° 46′
Surface temperature	250–320 K
Surface gravity (Earth = 1)	0·380

Orbit	
Semi-axis major	1·5236915 A.U. = 227·94 × 10^{6} km
Eccentricity	0·0934
Inclination to ecliptic	1° 50′ 59″
Sidereal period of revolution	686·98 d
Mean orbital velocity	24·13 km s^{-1}

mascon (acronym from *mass concentration*) A concentration of high mass below the surface of the Moon which causes anomalies in gravity. They were detected in the data obtained during the flight of Lunar Orbiter 5 in 1967. The mascons, of which at least a dozen are known, have diameters of 50–200 km and lie about 50 km below the surface. Their origin is as yet unexplained.

maser (acronym from *m*icrowave *a*mplification by *s*timulated *e*mission of *r*adiation) A microwave amplifier or oscillator whose energy is supplied by induced emission of electromagnetic radiation from molecules, atoms, or ions. If such a quantum-mechanical system is subjected to an alternating electromagnetic field whose frequency corresponds to the difference between two distinct energy levels of the system, then a photon will probably be absorbed if the system is at the lower energy level to start with; there will be induced emission of a photon if the system is at the higher energy level. The induced emission is coherent. The maser is an amplifier of extremely low noise which finds use in radio-astronomy laboratories and in satellite-tracking stations. The maser operates in the microwave region at about 10^{10} Hz. The same principle can be employed in the region of visible light of frequency about 10^{14} Hz, when the device is known as an optical maser or **laser**.

Maskelyne, Nevil (London, October 6, 1732 – Greenwich, February 9, 1811) An English astronomer who was appointed the fifth Astronomer Royal. He made a voyage to St Helena to observe the transit of Venus, and en route developed a method of determining longitude by lunar distances (1761). He made a journey to Scotland to determine the mean density of the Earth by deviation of a plumb-line at the mountain of Schiehallion. He is remembered for having founded *The Nautical Almanac*, the first edition of which appeared in 1767.

mass The quantity of matter present in a body. It is also the property of matter which characterizes its behaviour in a gravitational field and in changes in velocity. The unit of mass in the **SI** is the **kilogram**. Bodies have inertia and must be acted upon by a force in order to produce a change in velocity:

$$F = km_i$$

F is the force, k the constant of acceleration, and m_i the inertial mass. Bodies have weight, which is the force with which a body is acted upon by a gravitational field:

$$W = gm_g$$

W is the weight, g the acceleration of gravity, and m_g the gravitational mass. In the absence of any other forces k is the same for all bodies at a particular place, and g is constant for all bodies at a particular place. Hence the inertial mass is proportional to the gravitational mass. Equivalence of m_i and m_g is provided by the General Theory of Relativity. By the Special Theory of Relativity the energy in a body is given by the formula

$$E = mc^2$$

E is the energy, m the mass, and c the velocity of light. Furthermore, the mass of a body is not constant but increases with the velocity v of the body in accordance with the expression

$$m = m_o (1 - v^2/c^2)^{-\frac{1}{2}}$$

m_o is the rest mass of the body.

mass defect The difference between the mass of an atomic nucleus and the sum of the individual masses of its constituent nucleons. It is the loss of mass that takes place when light atomic nuclei combine to form heavier ones, and which is converted into energy in accordance with Einstein's equation $E = mc^2$, and radiated.

mass-luminosity relationship A relationship derived theoretically by A. S. Eddington in 1924. When plotted on a logarithmic scale, the luminosity (bolometric absolute magnitude) versus mass gives a curve which is almost a straight line. Thus

$$L \propto M^x$$

where L is the luminosity, M the mass, and x varies between 3·0 and 3·5. The relationship holds only for stars in the main sequence. It is not obeyed by white dwarfs. The relationship enables the mass of a star to be determined if its absolute magnitude is known.

mass number The total number of nucleons (protons and neutrons) in the nucleus of an atom.

mass of a planet or star The mass of a planet or other celestial body can be determined from its gravitational effect on the motion of nearby bodies, for example, when a planet has a satellite in orbit around it or, if it has no satellite as in the case of Venus, then by its effect on the motion of a space probe. In the case of a binary system of stars the ratio of their masses can be found from observations of their orbital motion. The mass of a single star can be determined from the gravitational red shift, if large enough, of the lines in its spectrum under the influence of a gravitational field. This has been done in the case of **white dwarfs**.

mass ratio The ratio of the masses of the components of a binary system.

Maunder, Edward Walter (London, April 12, 1851 – London, March 21, 1928) A British astronomer who was a pioneer in solar physics and is remembered for his **butterfly diagram**.

mean anomaly see **anomaly**

mean density of matter (universe) The total mass content of the universe is estimated at about 10^{23} solar masses, say 2×10^{56} g, and is contained in a sphere of radius about 2×10^{28} cm, assuming that the universe is bounded. Although the mean density of matter is extremely difficult to measure, it is the factor that will determine the future behaviour of the universe. If it is below a certain critical value then the universe will continue to expand; if it is above the critical density then the universe will eventually collapse upon itself. The critical density is thought to be in the region of 5×10^{-30} g cm^{-3}.

mean equinox The mean vernal **equinox**. The vernal equinox is subject to slight periodical displacements about its mean position on account of **nutation**.

mean motion A hypothetical concept used in celestial mechanics. It is the velocity a planet or other body would have if it moved in a circular orbit having a radius equal to its mean distance from the Sun in a period equal to its true period of revolution.

mean position The position of a star on the celestial sphere, as seen from the Sun, after all corrections have been made for aberration, annual parallax, atmospheric refraction, proper motion, etc., and referred to the mean equinox.

mean Sun A fictitious sun moving along the celestial equator at a constant velocity equal to the average velocity of the true Sun along the ecliptic. It completes its annual course with respect to the vernal equinox in exactly the same time as the true Sun.

mean time Time based on the motion of the **mean Sun**, which travels at a uniform rate along the celestial equator. The true Sun travels at a variable rate along the ecliptic, and cannot therefore give a uniform measure of time.

megaparsec (*symbol:* Mpc) A unit of distance equal to one million **parsecs**. 1 Mpc = $3{\cdot}2616 \times 10^6$ ly.

Melotte, Philibert Jacques (London, January 29, 1880 – Abinger, March 30, 1961) A British astronomer who was an expert in photographic astronomy: he is remembered for his work on the **Carte du Ciel** (Greenwich astrographic zone) and the **Franklin-Adams Charts**. He discovered satellite VIII of Jupiter (1908).

meniscus lens A concavo-convex lens, i.e. one thicker in the middle than elsewhere (a CONVERGING OR POSITIVE MENISCUS), or a convexo-concave lens, i.e. one thinner in the middle than elsewhere (a DIVERGING OR NEGATIVE MENISCUS).

meniscus-Schmidt telescope or **Schmidt-Maksutov telescope** A Maksutov telescope in which the **meniscus lens** is supplemented by an aspherical correcting plate of the type used in the conventional Schmidt camera.

Mercury The innermost planet of the solar system. It is difficult to observe because of its closeness to the Sun, and its small angular diameter makes identification of surface detail very difficult. Faint markings against a whitish background are observable on the disc using a telescope. Very little was thus known about surface conditions until the space-probe Mariner 10 provided some excellent pictures during three close approaches to the planet in 1974 and 1975. These showed the surface to be heavily cratered. Detailed photographs

A photomosaic of Mercury taken by Mariner 10 from a distance of just over 200,000 km.

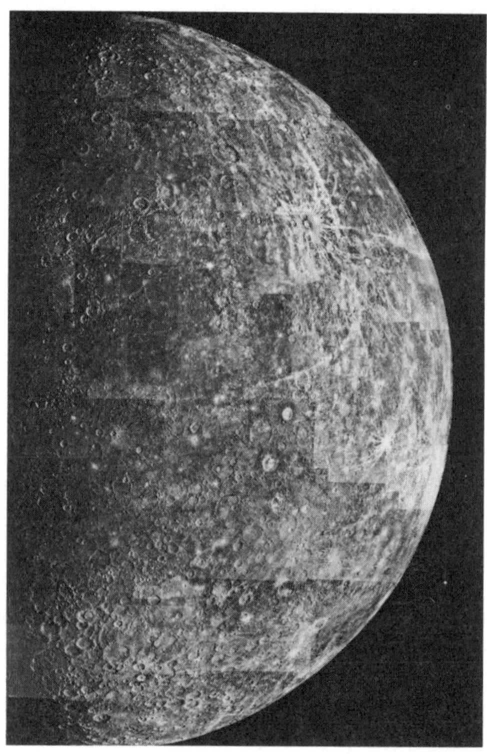

have enabled a map of the surface to be made. The maximum surface temperature recorded by Mariner 10 during the day was 460 K; it could reach over 700 K at perihelion. The temperature on the planet's dark side, away from the Sun, can drop as low as 90 K.

Helium and argon have been detected in the planet's very thin atmosphere and a weak magnetic field discovered. Mercury's orbital velocity is greater than that of any other planet; its orbital eccentricity is greater than that of any other planet except Pluto. A transit of Mercury across the Sun's disc occurs at intervals of either 7 or 13 years. The planet has no known satellite. The main data relating to Mercury are given in the table.

Globe	
Diameter (equatorial)	4880 km
Density (water = 1)	5·5 g cm^{-3}
Mass	$3·15 \times 10^{23}$ kg
Volume	$5·8 \times 10^{10}$ km^3
Sidereal period of axial rotation	58·7 d
Escape velocity	4·3 km s^{-1}
Albedo	0·06
Inclination of equator to orbit	0°
Daytime surface temperature	400–700 K
Surface gravity (Earth = 1)	0·38
Orbit	
Semi-axis major	0·387 A.U. = $57·91 \times 10^6$ km
Eccentricity	0·2056
Inclination to ecliptic	7° 00′ 16″
Sidereal period of revolution	87·97 d
Mean orbital velocity	47·87 km s^{-1}

Mercury project The first US manned space-flight programme for the purpose of establishing that man could travel into space and return safely to Earth. The first American to go into space was Alan Shepard, who was launched in Mercury 3 on May 5, 1961, and returned safely after a flight of 15 minutes. The last test in the series was on May 15, 1963, by Gordon Cooper, whose flight lasted 34 hours.

meridian see **celestial meridian**

meridian astronomy That branch of astronomy which deals with observations made only in the plane of the observer's meridian.

meridian circle A transit instrument. See **transit circle**.

meridian photometer A visual photometer developed by E. C. Pickering (see **Pickering, Edward**) to compare the apparent magnitudes of two stars at different altitudes near the meridian. Polaris is used as a reference for comparison: its image is reflected by a mirror into the telescope simultaneously with that of the star to be examined. The apparent brightness of the two images is then equalized. The amount of adjustment for equalization provides a measure of the difference in magnitude between the two stars.

meridian telescope Any astronomical instrument used in **meridian astronomy**, usually the **transit circle**.

meridian transit The passage of a star or any other celestial body across the observer's meridian as a result of the Earth's rotation.

mesopause That part of the Earth's atmosphere lying between the **mesosphere** and the thermosphere.

mesosphere That part of the Earth's **atmosphere** immediately above the stratopause and between about 45 and 80 km above ground level. The temperature falls with increasing altitude from about 270 K to 170 K. Between altitudes of about 50 and 90 km lies the D layer of the ionosphere, in which there is molecular ionization. This ionization almost disappears during the night.

Messier, Charles Joseph (Badonviller, June 26, 1730 – Paris, April 12, 1817) A French astronomer who compiled a catalogue of the fuzzy extended objects visible in the northern sky so that he should not confuse them with comets, of which he was an eager searcher. The list is still used for reference to the objects in question, which include nebulae, star clusters, and galaxies; they are indicated by the letter 'M' followed by a number, e.g. M29 is an open star cluster in Cygnus.

Name		*NGC*	*Description*
Andromeda Galaxy	31	244	Spiral galaxy in Andromeda
Black Eye Galaxy	64	4826	Spiral galaxy in Coma Berenices
Crab Nebula	1	1952	Remnant of supernova in Taurus
Dumb-Bell Nebula	27	6853	Planetary nebula in Vulpecula
Lagoon Nebula	8	6523	Emission nebula in Sagittarius
Omega (or Swan or Horseshoe) Nebula	17	6618	Emission nebula in Sagittarius
Orion Nebula	42	1976	Emission nebula in Orion
Owl Nebula	97	3587	Planetary nebula in Ursa Major
Praesepe (or Beehive)	44	2632	Open star cluster in Cancer
Pleiades	45	—	Open star cluster in Taurus
Ring Nebula	57	6720	Planetary nebula in Lyra
Sombrero Galaxy	104	4594	Edge-on spiral galaxy in Virgo
Trifid Nebula	20	6514	Emission nebula in Sagittarius
Whirlpool Galaxy	51	5194	Spiral galaxy in Canes Venatici
Wild Duck	11	6705	Open star cluster in Scutum

Messier objects The nebulae, star clusters and galaxies, numbering over 100, which appear in the Messier catalogue. The common names and NGC numbers of certain of the objects are given in the table above.

metagalaxy or **hypergalaxy** The whole universe, comprising all bodies known and unknown in it, as well as the space in which they are contained.

meteor A small solid particle or body entering the Earth's atmosphere at high speed from space. Friction causes the particle to become very hot and luminous. Many are completely consumed. The smallest survive and reach the Earth's surface as **micrometeorites**.

As the meteor passes through the atmosphere, atoms are removed from its surface by friction and heating (ablated). Collisions between these atoms and nearby air molecules result in the production of a visual streak of light in the sky, through excitation and de-excitation of the atoms. The name METEOR is also applied to this luminous streak, of which another more popular name is SHOOTING STAR. A FIREBALL is a meteor whose brightness approaches that of Venus. A BOLIDE is an exceptionally bright meteor whose brightness approaches that of the Full Moon. It may explode during its descent. See also **meteor shower**.

meteorite That part of a large meteoroid that survives passage through the Earth's atmosphere and reaches the ground. It is estimated that hundreds of tonnes of meteoritic matter of all kinds fall on the Earth daily. There are three main classes of meteorites: AEROLITES (or STONY METEORITES) are composed mainly of silicates of iron, magnesium, aluminium, calcium, and sodium; SIDERITES (or IRON METEORITES) are composed of iron and nickel; SIDEROLITE (or STONY IRON METEORITES) have an intermediate composition.

Meteorites have diameters from 0·1 mm upwards and enter the Earth's atmosphere with a velocity of 15–70 $km\,s^{-1}$. Occasionally a very large object falls to the ground: one fell in Arizona in the remote past; on February 12, 1947 a mass of iron-nickel estimated to weigh over 100 tonnes fell in the Sikhote-Alin Mountains in the USSR. The most recent recorded fall of a meteorite of any considerable size in England was on December 24, 1965 at Barwell in Leicestershire; it probably weighed about 450 kg before disintegrating. The age of meteorites has been estimated at about $4{\cdot}5 \times 10^9$ years. The origin of meteorites is unknown: it has been suggested that they are the debris from the disintegration of one or more minor planets.

meteoritic theory The impact theory of the origin of the surface features of the Moon.

meteoroid A small solid meteoritic body or particle that occurs in the solar system. A meteoroid that enters the Earth's atmosphere is known as a **meteor**.

meteor shower A heavy, profuse fall of meteors during a period of several hours, seeming to fall from a fixed point, known as the RADIANT, in the sky. They occur when the Earth passes through a **meteor stream**. The meteoroids in the stream have elliptical orbits so they are true members of the solar system; furthermore they are associated with comets (see **Leonids**). SPORADIC METEORS account for most of the meteors appearing during the year. They are not associated with cometary orbits but appear with fluctuating hourly and seasonal rates.

meteor stream A large quantity of meteoritic bodies circling the Sun in close orbit and associated with a decaying comet. A meteor shower occurs when the Earth in its passage around the Sun crosses a meteor stream.

methane A hydrocarbon gas of chemical formula CH_4, commonly known as firedamp or marsh gas. It occurs in the atmospheres of the giant planets.

Metonic Cycle A period of 19 solar years, almost equal to 235 lunar months. The mean phases of the Moon therefore recur on the same days of the month after 19 years. See also **Golden Number**.

Meton of Athens (*fl.* about 432 B.C.) A Greek astrologer and mathematician who made the first accurate solstitial observations in collaboration with Euctemon. He derived the **Metonic Cycle**.

metre (*symbol:* m) The unit of length in the **SI**. It is defined as the length equal to 1 650 763·73 wavelengths in vacuum of the radiation corresponding to the transition between the levels $2p_{10}$ and $5d_5$ of the krypton–86 atom.

Michelson, Albert Abraham (Strelno, Poland, December 19, 1852 – Pasadena, California, May 9, 1931) A Polish-born American physicist who determined the velocity of light and, in collaboration with E. W. Morley (1838–1923), designed the interferometer for measuring length in terms of wavelengths of light. This instrument was used in their famous experiment to measure the relative motion of the Earth through the aether. The result did much to destroy that concept, and contributed a foundation for the Theory of Relativity. Michelson also measured the angular diameters of several nearby giant stars and of Jupiter's satellites with his stellar interferometer. He showed that the Earth's core is molten.

micrometeorite A particle of meteoritic dust whose diameter is of the order of microns. Micrometeorites are continuously slowly falling through the atmosphere and settling on the Earth's surface; they are too small to be burnt up in the atmosphere. The quantity that falls greatly exceeds the total of all other meteorites. There are three main types: spheres of high density, irregular compacted particles, and fluffy non-compacted particles of lower density.

micrometer An instrument used to make an accurate measurement of small distances, or apparent diameters of objects subtending very small angles. The instrument was invented by W. Gascoigne (1612?–44). Micrometers are of various constructions. The CROSS-BAR MICROMETER consists of two bars set accurately at right angles to each other, and used to determine the position of an object with reference to a star of known position. The FILAR MICROMETER has two wires, separately movable, one of which is connected to a frame with a micrometer head for reading off the movement. The whole micrometer can be rotated, locked in position, and its angle read off on a circular scale. It is used for measuring the apparent separation of a double star or the diameter of a planet. The IMPERSONAL MICROMETER is designed so as to reduce to a minimum the observer's **personal equation** when making observations with the transit circle. Errors in timing have been reduced to the order of

0·02–0·06 second. The RING MICROMETER consists of a metal ring, or a glass plate on which a ring has been prepared, mounted in the focal plane of the telescope. The eyepiece can observe the ring and the image produced by the objective simultaneously. As in the case of the cross-bar micrometer, the times taken for the two objects under observation to cross certain points of the field are recorded. It is used for ascertaining the coordinates of a body.

micron (*symbol:* μ) A unit of length equal to 10^{-6} m (i.e. 10^4 Å), now usually referred to as the MICROMETRE.

microphotometer An instrument for measuring variations in density of images on a photographic plate. The relative brightness of the sources of these images can be deduced from such measurements.

microwave background radiation Weak microwave radiation of cosmic origin, first detected in 1965 by A. A. Penzias and R. W. Wilson of the Bell Telephone Laboratories. It has an almost equal intensity from all directions in space and is black-body radiation characteristic of a temperature of 3 K. It is considered to be the remnant of the radiation content of the very early universe. See **big bang theory**.

Milky Way The faint band of light visible on clear dark nights encircling the sky along the line of the galactic equator. At one point it is divided into two parts of unequal brightness. It is composed of an enormous number of stars, occurring singly or in clouds separated by obscuring portions of interstellar dust. It is part of the Milky Way System or **Galaxy**.

Milne, Edward Arthur (Hull, February 14, 1896 – Dublin, September 21, 1950) A British astrophysicist who made important contributions to the theory of stellar atmospheres, the system of kinematic relativity, and cosmology.

Mimas Satellite 1 of **Saturn**. See also **satellite**.

minor planets or **asteroids** A large group of small bodies moving in orbits around the Sun under its gravitational attraction. The majority move within a zone which lies about 2·17–3·3 A.U. from the Sun, between the orbits of Mars and Jupiter.

About 2000 have been catalogued and have had approximate orbits calculated; about 4000 have been observed once. The total possible number is estimated at 50,000 or more. The first minor planet to be discovered was Ceres on January 1, 1801 by G. Piazzi (see **Piazzi, Giuseppe**). It is the largest of the known minor planets, having a diameter of about 1000 km. The second to be discovered was Pallas in 1802 by H. W. M. Olbers (see **Olbers, Heinrich**). The third, Juno, was found in 1804 by K. L. Harding (1765–1834) and the fourth, Vesta, in 1807, also by H. W. M. Olbers. It was not until 1845 that the fifth minor planet, Astraea, was discovered by K. L. Hencke (1793–1866), since when they have been discovered in increasing numbers.

The origin of the minor planets is not known. It has been suggested that they are the remnants of a large planet that disintegrated; the estimated total mass of the minor planets, of the order of 1/1600 of the Earth's mass, makes this unlikely. More probably they are the remains of bodies which fragmented and scattered during the formation of the solar system. The larger minor planets are spherical, but most of them are irregularly shaped bodies. Their orbits are elliptical but more eccentric than those of the major planets, and the inclination to the ecliptic is greater.

Some minor planets are of special interest. The **Trojans** are interesting dynamically. Hidalgo, the outermost minor planet, has the largest known orbit. Icarus, the innermost asteroid, has one of the smallest orbits and highest eccentricities. It passes inside the orbit of Mercury and outside that of Mars. Eros came to within 0·15 A.U. of the Earth in 1975 and will do so again in 1982. The closest approach to Earth was made by Hermes on October 30, 1937 when it was within 6×10^5 km. The minor planet No. 532, Herculina, has been found to have a satellite. The exact nature of the extraordinary object, **Chiron**, recently discovered, has not been determined, although it has provisionally been included with the minor planets.

The first four minor planets having been named from classical mythology, the practice was continued until the number discovered caused problems in finding names. Several systems of identification have been proposed, and abandoned on account of the rapid rate of discovery, which had been accelerated by photographic methods. The present system of designating a minor planet is shown in the table.

Time of discovery	*Designation*
1979 January (1st half)	1979 AA, AB, AC, and so on
(2nd half)	1979 BA, BB, BC, and so on
1979 February (1st half)	1979 CA, CB, CC, and so on
(2nd half)	1979 DA, DB, DC, and so on

The first letter indicates the half of the month in which the discovery took place, and the second letter indicates the order of discovery within the period. This system allows for the discovery of minor planets at a rate approaching two a day. These are temporary designations. A permanent number is given when the orbit has been calculated.

minute of arc (*symbol:* ′) A unit of angular measure equal to one sixtieth of one degree.

Miranda Satellite V of **Uranus**. See also **satellite**.

Mira stars see **long-period variables**

mirror A reflecting surface which may be plane, spherical, elliptical, or parabolic and is used in many optical instruments. Mirrors may be of glass or quartz, in which case the surface is either silvered or metallized; alternatively they may be of polished metal. Reflecting telescopes embody a concave primary mirror of glass or quartz, which is aluminized in preference to being silvered.

MK system or **MKK system** Another name for the **spectral-luminosity classification**. The abbreviation is derived from the initial letters of Morgan, Keenan and Kellman, who devised the system.

mock Moon or **paraselene** An image of the Moon produced in an analogous way to a **mock Sun**.

mock Sun or **parhelion** An image of the Sun, coloured or white, which is more frequently seen in polar regions and is usually at the same elevation as the Sun. When the Sun is near the horizon the angular separation between it and the mock Sun is 22°. The phenomenon is caused by refraction of sunlight by ice crystals suspended in the upper atmosphere.

moldavites see **tektites**

monochromatic light Light of one colour, i.e. of electromagnetic vibrations of the same or nearly the same frequency.

monochromator An optical device or instrument used to produce a narrow band of monochromatic light from light comprising many wavelengths. The required line or band in the spectrum is isolated by means of a prism, diffraction grating, or **Lyot filter**.

month The period of revolution of the Moon around the Earth. As the motion of the Moon is complicated, the period of its revolution around the Earth varies according to the choice of point to which the motion is referred. Hence the following months are distinguished. The ANOMALISTIC MONTH is the period between two successive transits of the Moon through its **perigee**. It equals 27·554 55 days. The DRACONIC MONTH is the period between two successive transits of the Moon through its ascending node. It equals 27·212 22 days. The SIDEREAL MONTH is the period between two successive conjunctions of the Moon with the same fixed star. It equals 27·321 66 days. The SYNODIC MONTH is the period between two identical phases of the Moon. It equals 29·530 59 days. The TROPICAL MONTH is the period between two successive transits of the Moon through the meridian of the vernal equinox. It equals

27·321 58 days. In the above examples the periods are reckoned by mean solar time.

The CALENDAR MONTH is one of the 12 divisions of the Gregorian calendar (see **calendar**), and is approximately equal in length to one synodic month.

Moon The only natural satellite associated with the Earth. Apart from the Sun it is the brightest object in the sky because of its proximity, being at a mean distance of only 384,000 km. It shines by light reflected from the Sun; the Full Moon has a visual magnitude of −12·7; at Half Moon the visual magnitude is about −10. An observer on Earth always sees the same side of the Moon because the Moon's orbital period around the Earth is the same as its axial rotation. Measurements of brightness, albedo, and polarization lead to the conclusion that the Moon has a rocky surface with areas of volcanic dust; this has been confirmed by results obtained with lunar roving vehicles and from visits by Apollo astronauts. The darker areas seen on the Moon are expanses of iron-rich lava known as maria (see **mare**). The brighter areas are rugged highland regions which occur predominantly in the southern part of the Moon's nearside and over the entire farside. The dust-covered surface is heavily marked with craters of all sizes. Additional features are mountain peaks and chains, valleys, elongated depressions known as rilles, and bright rays of crater ejecta radiating from the youngest craters. Because of the Moon's closeness these features have been studied, named, and depicted on maps that are comparable with large scale maps of the Earth. By analogy with geographical practice the lunar features are designated as oceans (oceanus), seas (mare), bays (sinus), lakes (lacus), swamps (palus), mountains (mons). Space probes have photographed the farside of the Moon, of which maps with named features have since been prepared. The Soviet Luna 3 obtained the first photographs of the farside on October 7, 1959; the American Lunar Orbiters also obtained photographs subsequently.

The origin of the Moon's craters has been ascribed to volcanic action and to the impact of meteorites. It is now believed that the majority were formed by impacting bodies from space. The largest craters, usually called BASINS, were produced during the early history of the Moon: this was when the bombardment was at its heaviest. The basins were subsequently filled with up-welling lava to produce the maria, which in turn became cratered. The activity observed by N. A. Kozyrev on the night of November 2–3, 1958 in the crater Alphonsus, as well as in other craters at other times by various observers, suggests that pockets of gas trapped under the lunar surface may occasionally erupt.

Water is absent on the Moon. The atmosphere is extremely attenuated and not detectable by optical methods: its existence was first revealed by deflection of radio-frequency radiation from the Crab Nebula during an occultation by the Moon on January 24, 1956. Atmospheric pressure on the Moon is about 10^{-13} of that on Earth. Any atmosphere formerly present would have been lost because of the low value of gravitational attraction, and hence the low escape velocity.

The chemical composition of material brought back from the Moon has been found to consist mainly of silica, iron oxide, aluminium oxide, calcium oxide, titanium dioxide, magnesium oxide, and other substances in minor amounts. Lunar samples show a higher titanium content than do terrestrial samples. Lunar rocks are igneous and usually have a basaltic composition; the highlands consist largely of anorthosite. The Moon, as regards its origin, is now considered to be the agglomeration of pre-existing particles of matter between four and five thousand million years ago.

The first men to land on the Moon were the Americans Neil Armstrong and Edwin Aldrin, who on July 21, 1969 stepped from Eagle, the lunar module of Apollo 11, on to Mare Tranquillitatis. Of the six subsequent Apollo missions, five reached the lunar surface in the period November 1969 to December 1972. As the Apollo programme progressed, each mission explored a larger area of the Moon and returned with an increasing mass of rock samples.

For data relating to the Moon see **satellite**.

A lunar landscape at Hadley Rille in the Apennine Mountains.

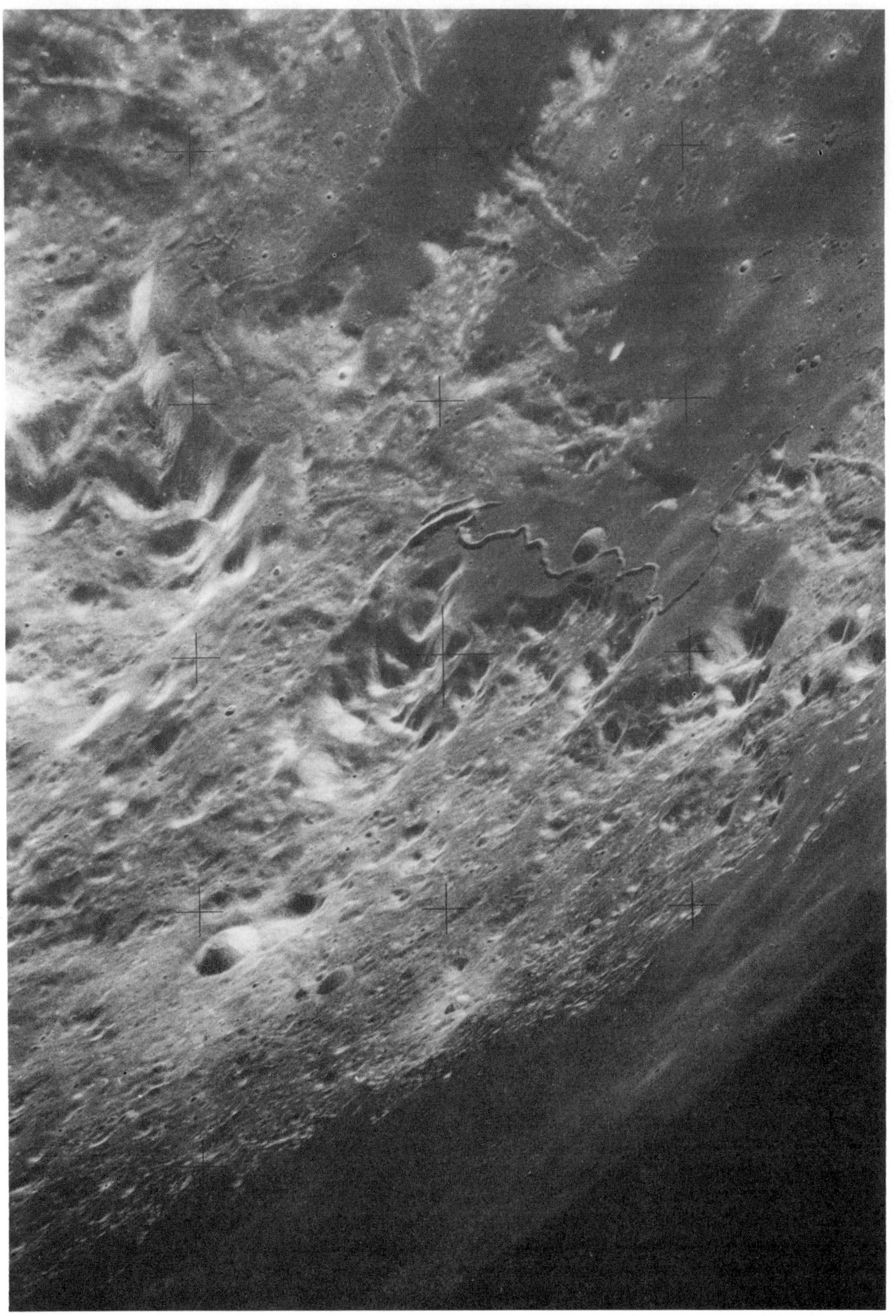

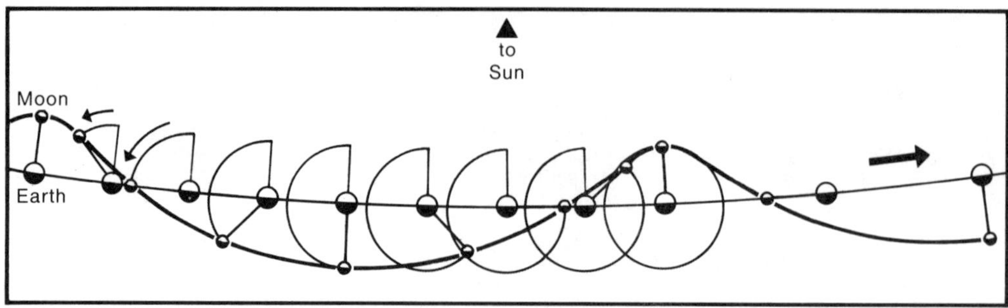

Motion of the Moon around the Earth and the Earth around the Sun: not to scale.

Moon, age of the The time that has elapsed since the last New Moon.

Moon, motion of the The Earth and Moon form a system in orbit around the Sun as shown in the diagram. The Moon's motion is an extremely complicated one on account of the perturbations to which it is subjected by the Sun, the planets, and other causes. Among the inequalities of the Moon's motion are the **equation of the centre, evection**, regression of the Moon's nodes, progression of the line of apsides (which gives the **anomalistic month**), **variation** and **secular acceleration** of the motion.

Moon, phases of the The succession of aspects of the Moon resulting from the changing amount of its surface illuminated by the Sun as the Earth and Moon orbit around the Sun. FULL MOON occurs when the Moon is at opposition (i.e. on the other side of the Earth from the Sun) and appears fully illuminated. NEW MOON occurs when the Moon is at conjunction (i.e. between the Earth and the Sun) and is apparently not illuminated. At intermediate positions the Moon appears as a crescent, gibbous, or as a Half Moon.

Morgan's classification A classification scheme for galaxies devised by W. W. Morgan (1906–). The spectral type (a, af, f, fg, g, gk, k) is written first and is followed by the form of galaxy as spiral (S), barred spiral (B), elliptical (E), irregular (I), elliptical showing dust absorption (Ep), rotational symmetry without any definite spiral or elliptical form (D), low surface brightness (L), small bright nucleus (N); and finally a number indicating the position of the plane of symmetry with respect to the line of sight, the number 1 indicating a face-on view and 7 an edge-on view.

morning star The planet Venus when it appears in the east before sunrise.

moving cluster see **cluster**

M stars see **stars, spectral classification of**

multiple star A star system which contains more than two stars and which behaves as a physical entity through the gravitational attractions of the components. Some of these systems comprise a binary system with an invisible third component, as with 61 Cygni.

mural circle or **mural quadrant** An early astronomical instrument used for measuring **Declination**. It consists of a graduated circle with a sighting arm and telescope, the whole firmly fixed to a wall in the plane of the meridian.

nadir The point on the celestial sphere vertically below the observer. It is diametrically opposite to the **zenith**.

naked-eye A term applied to observations made by the eye without the aid of any optical instrument.

NASA The National Aeronautics and Space Administration. The US organization responsible for all US nonmilitary space research. It was formed in October 1958.

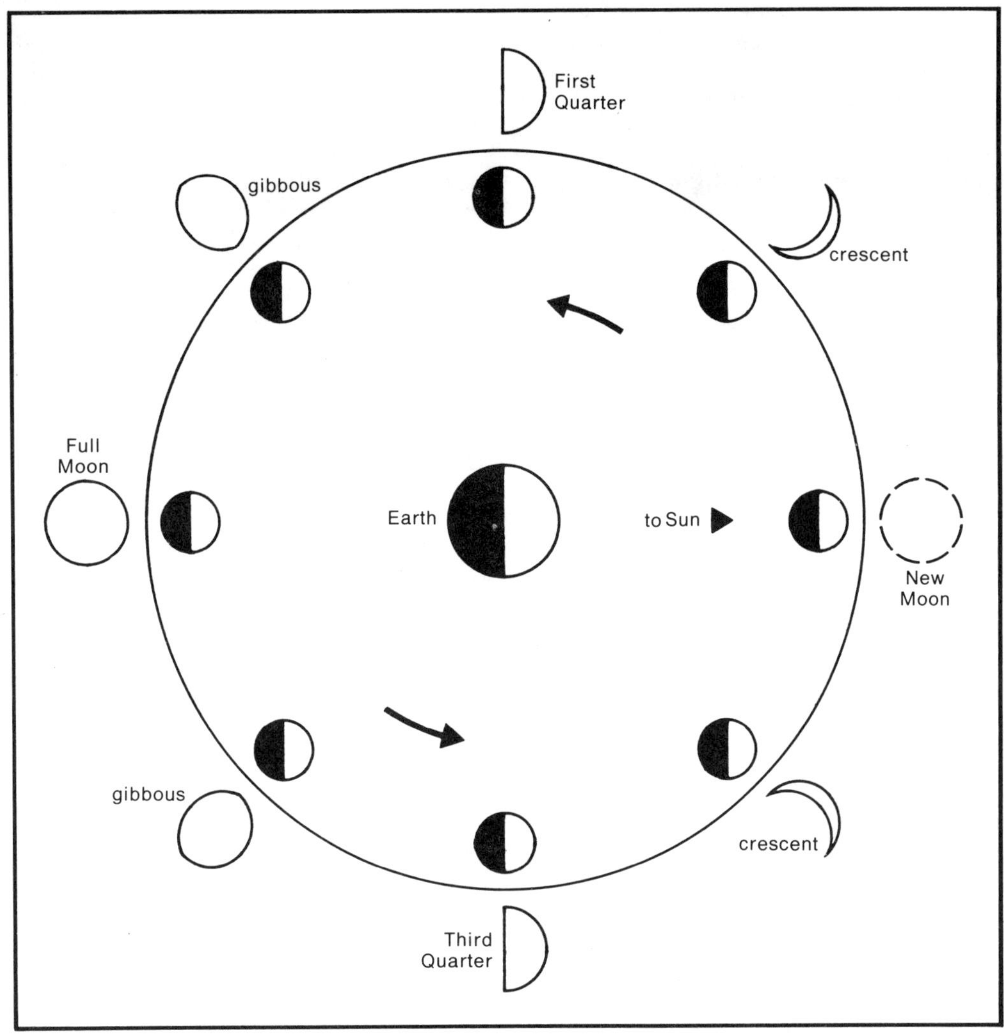

Phases of the Moon shown viewed from the visible hemisphere.

Nasmyth focus The prime focus in the **Newtonian-Cassegrain telescope**. It is so-called because it was used by James Nasmyth in his reflector.

Nasmyth, James (Edinburgh, August 19, 1808 – London, May 7, 1890) A British engineer and inventor. He was the first to observe the mottled appearance of the Sun's surface known as 'rice grains' or 'willow leaves' (1860). He made extensive observation of the Moon's surface features, and published in collaboration with J. Carpenter *The Moon Considered as a Planet, a World, and a Satellite* (1874). He was a supporter of the volcanic theory of the origin of the lunar features.

Nautical Almanac A special edition of *The Astronomical Ephemeris* published for navigational use. See **almanac**.

nautical astronomy or **navigational astronomy** That branch of practical astronomy which deals with the determination of longitude and latitude at sea (or in the air), i.e. determination of position by means of astronomical observations. With the development of space technology, navigation satellites have been placed in orbit to provide radio signals when bad weather prevents astronomical observations from being made.

nautical twilight see **twilight**

neap tide see **tides**

nebula A term formerly applied to any celestial body having a hazy cloudy appearance. It was subsequently found that some could be resolved into innumerable stars, i.e. were star clusters or galaxies, whilst others remained as diffuse masses of light. W. Herschel (see **Herschel, William**) divided them into six classes. The term is now restricted to objects which are contained within the Galaxy and are regions of interstellar gas and dust. There are three main classifications. EMISSION NEBULAE are bright diffuse nebulae which emit light and other radiation as a result of ionization and excitation of the gas atoms by ultraviolet radiation. The source of the ultraviolet is usually one or more hot stars. Recombination of gas ions and free electrons, and also forbidden transitions in excited atoms, leads to the emission of radiation and produces emission lines in the spectrum of such a region. Examples of emission nebulae include **H II regions** such as the **Orion Nebula, planetary nebulae** such as the Ring Nebula in Lyra, and **supernova remnants** such as the **Crab Nebula**.

In contrast, the brightness of REFLECTION NEBULAE results from the scattering by dust particles of light from nearby stars. DARK NEBULAE are not luminous. The interstellar gas and dust absorbs the light from the background stars and they appear as dark patches in the sky. This third class also comprises a group known as GLOBULES consisting of very dense absorption nebulae of nearly spherical shape. They are thought to be protostars.

The Horsehead Nebula in Orion.

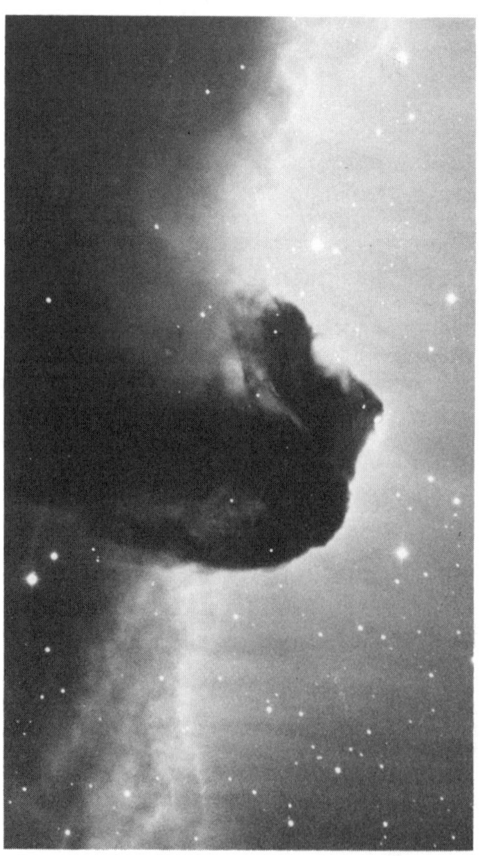

nebular hypothesis The theory that the solar system developed from a primeval nebula that was able to contract into a rotating disc. It has been put forward at various times in various ways: by I. Kant in 1755; by P. S. Laplace in 1796; and in modern times by C. F. von Weizsäcker, G. P. Kuiper and others. It is the basis of most recent ideas on the formation of the Sun and planets.

nebulium A hypothetical chemical element to whose existence the so-called **forbidden lines** in the spectra of certain nebulae were ascribed. The lines in question are now known to be produced by highly ionized gases, in particular, oxygen. The name was coined by W. Huggins (see **Huggins, William**).

Neptune The eighth planet in order of distance from the Sun and the most distant giant planet. It was first observed by J. G. Galle (see **Galle, Johann**) assisted by H. L. D'Arrest at Berlin on September 23, 1846, on the basis of predictions by U. J. J. Le Verrier (see **Le Verrier, Urbain**) and J. C. Adams (see

Adams, John) following investigations to account for perturbations in the orbital motion of the planet Uranus. Neptune is invisible to the naked eye. Observed through the telescope it appears as a small bluish-greenish disc on which very few details can be distinguished. Neptune has an atmosphere consisting mainly of methane, hydrogen, and ammonia. The globe of the planet is thought to consist either of a rocky core surrounded by a thick layer of ice under its atmosphere of methane and other gases, or of condensed hydrogen which is metallic in the central region. Neptune has two known **satellites**, Triton and Nereid. The main data relating to Neptune are given in the table.

Globe	
Diameter (equatorial)	50,000 km
Density (water = 1)	1·7 g cm^{-3}
Mass	1·03 × 10^{26} kg
Volume	6·1 × 10^{13} km^{3}
Sidereal period of axial rotation	15^{h} 48^{m}
Escape velocity	23·9 km s^{-1}
Albedo	0·84
Inclination of equator to orbit	28° 48′
Surface temperature	72 K maximum
Surface gravity (Earth = 1)	1·1
Orbit	
Semi-axis major	30·058 A.U. = 4496·7 × 10^{6} km
Eccentricity	0·008589
Inclination to ecliptic	1° 46′ 19″
Sidereal period of revolution	60,190·4 d
Mean orbital velocity	5·43 km s^{-1}

Nereid Satellite II of Neptune. See **satellite**.

Nernst Heat Theorem see **thermodynamics**

neutrino An elementary particle of rest mass zero and with no electrical charge. It is of importance for theories concerning the processes taking place in stars, as it is the particle that carries energy away during nuclear reactions.

neutron An electrically neutral elementary particle of mass 1·008665 amu, i.e. 1·6749 × 10^{-24} g. A free neutron decays into a proton, an electron, and an antineutrino. Neutrons and protons together constitute the atomic nucleus.

neutron star An extremely small dense star that in a late stage of evolution has undergone considerable **gravitational collapse** so that its component protons and electrons have been compressed into neutrons. Neutron stars have a diameter of only 10–20 km, a density of about 10^{13}–10^{15} g cm^{-3}, a core temperature of about 10^{9} K, and a strong magnetic field of about 10^{12} gauss. It was first suggested in the 1930s that when a **supernova** occurs, blasting off most if not all of the material in a star, any remaining stellar core would most likely be a neutron star. **Pulsars** are considered to be rotating neutron stars.

Newcomb, Simon (Wallace, Nova Scotia, March 12, 1835 – Washington, D.C., July 11, 1909) A Canadian-born American astronomer who prepared extremely accurate tables of the motions of the planets, Moon, and stars. He worked with A. A. Michelson (see **Michelson, Albert**) on the determination of the velocity of light.

New Moon The phase of the Moon when at conjunction. The dark side of the Moon then faces the Earth; the illuminated side is invisible to us.

newton (*symbol:* N) A derived unit in the **SI**. It is the unit of force, and is that force which gives an acceleration of 1 m s^{-2} to a mass of 1 kg.

Newtonian-Cassegrain telescope **1.** A modified form of **Cassegrain telescope** in which the light is reflected from the secondary mirror down the tube to a flat mirror inclined at 45°. This directs the rays to a Newtonian focus at the side of the tube near the primary mirror. Piercing of the primary mirror, as in the conventional Cassegrain construction, is thus avoided. **2.** A large reflector designed so that it can be used either as a Newtonian or Cassegrain instrument.

Newtonian focus The position towards the upper end of a telescope tube where the image formed by the optical system is viewed by an eyepiece placed at the side of the tube.

Newtonian telescope The simplest form of reflecting telescope, first made by Newton in 1668 and described by him in the *Philosophical Transactions* (1672) 7, No. 81, p. 4004. In its original form it consists of a primary concave spherical mirror of speculum metal which reflects the light up the tube of the telescope. A flat speculum mirror, placed at 45°, directs the light to the side of the tube where the image of the object being observed is seen through a small planoconvex eyepiece. C. Huygens (see **Huygens, Christiaan**) pointed out that better performance would result if the primary mirror were paraboloid; this has been the case since the technique of making such mirrors has been developed.

Newton, Isaac (Woolsthorp, December 25, 1642 – London, March 20, 1727) A brilliant English mathematician and natural philosopher. He is distinguished for his statement of the Laws of Motion and of Gravitation and for his discoveries in optics. He also developed fluxional calculus. His interest in the problem of gravitation led to his immortal work the *Philosophiae Naturalis Principia Mathematica* (1687). He built the first reflecting telescope.

Light path in a Newtonian telescope.

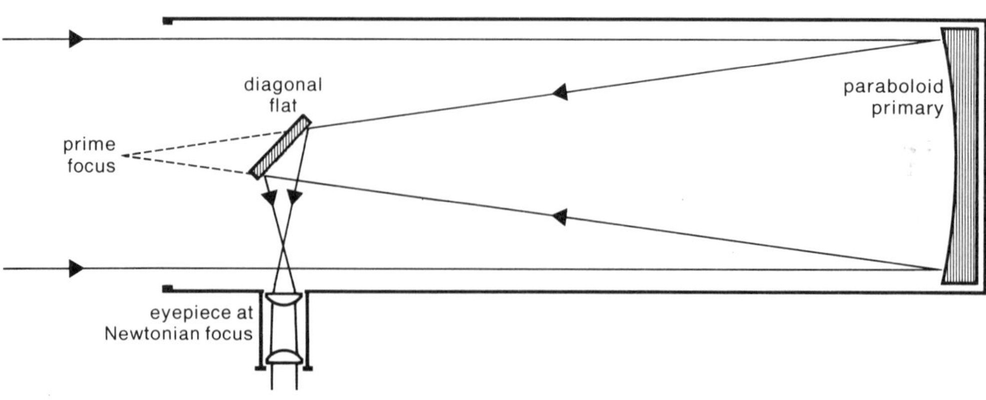

A facsimile of Newton's original reflecting telescope.

Newton's Law of Gravitation see **gravitation**

Newton's Laws of Motion The fundamental laws of classical mechanics formulated by Isaac Newton and set forth in his *Principia*. 1. LAW OF INERTIA. Every body continues in its state of rest, or of uniform motion in a straight line, until acted upon by some outside force. 2. LAW OF ACCELERATION. The acceleration of a body is directly proportional to the force acting upon it and is in the same direction of the straight line in which the force is acting. 3. LAW OF ACTION AND REACTION. To any action there is an equal and opposite reaction.

Sir Isaac Newton. National Portrait Gallery, London.

N galaxy A compact galaxy, intermediate in its properties between a **Seyfert galaxy** and a **quasar**. It has a small blue nucleus superimposed on a faint red background.

Nicholson, Seth Barnes (Springfield, Illinois, November 12, 1891 – Los Angeles, July 2, 1963) An American astronomer who specialized in solar and stellar astronomy. He discovered satellites IX, X, XI, and XII of Jupiter. With E. Pettit (1890–) he developed a thermocouple for measuring the surface temperature of planets, the Moon, and stars.

night-sky camera A fixed camera set to record the trails of stars resulting from the Earth's diurnal motion; the position of any interruption in the trails can be related to the time. It is used in connection with automated instruments to determine the reduction in effective exposures of astronomical photographs caused by cloud.

night-sky illumination The sky on a moonless night, although dark, yet receives weak illumination from various sources, namely **airglow, aurorae, galactic light, Gegenschein,** starlight (which is the main contributor), and **zodiacal light**.

night vision The diameter of the pupil of the eye depends on the intensity of the light to which it is exposed. On going from a place which is brightly illuminated into the darkness of night, the eye can take from five to fifteen minutes to adjust itself to maximum sensitivity so as to see the faintest stars that can be seen with the naked eye.

nimbus A bright ring surrounding a lunar surface feature.

nitrogen An inert gaseous chemical element which is an important constituent of the planetary atmospheres of Earth, Mars, and Venus and occurs as ammonia in the atmospheres of the four giant planets.

nitrogen sequence The spectral subclass WN of **Wolf-Rayet stars**. The spectra show emission lines of nitrogen.

N lines Two forbidden green lines produced by doubly ionized oxygen at 5007 Å and 4959 Å.

noctilucent clouds Clouds consisting of dust particles and thought to be of interplanetary origin. They appear long after sunset in the summer months, most frequently just after the solstice. They look like cirrus with a bluish-white to yellowish colour, occurring at a mean height of 80 km and moving usually from the north-east at 160 to 500 km per hour.

nodal line or **line of nodes** The line which joins the ascending and descending nodes of an orbit. It is the line of intersection of the planes of a planetary orbit and the ecliptic.

node The point in which one orbit cuts another. Specifically, one of the two points of the orbit of a planet or a comet in which it cuts the ecliptic, or in which the orbit of a satellite cuts that of its primary. The ASCENDING NODE is the node at which a celestial body passes to the north. The DESCENDING NODE is the node at which a celestial body passes to the south. The REGRESSION OF THE MOON'S NODES is the backward motion of the Moon's nodes around its orbit. They move slowly westward so that the nodal line makes one complete revolution of the orbit in 18·6 years. The regression is caused by the Sun's gravitational attraction.

nodical month or **draconic month** see **month**

North America Nebula A bright emission nebula (NGC 7000) in the constellation Cygnus. It is so called because of its slight resemblance to the North American continent.

Northern Cross The cross-shaped arrangement of the brightest stars in the constellation Cygnus.

Northern Hemisphere Observatory A new observatory for use by British and other astronomers, sited on La Palma in the Canary Islands. The Isaac Newton telescope will be moved there.

north point That point on the celestial sphere due north of the observer where the meridian intersects the horizon.

north polar distance (*abbrev.*: N.P.D.) The angular distance, measured in the meridian, between an object and the north celestial pole. It is equal to 90° minus the **Declination** of the object.

North Polar Sequence (*abbrev.*: N.P.S.) A series of accurately measured magnitudes of stars within 2° of the north celestial pole. It includes stars from magnitude 2 down to magnitude 20. The series was used to provide an arbitrary zero point on the magnitude scale. Comparison with the North Polar Sequence enables the magnitudes of other stars to be determined. At the present time, stellar standards of reference measured photometrically are preferred.

nova An existing star, usually quite faint, whose brightness suddenly increases for a short or a long period before slowly returning to its pre-nova state. The sudden surge of energy can increase the luminosity by as much as 14 magnitudes. The phenomenon is thought to be the result of some readjustment between the core of a star and its surface layers, which are ejected at speeds of the order of 1000 km s^{-1}. In a typical outburst the release of energy is about 10^{37} joules. Novae occur in close binary systems where one component is a cool red giant and the other is a hot smaller body, usually a **white dwarf**, which is in an unstable evolutionary condition. Matter transferred from the larger to the smaller star possibly triggers the eruption. Stars that undergo more than one outburst are known as RECURRENT NOVAE. Some bright novae that have been observed are Nova Persei (1901), Nova Aquilae (1918), Nova Herculis (1934), Nova Delphini (1967), and Nova Cygni (1975). See also **dwarf nova**; **supernova**.

Nubecula Major and **Nubecula Minor** The Large and Small **Magellanic Clouds** respectively.

nuclear reaction The process whereby one chemical element is transformed into another by the fusion or fission of atomic nuclei. The nuclei of lighter elements combine to form heavier ones. The mass of the resulting nucleus is less than the sum of the individual masses of the combining nuclei (see **mass defect**). The mass lost is emitted as radiant energy. The **carbon-nitrogen cycle** and the **proton-proton reaction** provide the sources of radiant energy of the Sun and similar stars.

nuclear time scale The time required for a star to convert all its available hydrogen into helium. In the case of the Sun it is 10^{10} years.

nucleon A general term for the proton and the neutron, i.e. for those elementary particles which constitute the composite nucleus, as well as their antiparticles.

nucleosynthesis The production of atomic nuclei through **nuclear reactions** in stellar interiors.

nucleus of a comet That part of a comet containing most of its mass, consisting of solid particles and frozen volatile matter, such as water, carbon dioxide, methane, ammonia, and cyanogen. The coma and tail of the comet are formed from this material and shine partly by reflection of sunlight from the solid matter and partly by fluorescence of the gaseous components.

nucleus of an atom The charged central part of an atom consisting of protons and neutrons. Electrons revolve around the nucleus in one or more shells. There are stable and radioactive nuclei. The radii of all nuclei are of the order of 10^{-12} cm, and are directly proportional to the cube root of their masses. As nearly all the mass of an atom is contained in such a tiny volume, the nuclear matter has a density of about $1{\cdot}4 \times 10^{14}$ g cm^{-3}.

nutation The periodic oscillation in the precessional motion of the Earth's axis of rotation, discovered by J. Bradley (see **Bradley, James**) in 1747. It is in effect a slight 'nodding' of the Earth's axis, as the term implies. It is caused by the combined effect of the gravitational attractions of the Sun and Moon upon the Earth: the effect is constantly varying as the relative positions of those bodies are continuously changing. The LUNAR NUTATION, which causes the Earth's axis to describe an ellipse, has a period of 18·6 years. The SOLAR NUTATION has a period of 0·5 of a tropical year. The FORTNIGHTLY NUTATION has a period of 15 days.

OB association see **stellar association**

Oberon Satellite IV of Uranus. See also **satellite**.

objective A light-gathering element of an optical instrument. In a refracting telescope it is the lens or lens system, also known as the OBJECT GLASS, that collects light from a celestial body and forms an image of that object. The term is also sometimes applied to the primary mirror of a reflecting telescope. The light-gathering power of a telescope objective increases as its area is increased, i.e. with the square of its aperture.

objective prism A large prism of small angle mounted in front of the objective of a refractor. The image of each star in the field of view is transformed into the image of its spectrum. From photographs taken by this means it is possible to make a rapid examination of the stars in a given area and to classify them in respect of spectrum or luminosity.

oblateness A measure of the deviation of a spheroid from a true sphere. (A spheroid is the solid produced by the revolution of a semi-ellipse about its semi-axis minor.) The polar flattening of a celestial object is given by the ratio of the difference between the equatorial and polar radii to the equatorial radius. It is an indication of the velocity with which a body is rotating.

oblate spheroid see **spheroid**

oblique ascension The geocentric longitude of a body measured along the ecliptic from the First Point of Aries, the Right Ascension being measured along the celestial equator. It is so called because the planes of the ecliptic and the equator are inclined to each other.

obliquity of the ecliptic The angle between the plane of the ecliptic and the plane of the celestial equator. Alternatively it is the angle between the axis of rotation of a planet and the pole of its orbit. The Earth's obliquity is about 23° 27′, and is decreasing by 0·47″ per annum. The angle will start to increase again after about 1500 years. The obliquity is responsible for the seasons of a planet. Its value represents the greatest angular distance that the Sun can lie north and south of the equator.

observatory A building or establishment arranged and fitted with the necessary instruments for making systematic observations of natural phenomena, e.g. astronomical, geophysical, meteorological, and seismological. The site for an astronomical observatory has to be chosen with care. When considering optical and infrared observatories, the vicinity of towns has now to be avoided because of atmospheric pollution and the glare produced by street lighting, which have spoilt conditions for good seeing. The best location for astronomical purposes is on a mountain or a high plateau. Even then meteorological conditions have to be considered, for the site must offer the likelihood of a reasonable number of cloudless or near cloudless nights for observational work. That is why the largest observatories are in South Africa, Australia, Chile, and on the west coast of America.

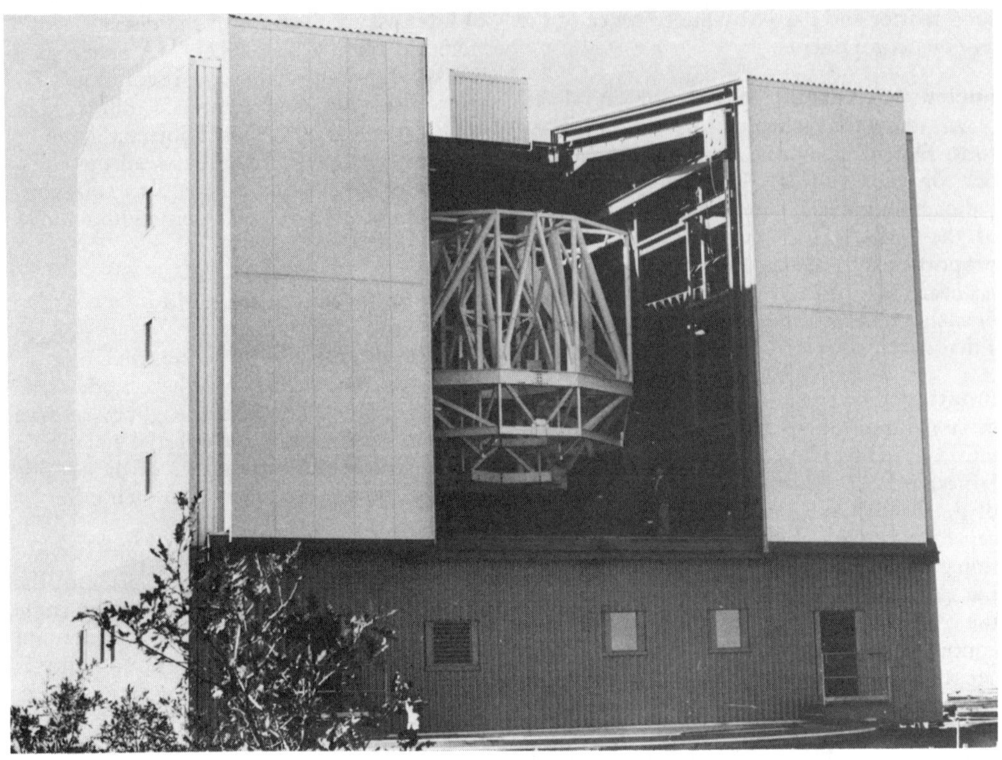

The multiple-mirror telescope at Mount Hopkins, Arizona, which is operated jointly by the Smithsonian Astrophysical Observatory and the University of Arizona.

Optical and infrared telescopes of any considerable size are installed in domed structures. The hemispherical dome is rotatable and has a slit, closed by sliding shutters when not in use, through which observations are made. In addition there have to be laboratories, photographic dark rooms, workshops, and living accommodation for personnel. There are over 200 professional optical observatories distributed throughout the world.

Many radio observatories have come into existence since 1945, subsequent to the development of radar in wartime. Their sites are less critical than those of optical observatories because radio measurements are not greatly affected by poor seeing, clouds, etc. Radio telescopes consisting of an array of antennae do however require a considerable area of land for their installation.

The accompanying lists, which are by no means complete, contain the names of certain well-known optical and radio observatories. They indicate briefly the location, date of foundation, and main interests.

The latest observatories are those being launched into Earth orbit, where the Earth's atmosphere has no absorbing or distorting effects. In addition to optical and radio studies these craft carry out measurements in the ultraviolet, X-ray, and gamma-ray region of the electromagnetic spectrum. Such measurements are not possible from ground-based observatories.

Optical observatories

Abastumani (Georgia, USSR). Altitude 1580 m. Founded 1937. Astrophysical Observatory of the Academy of Sciences. Spectroscopy, photometry, Sun and planets.

Allegheny (Pittsburgh, USA). Altitude 370 m. Founded 1860. Observatory of the University. Spectroscopy, positions of stars.

Alma-Ata (Kazakh, SSR). Altitude 2608 m. Founded 1948. Mountain observatory of the

Academy of Sciences. Spectroscopy, photometry, theoretical astrophysics.

Armagh (Northern Ireland). Altitude 64 m. Founded 1790. Armagh Observatory. Star fields. There is a planetarium in the grounds of the observatory.

Berlin-Babelsberg (East Germany). Altitude 82 m. Founded 1705. Observatory of the German Academy of Sciences. Photometry, positions of stars.

Budapest (Hungary). Altitude 474 m. Founded 1926. Konkoly Observatory. Spectroscopy, photometry.

Burakan (Armenia, USSR). Founded 1952. Astronomical Observatory of the Academy of Sciences. Faint objects.

Cambridge (Great Britain). Altitude 28 m. Founded 1820. University Observatory. Spectroscopy, photometry, Sun.

Cambridge (Massachusetts, USA). Altitude 24 m. Founded 1830. Harvard College Observatory. Photometry, faint objects.

Canberra (Australia). Altitude 768 m. Founded 1924. Mount Stromlo Commonwealth Observatory. Spectroscopy, photometry, faint objects, positions.

U.S. Navy telescope at Flagstaff, Arizona.

Cape of Good Hope (South Africa). Altitude 10 m. Founded 1820. South African Astronomical Observatory. Photometry, Sun, positions.

Castel Gandolfo (Italy). Altitude 45 m. Founded 1578. Vatican Observatory. Spectroscopy, photometry, positions.

Córdoba (Argentina). Altitude 434 m. Founded 1870. National Observatory of Argentina. Photometry, spectroscopy.

Dublin (Ireland). Altitude 86 m. Founded 1785. Dunsink Observatory. Photometry, Sun.

Edinburgh (Scotland). Altitude 146 m. Founded 1818. Royal Observatory. Photometry, spectroscopy, tracking.

Flagstaff (Arizona, USA). Altitude 2210 m. Founded 1894. Lowell Observatory. Spectroscopy, faint objects, planets, positions.

Fort Davis (Arizona, USA). Altitude 2081 m. McDonald Observatory.

Groningen (Netherlands). Altitude 4 m. Founded 1962. Kapteyn Astronomical Observatory. Stellar astronomy.

Hamburg-Bergedorf (West Germany). Altitude 41 m. Founded 1823. Hamburg Observatory. Photometry, spectroscopy, faint objects, Sun, positions.

Heidelberg-Königstuhl (West Germany). Altitude 570 m. Founded 1775. State Observatory. Spectroscopy, photometry, positions.

Helwan (Cairo, Egypt). Altitude 115 m. Founded 1905. Spectroscopy, photometry.

Herstmonceux (England). Altitude 34 m. Founded 1675. Royal Greenwich Observatory. Spectroscopy, photometry, faint objects, Sun, positions. This is the successor of the Royal Observatory, Greenwich, whose longitude was adopted in 1884 as the zero of longitude.

Jena (East Germany). Altitude 164 m. Founded 1955. University Observatory. Photometry, theoretical astrophysics.

Kodaikanal (India). Altitude 2343 m. Founded 1792. Astrophysical Observatory. Solar physics, positions.

Leiden (Netherlands). Altitude 6 m. University Observatory. Theoretical astronomy and astrophysics.

Lembang (Indonesia). Altitude 1300 m. Founded 1926. Photometry, spectroscopy, positions.

Meudon (Paris, France). Altitude 162 m. Founded 1670. Observatory of Physical Astronomy. Spectroscopy, photometry, solar physics, positions.

Mount Hamilton (California, USA). Altitude 1283 m. Founded 1875. Lick Observatory of the

University of California. Photometry, faint objects, positions.

Mount Wilson (California, USA). Altitude 1742 m. Founded 1904. Hale Observatory of the Carnegie Institution of Washington. Faint objects, solar physics.

Nanking (China). Altitude 367 m. Founded 1934. Purple Mountain Observatory. Photometry, solar physics, positions.

Narrabri (New South Wales, Australia). Founded 1963. University Observatory. Interferometry for stellar diameters.

Ottawa (Ontario, Canada). Altitude 58 m. Founded 1910. Ottawa River Solar Observatory. Solar physics, positions.

Oxford (England). Altitude 64 m. Founded 1868. University Observatory. Solar physics.

Palomar Mountain (California, USA). Altitude 1706 m. Founded 1948. Hale Observatory of the Carnegie Institution of Washington and the California Institute of Technology. Photometry, faint objects.

Pic du Midi (France). Altitude 2862 m. Founded 1930. Observatory of the University of Toulouse. Spectroscopy, solar physics, positions.

Potsdam (East Germany). Altitude 107 m. Founded 1878. Astrophysical Observatory of the Academy of Sciences. Photometry, spectroscopy, solar physics, positions.

Pretoria (South Africa). The Radcliffe Observatory (Oxford, England), founded in 1772, was moved to Pretoria in 1948. Spectroscopy, photometry.

Pulkovo (Leningrad, USSR). Altitude 75 m. Founded 1839. Astronomical Observatory of the Academy of Sciences. Spectroscopy, photometry, solar physics, faint objects, positions.

St Michel (France). Altitude 651 m. Founded 1940. Observatory of Haute-Provence. Spectroscopy, photometry, solar physics.

Saltsjöbaden (Sweden). Altitude 55 m. Founded 1748. Stockholm Observatory of the Royal Swedish Academy of Sciences. Spectroscopy, photometry, positions.

La Serena (Chile). Altitude 2399 m. Founded 1962. Cerro Tololo Interamerican Observatory. Spectroscopy, photometry, planets.

La Serena (Chile). Altitude 2282 m. Las Campanas Observatory, one of the Hale Observatories of the Carnegie Institution of Washington. Spectroscopy, photometry, faint objects.

Siding Spring (New South Wales, Australia). Altitude 1165 m. Siding Spring Observatory housing the Anglo-Australian Telescope and the UK Schmidt telescope. Stars, galaxies, spectroscopy, photography.

La Silla (Chile). Altitude 2400 m. Founded 1962. European Southern Observatory. Photography, faint objects, positions, spectroscopy, photometry.

Simeis (Ukraine, USSR). Altitude 676 m. Founded 1948. Crimean Astrophysical Observatory. Spectroscopy, photometry, solar physics, theoretical astrophysics.

Skalnate Pleso (Czechoslovakia). Altitude 1783 m. Founded 1943. Observatory of the Slovak Academy of Sciences. Photometry, solar physics, positions.

Tartu (Estonia, USSR). Altitude 67 m. Founded 1810. Astronomical Observatory of the Academy of Sciences. Spectroscopy, photometry.

Tautenberg (near Jena, East Germany). Altitude 331 m. Founded 1960. Karl Schwarzschild Observatory of the Academy of Sciences. Faint objects.

Tokyo (Japan). Altitude 59 m. Founded 1920. Tokyo Astronomical Observatory at Mitaka. Spectroscopy, photometry, solar physics, tracking.

Tonantzintla (Mexico). Altitude 2150 m. Founded 1942. National Astrophysical Observatory. Spectroscopy, photometry, faint objects.

Tucson (Arizona, USA). Altitude 2064 m. Founded 1959. Kitt Peak National Observatory. Photometry, polarization, faint objects, solar physics.

Turku (Finland). Altitude 28 m. Founded 1934. University Observatory. Photometry.

Uccle (Brussels, Belgium). Altitude 105 m. Founded 1887. Royal Observatory. Spectroscopy, photometry, solar physics, positions.

Uppsala (Sweden). Altitude 21 m. Founded 1730. University Astronomical Observatory. Faint objects.

Victoria (British Columbia, Canada). Founded 1917. Dominion Astrophysical Observatory. Spectroscopy.

Washington (D.C., USA). Altitude 84 m. Founded 1830. U.S. Naval Observatory. Planets, positions, time.

Williams Bay (Wisconsin, USA). Altitude 834 m. Founded 1897. Yerkes Observatory. Spectroscopy, photometry, polarization, positions. The observatory's 40-inch refractor is largest ever made.

Zelenchukskaya Stanitsa (Northern Caucasus, USSR). Altitude 2100 m. Special Astrophysical Observatory of the Academy of Sciences. This observatory erected on Mount Pastukhov houses the world's largest reflector, whose mirror has a diameter of 6 m.

Zürich (Switzerland). Altitude 2050 m. Founded 1855. Federal Observatory. Solar physics.

Radio observatories

Arecibo (Puerto Rico). Altitude 496 m. Founded 1963. Arecibo Observatory of Cornell University. Largest radio telescope in use: the dish has a diameter of 1000 feet. Planets, radar, interstellar gas.

Berlin-Adlershof (East Germany). Altitude 50 m. Founded 1958. Heinrich Hertz Institut. Sun, interstellar gas.

Cambridge (England). Altitude 26 m. Founded 1945. Mullard Radio Astronomy Observatory of the University of Cambridge. Sun, planets, radio sources.

Crimea (Ukraine, USSR). Altitude 550 m. Crimean Astrophysical Observatory. Sun, planets, interstellar gas.

Dwingeloo (Netherlands). Altitude 25 m. Founded 1955. University of Leiden, Foundation for Radio Astronomy. Interstellar gas.

Effelsberg (near Bonn, West Germany). Altitude 366 m. Max Planck Institute for Radio Astronomy.

Goldstone (California, USA). Altitude 1038 m. Founded 1958. Jet Propulsion Laboratory. Planets, satellites, radar.

Green Bank (West Virginia, USA). Altitude 823 m. National Radio Astronomy Observatory.

Hoskinstown (New South Wales, Australia). Altitude 732 m. Molonglo Radio Observatory.

Jodrell Bank (Cheshire, England). Altitude 70 m. Founded 1949. Nuffield Radio Astronomy Laboratories of the University of Manchester. Planets, radar, interstellar gas, radio sources.

Kitt Peak (Arizona, USA). Altitude 1920 m. Founded 1968. National Radio Astronomy Observatory. Planets, interstellar gas, radio sources.

Malvern (England). Altitude 20 m. Founded 1948. Royal Radar Establishment. Planets, meteors, radar, radio sources.

Parkes (New South Wales, Australia). Altitude 392 m. Founded 1952. CSIRO, National Radio Astronomy Observatory. Interstellar gas, radio sources. CSIRO has a Solar Observatory at Culgoora, N.S.W.

Pulkovo (Leningrad, USSR). Altitude 70 m. Founded 1957. Astronomical Observatory of the Academy of Sciences. Sun, planets, polarization, radio sources.

Stanford (California, USA). Altitude 80 m. Founded 1959. Radio Astronomy Institute of Stanford University. Sun, Moon, planets, radar, radio sources.

Zelenchukskaya Stanitsa (Northern Caucasus, USSR) Altitude 2100 m. Academy of Sciences Astrophysical Observatory housing the 600-m RATAN.

occultation The temporary cutting off of the light from a celestial body by the passage of another over its face. For example, the Moon may pass in front of a star; a planet may do the same and may also occult its own satellites, as in the case of Jupiter. Strictly speaking, a solar eclipse is an occultation of the Sun by the Moon. Occultations of stars, whose positions are known accurately, are valuable for determining accurately the position of the Moon by timing the moment of those events.

occulting bar A bar placed in the focal plane of the eyepiece of a telescope for the purpose of obscuring an object in the field of view. It is of particular use when a bright object has to be obscured so that a nearby faint one can be seen.

occulting disc An alternative to the occulting bar, differing only in shape.

ocular see **eyepiece**

off-axis telescope A modified Newtonian telescope in which only one half of the primary mirror is used. The light falling on the mirror is slightly off-axis so that the diagonal mirror can be moved out of the path of light coming from the object under observation. See illus. on page 120.

Olbers, Heinrich Wilhelm Matthäus (Arbergen, October 11, 1758 – Bremen, March 2, 1840) A German astronomer and physician who is remembered chiefly for what is now known as **Olbers' paradox**. He discovered the minor planets Pallas (1802) and Vesta (1807). He put forward the hypothesis that minor planets between Mars and Jupiter were the fragmented remains of a former planet. He also proposed the theory that the tails of comets are highly rarified

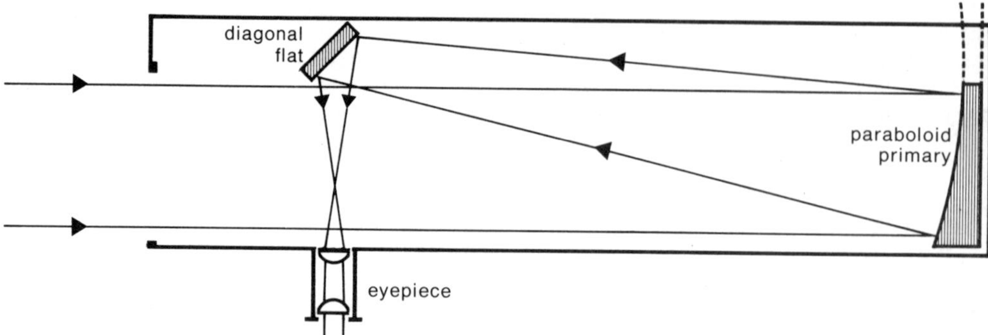

Light path in an off-axis telescope.

matter expelled from the head of the comet by pressure of some kind from the Sun, thus anticipating the concept of radiation pressure.

Olbers' paradox The paradox that arises in attempting to answer the question: why is the sky dark at night? H. W. M. Olbers was not the first to discuss this problem. E. Halley (see **Halley, Edmond**) had done so, besides others, before Olbers published his paper on the subject in 1823. If the number of stars is infinite and space is infinite, then, according to Olbers, the cumulative effect of all this radiation would be to make the night sky intensely bright and to produce on Earth a surface temperature of more than 5000 K. To account for the observed appearance of the night sky he put forward the hypothesis that light is absorbed by the homogeneous medium through which it passes. The various assumptions made by Olbers in trying to resolve the problem are either inaccurate or incorrect. It is now believed that the night sky is dark because of the **red shift** of light emitted by receding galaxies, which causes a diminution in the intensity of the light as observed, and because of the finite age of the universe. The conclusion is that the night sky is dark because the universe is expanding.

Oort cloud A reservoir of a 'cloud' of comets in orbit around the Sun at a distance of about 10^5 A.U., where they are loosely bound gravitationally. If one of these comets is perturbed for any reason and is displaced towards the Sun, then it may be further perturbed by a planet (such as Jupiter) and thus driven into a smaller orbit. The concept was postulated by J. H. Oort.

Oort, Jan Hendrik (Franeker, April 28, 1900–   ) A Dutch astronomer distinguished for his researches on the structure and dynamics of stellar systems, especially the Galaxy, the rotation of which he discovered. He was also a pioneer of radio astronomy, in particular with respect to the 21-cm radiation of interstellar hydrogen. In 1950 he proposed the existence of the so-called Oort 'cloud' of comets.

Oort's constants The constants A and B in the formulae derived by J. H. Oort for the differential rotation of the Galaxy.

$A = 15\,\mathrm{km\,s^{-1}\,kpc^{-1}}$; $B = -10\,\mathrm{km\,s^{-1}\,kpc^{-1}}$.

open cluster see **cluster**

opposition The position of a body in the solar system when it lies exactly opposite the Sun in the sky. It is the most favourable position for observing the body. See also **aspect**.

optical axis The central line through the mid-points of all the elements, such as lenses and mirrors, in an optical system. It is not necessarily the light-path, which may be inclined to the optical axis.

optical double Two stars that appear to be close together, but are not components of a single system and are in fact separated by great distances. They appear close together because they happen to be in almost the same line of sight.

optical figuring The final stage of fine-grinding/polishing of the surface of glass components of optical instruments, such as the lens or mirror of a telescope, to achieve the required form.

optical flat A plane surface of glass which for astronomical use has been carefully polished under controlled conditions so that deviations from a true plane are not in excess of one-tenth of a wavelength of light.

optical glass Glass developed specifically for use in optical instruments. It differs from ordinary glass mainly in respect of the high purity of the materials used in manufacture, especially freedom from iron which imparts a yellowish-green colour. Optical glass must be free from inclusions of air or other substances, show no streaks, have a high transmission for light, and not be affected by atmospheric conditions. CROWN GLASS contains a high proportion of silica and has a higher refractive index than FLINT GLASS, which contains much less silica with a high percentage of lead oxide and has a higher refractive index. Combinations of these two glasses make it possible to reduce undesirable optical effects such as chromatic aberration. Borosilicate glasses, of which PYREX GLASS is typical, have a high percentage of silica with a significant amount of boron oxide. They have a low coefficient of thermal expansion and have found wide application, particularly for the mirrors of reflecting telescopes.

optical temperature The temperature of a celestial body as calculated from its optical radiation.

optical wedge A plate of glass or quartz bearing an obscuring film whose density increases continuously from one end to the other so that the transmission of light is a maximum at one end and zero at the other. Wedges are used in conjunction with various apparatus when it is required to reduce the image of a very bright object.

optical window The radiation ranging from near ultraviolet through visible to near infrared wavelengths that is almost unaffected by atmospheric absorption and can therefore reach the Earth's surface. It covers the range of wavelengths between about 3000 Å and 9000 Å. Lower-wavelength radiation is absorbed by the atmosphere. Most higher wavelengths, with the exception of the **radio window** and some narrow infrared windows, are also absorbed.

optics That branch of knowledge which treats of the properties of light. GEOMETRICAL OPTICS, based upon the laws of reflection and refraction and the rectilinear propagation of light, is concerned with the formation of images in optical instruments.

orbit The path of a celestial body under the influence of gravitation. The path is usually a closed one about the focus of the system to which it belongs, as with those of the planets of the solar system about the Sun, or of the components of a binary system about their common centre of mass. Most orbits are elliptical, though the eccentricities can vary greatly. It is rare for an orbit to be parabolic or hyperbolic. See illus. on page 122.

In order to define the size, shape, and orientation of the orbit of a celestial object in space, seven quantities must be determined by observation. These are known as ORBITAL ELEMENTS. In the case of a planetary orbit (see illus. on page 122), only six are needed. They are as follows:

- a the semi-axis major, in A.U.
- e the eccentricity
- i the inclination of the orbital plane to the plane of the ecliptic
- Ω the longitude of the ascending node
- $\tilde{\omega}$ the longitude of perihelion
- T the epoch (the time of perihelion passage)

In the case of the orbit of a binary star system a seventh element is needed, if the mass is not known; this is the period P, or alternatively the mean motion n.

In general, the same considerations apply in determining the orbit of a satellite as in the case of determining the orbit of a planet, except that the inclination is usually referred to the equatorial plane of the primary planet, instead of to the plane of the ecliptic.

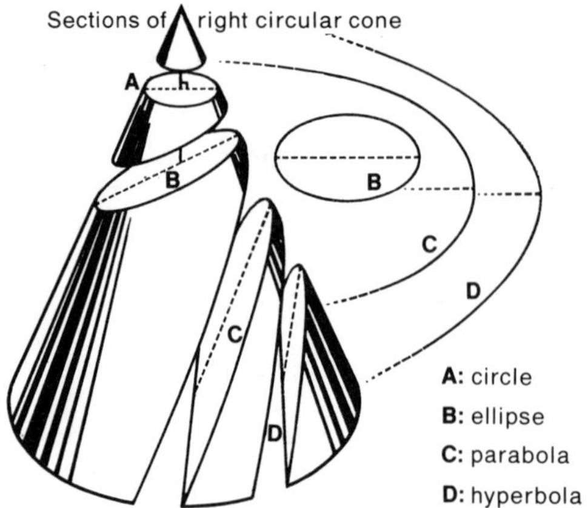

Orbits: their relations to conic sections.

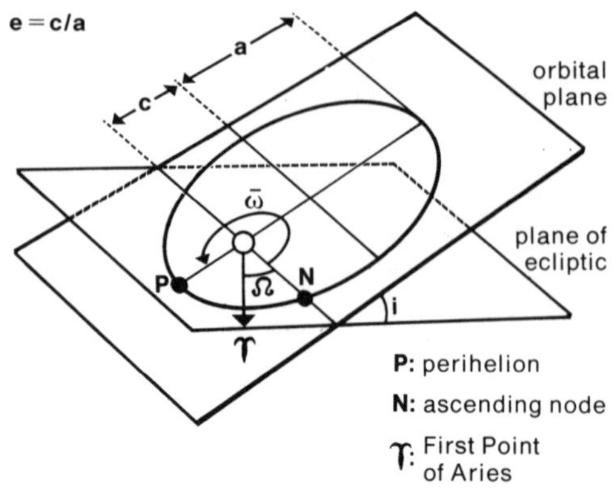

Orbital elements of a planetary orbit.

orbital element One of the numerical quantities needed to define an orbit, or the position of a body in its orbit at a given time. See **orbit**.

orbital inclination 1. OF A PLANET. The angle between the orbital plane of the planet and the plane of the ecliptic. **2.** OF A SATELLITE. The angle between the orbital plane of the satellite and the equatorial plane of the planet. **3.** OF A DOUBLE STAR. The angle between the orbital plane of the double star and the plane which is tangential to the celestial sphere in the direction of the stars. The angle is zero if the orbit is seen in plan, and 90° if seen in profile.

orbital velocity The linear velocity of a body in its orbit, usually expressed as the mean orbital velocity, which ignores variations in velocity resulting from the eccentricity of an elliptical orbit. The mean orbital velocities of the planets are given in the data relating to the individual planets.

Orbiter One of the components of the Space Shuttle System. See also **Lunar Orbiter**.

Orbiting Astronomical Observatory (*abbrev.:* OAO) The name of a series of American satellites equipped with telescopes and other scientific apparatus for the detection of ultraviolet and X-ray radiation. OAO 2 was launched on December 7, 1968 and continued to function until February 1973. OAO 3, launched on August 21, 1972, was named *Copernicus* to commemorate the quincentenary in 1973 of the birth of the Polish astronomer.

Orbiting Geophysical Observatory (*abbrev.:* OGO) The name of a series of American satellites destined for geophysical studies of the ionosphere, Van Allen belts, etc. OGO 1 was launched on September 5, 1964.

Orbiting Solar Observatory (*abbrev.:* OSO) The name of a series of American satellites placed in orbit round the Earth for the purpose of solar studies. The first was launched on March 7, 1962. Its equipment included an X-ray spectrometer for the study of radiation emitted during solar activity.

Orion An impressive constellation in the equatorial zone of the sky containing as principal stars **Betelgeuse, Rigel, Bellatrix,** and **Saiph**. These form the great quadrilateral that marks the shoulders and feet of Orion, the Hunter. Between them are three stars called ORION'S BELT. Below the Belt are three stars that mark ORION'S SWORD. The constellation provides a fine example of a multiple star (θ Orionis, the Trapezium), and examples of double stars, irregular variable stars, and clusters. The whole area of the constellation is covered by a bright diffuse

The Orion Nebula or Great Nebula in Orion.

nebulosity, which includes the famous **Orion Nebula** and also the **Horsehead Nebula**. In the region of the Trapezium are radio sources and molecular clouds containing infrared sources.

Orion arm or **local arm** The spiral arm of the Galaxy, about 10 kpc from the galactic centre. The Sun is located on the inner side of that part of the arm known as the ORION SPUR.

Orion association see **Orion Nebula**

Orionids A meteor shower whose radiant is in the north part of Orion. It occurs between October 16 and 26, and results from the Earth passing close to the orbit of Halley's Comet when debris from the comet is encountered. The Eta Aquarids also are associated with this comet.

Orion Molecular Clouds Two dense interstellar clouds of molecules in the constellation Orion. CLOUD 1 is an infrared source near the Trapezium. It contains the **BN object** and the **KL nebula**. CLOUD 2 is another infrared source situated near the Trapezium.

Orion Nebula (M42, NGC 1976) or **Great Nebula in Orion** A gaseous emission nebula, known from antiquity, visible to the naked eye as a faint diffuse greenish illumination around θ Orionis, the middle star of Orion's Sword. It consists of interstellar dust and gas (mainly hydrogen). The luminosity results from the glowing gaseous matter and also from the reflection of light by dust particles. It is an **HII region** and most probably one where new stars are being formed. It is also a source of X-rays. The nebula contains many high luminosity stars of spectral type O, loosely grouped, which are known as the ORION ASSOCIATION. See picture on page 123.

Orion's Belt see **Orion**

Orion spur see **Orion arm**

Orion-type stars Stars of spectral type B, so called because several major stars of Orion are of this kind.

orrery A mechanical machine for demonstrating the relative motions of heavenly bodies. Orreries vary from simple models showing only the Earth-Moon-Sun system to highly complicated ones of the whole solar system, where the planets with their satellites not only revolve in orbit but also rotate on their axes. The machine is named after Charles Boyle, fourth Earl of Orrery, who bore the expense of one constructed by John Rowley in 1715, after a pattern by George Graham. The original orrery made for Rowley's patron was acquired by the Science Museum, London, in 1974.

oscillating universe A variant of the **big-bang theory** in which it is suggested that the universe is continuously passing through cycles of expansion and contraction (or col-

An example of an orrery.

lapse), so that there can be regular occurrences of a 'big bang'.

osculating orbit The path that an orbiting body, such as a planet or a comet, would follow if it were not subject to any perturbations, but were subject only to the inverse square law of attraction by the Sun or other central body. The orbital elements, in this case, are called OSCULATING ELEMENTS, and are used in calculating perturbations. The starting point is called the OSCULATING EPOCH.

outer planets The distant planets Jupiter, Saturn, Uranus, Neptune, and Pluto, which lie beyond the belt of minor planets.

oxygen A chemical element which is an important constituent of the Earth's **atmosphere**, and the most abundant element in the Earth's crust, combined with other elements to form rocks and water. It is not present in the atmospheres of the other planets except in very small amounts. It occurs in the Sun and in stars of all spectral types; in types O and B it is ionized. Titanium oxide has been detected in the spectra of K and M type stars.

Ozma project A project established by the National Radio Astronomy Observatory, Green Bank, West Virginia, under the superintendence of F. D. Drake in 1960 to try and detect interstellar radio signals from possible intelligent life in other planetary systems. It was so named 'for the queen of the imaginary land of Oz – a place very far away, difficult to reach, and populated by exotic beings'. The project obtained no positive results and was dropped after some months.

ozone A gaseous molecule comprising three atoms of oxygen singly linked to form a ring. It occurs in the Earth's **atmosphere** about 20 to 60 km above the surface. It absorbs the ultraviolet radiation shorter than 2950 Å which comes from space.

Pallas The second minor planet to be discovered. It was found by H. W. M. Olbers on March 28, 1802. It is the second largest minor planet, the diameter being about 608 km and the mass about $2{\cdot}6 \times 10^{19}$ kg.

Palomar Sky Atlas Large-scale photographic star charts (totalling nearly 2000) of the entire sky visible from the Palomar Observatory in southern California. Each region was photographed twice (in red and blue light) with the 48-inch Schmidt telescope, and includes objects down to the 20th magnitude. This vast undertaking was sponsored and financed by the National Geographic Society, Washington, D.C.

Pan The proposed name for satellite XI of **Jupiter**. The name Carme was ultimately adopted.

parabola A conic section whose eccentricity is unity. It is a plane curve, being the locus of a point which is equidistant from a fixed point and from a fixed given straight line. It is of interest in astronomy with reference to cometary orbits and mirrors for telescopes. The figure of telescope mirrors, if more than about 25 cm in diameter, is usually a parabola in order to obtain a point image of the beam of light parallel to the axis of the telescope.

parabolic comet Any of at least half of all comets observed that have an orbital eccentricity of approximately unity. As such comets are observable over only a short arc of their orbits near perihelion, it is not possible to determine the orbital elements sufficiently accurately to distinguish between exact parabolic orbits and extremely elongated elliptical orbits.

parabolic velocity The velocity that a body must acquire to achieve a parabolic orbit. If the velocity of a body moving in a circular orbit be increased by a factor of $\sqrt{2}$, then its orbit will become parabolic. Thus, the Earth would forsake its orbit around the Sun if its mean orbital velocity increased from $29{\cdot}8\,\mathrm{km\,s^{-1}}$ to $42\,\mathrm{km\,s^{-1}}$.

parallactic angle or **angle of situation** The angle formed at a celestial body in the spherical triangle whose vertices are at the celestial pole, the zenith, and the object under observation.

parallactic displacement The apparent change in the observed position of a star as a result of a change in the observer's position on the Earth or with the motion of the Earth in its orbit.

parallactic ellipse The elliptical orbit in which a star between the ecliptic and the pole of the ecliptic will appear to move, as observed against the background of more distant stars. Its parallactic displacement is altered by the change in position of the observer as a result of the Earth's motion in its orbit. The semi-axis major of the ellipse is equal to its **parallax**.

parallactic inequality A periodic inequality in the motion of the Moon during every lunation, caused by the perturbing effect of the Sun. The amount of the inequality is inversely proportional to the distance between the Moon and the Sun. This perturbation has been of use in determining the distance of the Sun and its mass.

parallactic mounting see **telescope, mounting of**

parallax The angle subtended at a celestial body by a baseline of known length. As this angle varies with the distance of the object in question, it provides a means of expressing and determining its distance from the observer. The choice of baseline is made according to the remoteness of the object. For fairly close objects, such as members of the solar system, the Earth's equatorial radius is used: this gives a **geocentric parallax**. For more remote objects the Earth's semi-axis major is used: this gives a **heliocentric parallax**. If the parallax of a star is determined by direct means, using the principle of triangulation and a known baseline, it is known as **trigonometrical parallax**. The first star to be measured by this method was 61 Cygni by F. W. Bessel (see **Bessel, Friedrich**) in 1838. If the parallax is deduced from examination of a star's spectrum, it is known as **spectroscopic parallax**. If it is determined from the absolute and apparent magnitudes, it is known as PHOTOMETRIC PARALLAX.

paraselene see **mock Moon**

parhelion see **mock Sun**

parsec (*symbol:* pc) A fundamental astronomical unit of length for large distances. It is the distance at which one astronomical unit (i.e. the semi-axis major of the Earth's orbit round the Sun) subtends an angle of one second of arc. The name was proposed by H. H. Turner (see **Turner, Herbert**) in 1913. 1 pc = 206 265 A.U. = $3{\cdot}086 \times 10^{13}$ km = 3·26 ly. The term is derived from *par*allax of one *sec*ond.

Parsons, Lawrence (fourth Earl of Rosse; Parsonstown, November 17, 1840 – Birr Castle, August 30, 1908) A British astronomer who continued his father's work in astrophysics at Birr Castle, Parsonstown. He made extensive observations on the Orion Nebula and also investigated the thermal radiation from the Moon.

Parsons, William (third Earl of Rosse; York, June 17, 1800 – Parsonstown, October 31, 1867) A British astronomer who established a private observatory at Birr Castle, Parsonstown, Ireland. Here, in 1854, he mounted a 72-inch reflector, which was then the largest telescope in the world. The tube is still to be seen at Birr Castle. He studied in particular the hazy cloud-like objects then known collectively as **nebulae**. One object closely observed was the Orion Nebula. He discovered spiral nebulae (later recognized as galaxies) and also the annular structure in many planetary nebulae. He was able to resolve certain nebulae into groups of stars.

partial eclipse see **lunar eclipse; solar eclipse**

pascal (*symbol:* Pa) A derived unit in the **SI**. It is the unit of pressure and equals one **newton** per square metre.

$$1\,\text{Pa} = 1\,\text{N}\,\text{m}^{-2} = 10\ \text{dynes}\ \text{cm}^{-2}.$$

Paschen series The series of spectral lines in the infrared associated with the third energy level of the hydrogen atom. It was investigated by F. L. C. H. Paschen (1865–1947).

Pasiphaë Satellite VIII of **Jupiter**. See also **satellite**.

pencil A term used in optics to denote a beam of rays of light converging to or diverging from a single point so that the spread of the light is minimal.

pendulum clock A clock in which the mechanism for indicating time is controlled by the period of swing of a pendulum. The measurement of intervals of time being of the utmost importance in astronomy, much attention has been given to improving the accuracy and constancy of astronomical clocks as instruments of precision. The SHORTT CLOCK made by the Synchronome Company was of the free-pendulum type. It held the field for astronomcal purposes until superseded by the quartz clock and the atomic clock.

penumbra **1.** When an object obstructs the path of light from an extended source the dark shadow which it casts is bordered by a less dark zone. The dark middle part is called the umbra; the less dark zone which borders it is called the penumbra, an example of which is the region surrounding the shadow-cone of the Moon during a solar eclipse. **2.** The lighter zone surrounding the dark central part (umbra) of a sunspot.

penumbral eclipse see **lunar eclipse**

periastron The point of nearest approach of the components of a binary star in their orbit.

perigee The point of nearest approach to Earth by another body in orbit around the Earth.

perihelion The point in the orbit of a planet or of a comet which is nearest to the Sun.

perijove The point in the orbit of any one of the satellites of Jupiter at which it is nearest to the planet.

periodic comet A common term for a short-period comet, i.e. one whose period is less than 200 years.

period-luminosity law A relationship between the period of light variation and the mean absolute magnitude (luminosity) of a **Cepheid**. It was discovered by H. S. Leavitt (see **Leavitt, Henrietta**) in 1912. The absolute magnitude increases as the period increases. If the period of a Cepheid in an extragalactic system is known, then its absolute magnitude can be deduced from the period-luminosity law. Furthermore, this absolute magnitude taken in conjunction with the apparent magnitude enables the distance of the Cepheid, and the extragalactic system, to be calculated (see **apparent magnitude**).

Perrine, Charles Dillon (Steubenville, Ohio, 1861 – June 21, 1951) An American astronomer who became Director of the National Observatory, Córdoba, Argentina. He is noted for his observation of and discovery of comets. He discovered satellites VI and VII of Jupiter.

Perseids A spectacular meteor shower which occurs between July 25 and August 18. Its radiant is in the north of the constellation Perseus. The shower is in the same orbit as Comet P/Swift-Tuttle 1862 III, and is believed to be debris of that comet. In the late Middle Ages this meteor shower was known as the 'Tears of Saint Laurence', whose martyrdom took place on August 10, 258.

persistence of vision The phenomenon whereby, when the retina of the eye has been excited by radiation of visible wavelengths, the sensation of light as interpreted by the brain does not cease immediately on removal of the stimulus. Depending upon the intensity of the radiation, it takes from one eighth to one tenth of a second for the sensation to subside. When the retina receives a succession of images sufficiently rapidly, a continuous impression is experienced. Use of the **comparator** depends on this physiological effect.

personal equation The time-lag peculiar to any individual in perceiving the occurrence of an event and recording it. The accuracy of a person's observation is consequently subject to systematic errors. The amount of the error varies for individual observers. An observer's personal equation cannot be measured directly, but in any programme of work being carried out by several observers a correction can be derived on a statistical basis from the mean of all the results. The error introduced by the personal equation is considerably reduced by using photographic or photoelectric methods instead of visual observation.

perturbation A deviation in the orbital motion of a celestial body brought about by the gravitational attraction of other bodies. PERIODIC PERTURBATIONS are small periodic oscillations whose effect disappears at regular intervals of time. SECULAR PERTURBATIONS are slow continuous changes in orbital motion in the same direction and which if continued for a long enough time will cause the system to depart from its equilibrium state.

Pfund series The series of spectral lines, far in the infrared, associated with the fifth energy level of the hydrogen atom.

phase The visible aspect of a nonluminous celestial body (the Moon or a planet) when only part of the surface appears illuminated by the Sun as seen from Earth. The changes in the shape of the phase are caused by the relative positions of the Sun, the Earth, and the illuminated body. In the case of the Moon all phases from New Moon to Full Moon are observable with the naked eye. The inferior planets, Mercury and Venus, show phases from a slender crescent to a fully illuminated disc when observed through a telescope. The superior planets, which do not come between the Sun and Earth whilst revolving in orbit, can exhibit only a partial cycle of phases, or none at all in the case of those beyond Jupiter.

Phillips, Theodore Evelyn Reece (Kibworth, March 28, 1868 – Headley, May 13, 1942) An English clergyman and distinguished amateur astronomer. He is known for the long series of observations of Jupiter over a period of nearly 50 years, besides observations of Mars and Saturn, made in his private observatory. He also studied double stars.

Phobos Satellite I of **Mars**. See also **satellite**.

Phoebe Satellite IX of **Saturn**. See also **satellite**.

phonic motor A small synchronous motor driven by an alternating current of constant frequency. It is used in certain astronomical instruments where a driving speed of great constancy is required, as for the **photographic zenith tube**.

Phosphor or **Phosphorus** Venus when it appears as a morning star.

photocell or **photoelectric cell** A device which converts a light signal into an equivalent electrical signal by utilizing the **photoelectric effect**. It consists of two electrodes, a CATHODE, coated with a photosensitive material, and an ANODE, which is maintained at a positive potential with respect to the cathode. The two electrodes are contained within a glass bulb which is either evacuated or, if greater sensitivity is required, contains gas at a low pressure. When light falls on the cathode, the emitted electrons are attracted to the anode and can be detected as a current in the external circuit. The proportionality between intensity of light and the photoelectric current, coupled with the great sensitivity of the device, make it greatly suitable for use in astronomy. More recent photocells use a semiconductor sandwiched between two electrical contacts.

photoelectric effect The liberation of electrons from a surface when illuminated by light or ultraviolet radiation. It was discovered by W. L. F. Hallwachs (1859–1922) in 1888. Each photon of radiation on striking a surface can release one electron, if the conditions are right. The number of electrons that are released in one second is proportional to the intensity of the incident radiation. The photoelectric effect is used in astronomy in the photoelectric photometer.

photoelectric magnitude The magnitude of a body measured by a photoelectric photometer attached to the telescope. The measurement of brightness is made through a number of filters in turn; three or more filters are used according to requirements. See also **stellar magnitude, determination of**.

photoelectric photometer An instrument for measuring the intensity of radiation using a **photocell**. It has enabled measurements to be made on far more distant stars.

photographic magnitude The magnitude of an object measured from photographs taken on photographic plates which have emulsions more sensitive to blue light than the human eye. The wavelength region used is 4000 to 5000 Å. See also **stellar magnitude, determination of**.

photographic zenith tube A telescope constructed for the particular purpose of photographing those stars which cross the field of the instrument when directed to the zenith. It is used to determine the time of transit of fundamental stars, and so to ascertain the clock error. It is used also to determine the zenith distance of those stars at transit. The instrument has a much greater degree of accuracy than the transit circle for observations in connection with time-service and variation of latitude.

photography The production of permanent pictures by the action of light on certain sensitive materials. The astronomer J. F. W. Herschel (see **Herschel, John**) played an important part in the early development of photography in that he discovered the use of sodium thiosulphate as the fixing agent for sensitized paper as applied to photography. The invention of this art was soon applied to astronomy and brought about great advances in the subject. Astrophotography has superseded visual observation in many cases, especially where large telescopes are involved. Photography has made it possible to obtain a permanent record of surface features of members of the solar system, and pictures of galaxies, nebulae, and other remote objects invisible to the naked eye. Spectroscopic observations are almost invariably carried out by photography. In recent years colour photography has found application. The great advantage conferred on astronomy by photography is the facility of studying pictures of observations at leisure at any subsequent time.

photoheliograph An instrument for taking photographs of the Sun (called PHOTOHELIOGRAMS). The image formed by the objective of the telescope is enlarged and thrown on to a photographic plate. The photoheliograms are measured for sunspots and bright faculae. Any other interesting features or changes are noted. Some observatories take photographs of the Sun daily. The Royal Greenwich Observatory has a very long record of these daily observations.

photometer An instrument used to measure the intensity of light from some source. For astronomical purposes the **photoelectric photometer** is used.

photometric binary see **binary**

photometric parallax see **parallax**

photomultiplier A **photocell** in which amplification of the photoelectric current takes place within the cell itself. A very low intensity light signal can therefore be converted into a detectable electrical signal. There are various designs of this kind of cell, which is more sensitive and efficient than the best photographic emulsion.

photon A quantum of electromagnetic radiation, and consequently of light. The energy (E) of a photon is equal to the product of the frequency (ν) of the radiation and Planck's constant: thus $E = h\nu$. To this energy E there corresponds a mass m (quite apart from the fact that the photon has a rest mass of zero) and a momentum p, whose values are $m = E/c^2$ and $p = h\nu/c$, where c is the velocity of light.

photosphere The visible disc of the **Sun**. It is the boundary of a spherical shell of highly luminous gas and has a temperature of about 6000 K.

photovisual magnitude The magnitude of an object determined from photographs taken on photographic plates using emulsions and filters transmitting those wavelengths of the spectrum to which the human eye is sensitive. See also **stellar magnitude, determination of**.

Piazzi, Giuseppe (Ponte, July 16, 1746 – Naples, July 22, 1826) An Italian astronomer who discovered the first minor planet (**Ceres**) on January 1, 1801. He published a catalogue of fixed stars (1803, 1814).

Picard, Jean (La Flêche, July 21, 1620 – Paris, October 12, 1682) A French astronomer who established the *Connaissance des Temps* (the French annual astronomical ephemeris) in 1679: it is still published. He made the first accurate measurement of a degree of meridian; the result was used by Newton in his work on gravitation. He invented several instruments including the transit instrument and the micrometer.

Pickering, Edward Charles (Boston, July 19, 1846 – Cambridge, Massachusetts, February 8, 1919) An American astronomer who became Director of the Harvard College Observatory. He is distinguished for his work in stellar photometry and spectroscopy, and for his discovery of spectroscopic binary systems and variable stars in globular clusters. He devised a meridian photometer. He published *The Henry Draper Catalogue*, the *Revised Harvard Photometry*, as well as many other contributions to astronomical literature.

Pickering series The series of spectral lines produced by singly ionized helium (He^+). Every other line is almost coincident with a line in the Balmer series for hydrogen. It was discovered by E. C. Pickering in the spectrum of ζ Puppis, and at first wrongly ascribed to hydrogen.

Pickering, William Henry (Boston, February 15, 1858 – January 17, 1938) An American astronomer known for his lunar and planetary observational work. He discovered the ninth satellite of Saturn (1898).

Pioneer The name given to a series of space probes launched by the USA for the purpose of developing techniques for launching and guiding the probes to be used in subsequent programmes of space research. They have explored various regions of the solar system. Pioneers 1–3 were intended as lunar probes but they did not reach their target. Pioneer 1 was launched on October 11, 1958, and its payload fell back into the Earth's atmosphere after a flight lasting a little over 43 hours. Pioneer 5, launched on March 11, 1960, was successfully put into a heliocentric orbit and sent back important new data on solar and cosmic radiation and magnetic fields. Pioneer 6, launched on December 16, 1965, went into a heliocentric orbit as a Sun probe. Pioneers 7 to 9 were also sent into orbits around the Sun for the purpose of studying the solar wind, magnetic fields, and other interplanetary phenomena.

Pioneer 10 was launched on March 2, 1972, travelled through the asteroid belt, and passed Jupiter in December 1973 at a distance of about 131,000 km after making observations of the atmosphere and magnetosphere of the planet and taking photographs. It will cross Pluto's orbit some time in 1987 and then leave the solar system. Pioneer 11 was launched on April 3, 1973 and made similar observations of Jupiter in December 1974 at a distance of about 46,000 km. It passed Saturn in September 1979 and will eventually leave the solar system.

Pioneer Venus 1 was launched on May 20, 1978 and in December 1978 went into orbit around Venus, approaching to within 150 km of the planet and sending back information. Pioneer Venus 2 was launched on August 8, 1978 and landed several probes on the planet in December 1978.

Planck, Max Karl Ernst Ludwig (Kiel, April 23, 1858 – Göttingen, October 4, 1947) A German physicist distinguished for his work on black-body radiation, which assisted the development of the **quantum theory**, and for his studies in thermodynamics. He is commemorated in the universal constant, h, which is called after him.

Planck's constant (*symbol: h*) The universal constant which relates the frequency of radiation (ν) with its quantum of energy (E). Thus $E = h\nu$, where h is equal to $6{\cdot}626196 \times 10^{-34}$ J s (or $6{\cdot}626196 \times 10^{-27}$ erg s). The constant has the dimensions of action, i.e. energy times time.

Planck's radiation law A mathematical expression which relates the emissivity of radiation to wavelength and temperature, under conditions of thermal equilibrium. The effective temperature, or the surface temperature, of a star can be ascertained by means of this law. See also **black-body radiation**.

planet A large celestial body in orbit around a star. Planets shine by reflecting the light from their primary; they are not self-luminous. The solar system contains nine MAJOR PLANETS: the innermost planets, Mercury and Venus, are called INFERIOR PLANETS because they are nearer to the Sun than Earth; the others, Mars, Jupiter, Saturn, Uranus, Neptune, and Pluto, given in order of distance from the Sun, are called SUPERIOR PLANETS 2because they are beyond the Earth's orbit. Between Mars and Jupiter there is a very large number of small bodies known as ASTEROIDS or MINOR PLANETS. Ceres, the largest, has a diameter of only about 1000 km. In addition to these natural objects there are now very many man-made objects which have been launched into space since the development of space-flight, and remain in planetary orbit. It is believed that other stars, apart from our Sun, may have planets associated with them. One example may be **Barnard's star**. See also **Planet X**; **planetary atmosphere**.

planetarium In its original sense, another term for **orrery**. It is now used to describe a far more complicated mechanism and is also applied to the building in which the instrument is housed. The positions and motions of celestial bodies can be projected on to a hemispherical dome for an audience seated below. Planetaria are to be found in certain large cities. The London Planetarium was opened in 1958.

planetarium, projector for The optical projector used in a modern planetarium was first designed by Walther Bauersfeld of the optical manufacturing company of Carl Zeiss, Jena, about 1923–24, and installed in the Deutsches Museum, Munich. The projector is of dumb-bell shape, consisting of two spheres connected by a girderwork. The spheres contain the powerful lamps for projecting the required star-field on to the dome of the planetarium building. The girderwork contains within it other special projectors for planets, comets, aurorae, etc. The dumb-bell is mounted so that it can be driven at various speeds about three axes of rotation. With such an instrument it is possible to demonstrate the precession of the equinoxes, the changing pattern of the heavens as a result of the diurnal east-west motion, as well as the aspect from any latitude on Earth. All these features can be accelerated or slowed down, and adjusted to show them as they were in the past or will be in the future. Projectors of this type are made also by a Tokyo company, and one of them is installed in the Armagh Planetarium, Northern Ireland. Smaller instruments having a restricted range of facilities are made as educational aids in the teaching of astronomy.

planetary atmosphere The nature of the atmospheres present on the planetary bodies of the solar system are as follows.

Mercury. Observations show that this planet cannot have more than an extremely tenuous atmosphere. The velocity of escape is low and the surface temperature is high. Minute traces of helium and argon have, however, been detected by the space probe Mariner 10. Atmospheric pressure on the planet can only be a minute fraction of that on Earth.

Venus. This planet has a dense atmosphere consisting of over 95% carbon dioxide, about 3% nitrogen, and some few parts per million of helium, neon, and argon. In the lower atmosphere there are trace amounts of free oxygen and water vapour. The thick layers of cloud are composed of sulphuric acid droplets and, at lower altitudes, sulphur particles. Atmospheric pressure at the surface is estimated to be about 90 atmospheres.

Earth. See **atmosphere**.

Mars. The thin Martian atmosphere consists of about 95% carbon dioxide, less than 3% nitrogen and 2% argon, some oxygen, and trace amounts of carbon monoxide and water vapour. Atmospheric pressure at the surface is about 0·7% of that on Earth.

The giant planets are constituted differently from the smaller rocky planets. They are thought to consist of a mixture of gases, primarily hydrogen and helium, whose density increases with depth.

Jupiter. That part of the planet's atmosphere above the layer of swirling cloud contains hydrogen, methane, ammonia, and traces of other molecules. Helium may be present. Any water present in the atmosphere will have been frozen and fallen below the clouds. The yellowish cloud zones apparently consist mainly of ammonia crystals while the lower darker cloud belts may contain sulphur compounds and other complex molecules. The Red Spot is probably coloured red by traces of phosphorus.

Saturn. This planet is similar to Jupiter. It, too, probably consists mainly of hydrogen and helium. Methane seems to be present to a greater extent than on Jupiter and ammonia to a less extent. Ethane has also been detected.

Uranus. Only methane and hydrogen have been definitely confirmed as existing on this planet. Any ammonia will be frozen and possibly exists in clouds under the methane layer.

Neptune. Only methane and hydrogen have been confirmed as existing on this planet.

Pluto. This planet is so remote that information about atmospheric conditions is sparse. It is too cold and small to have retained much of an atmosphere. Measurements indicate that methane ice might cover parts of the surface.

Planetary satellites. The Moon, being close to Earth, has been attentively observed for a very long time. An extremely thin atmosphere has been revealed, later found by Apollo instruments to consist of rare gases. The escape velocity is such that any gaseous atmosphere formerly present, or gases still being produced from rocks by radioactive decay, will be quickly dissipated. There is no evidence that the satellites of Jupiter possess any significant atmosphere. Io has a tenuous atmosphere, and the presence of a cloud of sodium has been detected. Titan, a satellite of Saturn, and Triton, a satellite of Neptune, have been shown by G. P. Kuiper (see **Kuiper, Gerard**) in 1944 to have some atmosphere containing methane.

Planetary Co-ordinates The title of three volumes of heliocentric, spherical, and rectangular co-ordinates, referred to the equinox of 1950.0, of the planets. They are used mainly for calculating perturbations of comets and minor planets. The volumes are published by H.M. Stationery Office, London.

planetary motion The motion of a planet in its orbit around the parent star, and of a satellite around its primary. Such motion is governed by Newton's Law of Gravitation (see **gravitation**) and by **Kepler's Laws**. The MEAN MOTION OF THE PLANET is the orbit as defined by those laws, but it is modified by the perturbations caused by the gravitational attractions of any other large bodies in the system. When these have been allowed for by calculations, the TRUE MOTION of the planet is obtained. Seven orbital elements are needed to define completely the **orbit** of a planet.

planetary nebula (A misnomer: the objects are not planets nor true nebulae. They were so named because when viewed in small telescopes they give the impression of planetary discs.) A hot central star at a temperature of 50,000 – 100,000 K surrounded by expanding shells of rarified ionized gases, mainly hydrogen plus helium but containing some nitrogen and oxygen. The gaseous envelope emits a fluorescent glow as a result of the ionizing ultraviolet radiation received from the central star. It is thought that a star having a carbon core and a mass between 0·6 and 4 solar masses will become a planetary nebula through contraction of the core during the final stage in the life of a red giant, when the shell of burning helium separates at high velocity from the central core and leaves behind a white dwarf. The Ring Nebula (M57) in the constellation Lyra is a well-known example of this type of object.

planetary precession The secondary component of the general **precession** of the equinoxes, it results from the gravitational effects of other planets upon the Earth. The consequence from planetary precession is movement of the equinox eastward by about 11″ per century and a decrease in the angle between the ecliptic and the equator by about 47″ per century.

planetesimals Very small planetary bodies which in the theories of T. C. Chamberlin (1901) and F. R. Moulton (1905) condensed from the matter erupted from the Sun as a result of a star in hyperbolic orbit passing close to it. The authors were the first to propose that planets were formed from the agglomeration of cold bodies.

planetocentric co-ordinates Co-ordinates referred to the centre of a planet. They are used in theoretical work.

planetographic co-ordinates Co-ordinates referred to the mean surface of a planet. The planetographic co-ordinates of a point on the surface of a planet will differ from the planetocentric co-ordinates if the figure of the planet departs to any great extent from a true spheroid. The differences between the two systems of co-ordinates is negligible in practice, except in the case of Jupiter.

planetoid Alternative name for **minor planet**. The term has also been applied to spacecraft which after being launched from Earth have become tiny planets revolving around the Sun.

Planet X A possible tenth planet orbiting the Sun beyond Pluto. The orbit of **Halley's Comet** as derived from observations at a given apparition does not agree exactly with those of the previous and subsequent apparitions. It has been calculated that an orbit which satisfies the observations of all apparitions since 1456 can be obtained by assuming the existence of a trans-Plutonian planet, situated at twice the distance of Neptune, having a mass three times that of Saturn, and moving in an orbit inclined at 120° to the ecliptic. Planet X, which would be of about magnitude 14, has not been discovered, in spite of systematic search. However, there are other possible theoretical explanations for the variations between successive orbits of Halley's Comet.

planisphere A contrivance in which plane surfaces move over one another to fulfil the purposes of a celestial globe. It consists of a map of the sky on a polar projection, over which is a rotatable disc with a suitable aperture. When the planisphere has been set for the appropriate date and hour the corresponding aspect of the sky is exhibited.

planoconcave lens A lens that is bounded by one plane and one concave surface. The lens is diverging.

planoconvex lens A lens that is bounded by one plane and one convex surface. The lens is converging.

Plaskett's star BD 6° 1309 (HD 47129) in the constellation Monoceros, identified by J. S. Plaskett (1865–1941) as a spectroscopic binary in 1922. It is a supergiant. For the brighter component Plaskett deduced a mass greater than 86 solar masses, and for the fainter component a mass of 72 solar masses. It is the most massive star known.

plasma Completely ionized gas consisting of free electrons and free atomic nuclei at an extremely high temperature. It is sometimes called the fourth state of matter. Plasma is almost neutral electrically as the numbers of positive and negative particles are approximately equal; it therefore becomes highly conducting. The Sun and other stars are in a plasma state.

plasma clouds Clouds of electrically charged particles (protons and electrons) having diameters of 10,000 to 100,000 km, ejected with the solar wind at a velocity of more than 1000 km s^{-1} during intense solar activity. The Earth's magnetic field is greatly disturbed when they encounter the Earth.

plasma oscillation An oscillation set up in plasma when the positive and negative electrical charges become separated through some perturbation, such as collision between interstellar clouds. It is believed that these oscillations may lead to the emission of radio-frequency radiation.

plate centre In photographic astrometry, the positions of star images on a photographic plate are measured in rectangular co-ordinates, whose origin is the centre of the plate. The centre ideally coincides with the optical axis of the telescope during exposure of the plate. The two axes are drawn on the plate: the ζ-axis is perpendicular to the meridian, the η-axis parallel to the meridian.

plate constants The small quantities (a, b, c, d, e, f) in the equations

$$\zeta - x = a\zeta + b\eta + c$$
$$\eta - y = d\zeta + e\eta + f$$

from which the standard co-ordinates (ζ, η)

of a star are derived from the co-ordinates (x, y) of the star's image as measured on the photographic plate. The plate constants depend upon various errors connected with the plate and the measuring machine.

plate measuring machine An instrument for measuring the relative positions of images of stars on photographic plates. It is essentially a microscope in which the stage carrying a photographic plate can be moved in two directions perpendicular to each other. The image of a star whose position is required is viewed exactly on the cross-wires of the microscope, which are then moved and set on stars whose positions are already accurately known. From the amount of movement of the photographic plate as shown by scales on the stage, it is possible to calculate the position of the star on the celestial sphere. Such a machine can have an accuracy of about 10^{-3} mm.

plate scale The ratio of the angular distance between two stars (measured in a plane tangential to the celestial sphere) to the linear distance between the images of the same two stars as measured on a photographic plate. It is a function of the focal length of the telescope's objective.

Platonic year see **year**

Pleiades (M45) An open cluster of stars in the constellation Taurus, known since antiquity and often called the SEVEN SISTERS (the seven daughters of Atlas and Pleione). Although only six or seven stars are visible to the naked eye, there are in fact several hundred embedded in nebulous matter consisting of gas and dust clouds which shine by reflected light. The brightest star of this cluster is Alcyone, which is about 1000 times more luminous than the Sun.

Plough or (U.S.) **Big Dipper** The seven bright stars in the constellation **Ursa Major**.

P-L relationship see **period-luminosity law**

Pluto The smallest and usually the outermost known planet of the solar system.

Dust clouds made visible by the light of stars in the Pleiades.

Unexplained irregularities in the motion of Uranus still remained after the discovery of **Neptune**. P. Lowell concluded that a trans-Neptunian planet was responsible, and in 1905 instituted a search for Planet X (as it was then called). The planet was eventually located within 5° of the predicted position by C. W. Tombaugh (see **Tombaugh, Clyde**) on photographic plates taken in January and February 1910: the discovery was publicly announced on March 13, 1930. The planet will not return to that point in its orbit where it was first located until the year 2178. The appropriate name Pluto (son of Saturn and brother of Jupiter and Neptune) provides the fitting monogram ♇ for the planet, as these two letters are the initials of Percival Lowell.

Pluto is very different from any other planet. It has the greatest orbital eccentricity and inclination to the ecliptic of all the planets. It does in fact approach closer to the

Sun than Neptune for part of its orbital period: it will be in this position during the period 1979–99.

Theory suggests that Pluto was originally a satellite of Neptune, and that a very close approach to the satellite Triton, whose motion was direct, resulted in Pluto being accelerated to escape velocity, thereby becoming a separate planet. Triton acquired retrograde motion and entered an orbit further removed from Neptune. An alternative suggestion is that Pluto is the first of a belt of small planets outside Neptune's orbit, similar to the belt of minor planets between Mars and Jupiter.

Studies of photographs of Pluto made by J. Christy of the US Naval Observatory and others in 1978 revealed that Pluto has a large and nearby satellite. This meant that for the first time an accurate determination of the planet's mass could be made. Thus it is now known that Pluto is a small low-density body, whose surface is probably partially or completely covered in a layer of frozen methane. It is too remote and small for surface features to be determined. The planet has no atmosphere. The main data relating to Pluto are given in the table.

Globe	
Diameter	2500–3000 km
Density (water = 1)	$1{\cdot}3\,g\,cm^{-3}$(?)
Mass	$1{\cdot}2 \times 10^{22}$ kg
Volume	$0{\cdot}9 \times 10^{10}\,km^3$
Sidereal period of axial rotation	$6^{d}{\cdot}39$
Escape velocity	$1{\cdot}1\,km\,s^{-1}$
Albedo	0·9
Inclination of equator to orbit	?
Surface temperature	63 K (maximum)
Orbit	
Semi-axis major	39·46 A.U. = $5{\cdot}9 \times 10^{9}$ km
Eccentricity	0·248
Inclination to ecliptic	17°
Sidereal period of revolution	248·4 y
Mean orbital velocity	$4{\cdot}74\,km\,s^{-1}$

Pogson, Norman Robert (Nottingham, March 23, 1829 – Madras, June 23, 1891) An English astronomer who was Government astronomer at Madras from 1860–91. He discovered several minor planets and variable stars but is remembered particularly for his work on stellar magnitudes.

Pogson's ratio The ratio of the fifth root of 100:1, i.e. 2·512. It gives the ratio between the brightness of two stars that differ by one magnitude. It was proposed by N. R. Pogson. See **apparent magnitude**.

Poincaré, Jules Henri (Nancy, April 29, 1854 – Paris, July 17, 1912) A French mathematician and physicist who worked on the theory of astronomical orbits and the three-body problem. He recommended Einstein for a post at the Polytechnic in Zürich.

Pointers The stars α and β in Ursa Major, which point towards Polaris, the Pole Star and brightest star in Ursa Minor.

polar caps The bright areas visible around the poles of Mars.

polar distance The complement of **Declination**. It is the angular distance of a celestial body from the celestial pole.

Polaris or **Pole Star** The brightest star (Alpha) in Ursa Minor, lying within 1° of the north celestial pole. On account of **precession** the star is getting closer to the pole and will continue to do so until the year A.D. 2100, when it will move away from the pole. In the year A.D. 14,000 Vega will be the Pole Star. Polaris is a supergiant visual binary whose primary, a Cepheid, is itself a variable spectroscopic binary. The southern hemisphere contains no bright equivalent of the Pole Star.

polarization A phenomenon which can be exhibited by waves of electromagnetic radiation, such as light. The waves normally vibrate in all planes perpendicular to the direction of propagation, but if the vibrations occur mainly in one particular plane the radiation is said to be plane polarized. Polarization can be brought about in various ways, for example, by reflection, refraction, transmission through substances having special optical properties, or scattering by small particles. The light and radio waves received from many celestial bodies are polarized. Study of this polarized radiation provides information as to the nature of its source, and is an important aid to astrophysical research.

polar lights see **aurora**

Polar Sequence see **North Polar Sequence**

polar telescope A type of instrument which provides a fixed position from which observations can be made. It consists of a straightforward refractor installed as the polar axis of its equatorial mounting, having at its lower end an equatorially mounted clock-driven flat mirror which reflects the light from any required part of the sky into the telescope.

polar variation A periodic change in the position of the poles of rotation of the Earth relative to the geoid. See **latitude, variation of**.

pole That point on a sphere which is at an angular distance of 90° from all points on some specified great circle of the sphere. The equatorial plane is normally taken as the plane of reference. The north and south CELESTIAL POLES are 90° north and south respectively of the celestial equator. Similarly, the ECLIPTIC POLES are 90° from the plane of the ecliptic, and the GALACTIC POLES are 90° from the galactic plane. In the case of a planet the line joining the poles coincides with the axis of rotation, which is perpendicular to the equatorial plane. The position of the celestial poles can be found by observation, as they are the points where the Earth's axis of rotation intersects the celestial sphere, and do not participate in the diurnal motion of the rest of the celestial sphere about the celestial axis.

Pole Star see **Polaris**

Pond, John (London, 1767 – Greenwich, September 7, 1836) An English astronomer who was appointed sixth Astronomer Royal. He is known for the accuracy of his work in positional astronomy, and published (1833) a catalogue of 1113 stars. He instituted improvements in methods of observing.

Population I stars A classification introduced by W. Baade (see **Baade, Walter**) in 1944. They are relatively young stars containing a high abundance of metals, such as O and B stars, supergiants, and classical Cepheids, and are found especially in the spiral arms of galaxies. They are sometimes called ARM POPULATION STARS. The Sun is a Population I star.

Population II stars A classification introduced by W. Baade in 1944. They are old stars containing a lower abundance of metals than Population I stars. They include RR Lyrae stars and subdwarfs, and are found in the halo of our Galaxy, especially in globular clusters, and in the nucleus of spiral galaxies. The oldest of them are sometimes called HALO POPULATION STARS.

pore A very small sunspot.

Poseidon The name proposed for satellite VIII of Jupiter. The name Pasiphaë was ultimately adopted.

positional astronomy see **astrometry**

position angle The angular distance, measured from 0° to 360° from north through east, between the direction of the north celestial pole and the direction of the line joining two stars. In the case of a binary system it is the angle between the direction of the celestial north pole and the direction of the line joining the primary and secondary components.

post-nova A **nova** after its outburst and the return to its original magnitude.

Poynting-Robertson effect The effect produced by the pressure of solar radiation on very finely divided particles of matter in orbit around the Sun, which finally causes the particles to spiral inwards towards the Sun. When the particles are close enough to the Sun, vaporization sets in, they are reduced in size, and then the remnants are repelled by the radiation pressure. The inner part of the solar system has by now probably been cleared of all small particles, except in so far as they are renewed by the disintegration of comets and minor planets. The effect is named after J. H. Poynting (1852–1914), who first described it in 1903, and H. P. Robertson, who proved its reality in 1937 by derivation from the theory of relativity.

preceding Denoting any object that goes before another in order of time. If the object that culminates earlier has a more northerly declination than the following one, it is said to be 'north preceding'; if more southerly, then it is 'south preceding'.

precession In astronomy the term usually refers to the precession of the equinoxes, i.e. the movement of the equinoctial points (the points of intersection of the celestial equator and the ecliptic) westward with respect to the background of stars. This westward movement amounts to 50 seconds of arc per year. It is caused by the celestial pole being made to describe a circle of radius 23½° on the sky, centred on the pole of the ecliptic, in a period of 25,800 years. The phenomenon results from the gravitational attraction of the Sun and Moon on the Earth's equatorial bulge, which produces an oscillation of the Earth's axis around the pole of the ecliptic. This slow periodic change in the direction of the Earth's axis is termed the precession of the Earth's axis.

The gravitational attraction by the Sun and Moon constitutes the main component of precession and is called the LUNI-SOLAR PRECESSION. The smaller similar effect caused by the gravitational influence of the other planets is called the PLANETARY PRECESSION. The total effect, the sum of these two, is called the GENERAL PRECESSION. The consequence from precession is that the position of the Pole Star and the positions of all other stars in Right Ascension and Declination are continuously changing. Star catalogues which give these data for a given epoch, e.g. 1850, 1900, 1950, must be corrected for precession when the data are required at any other time.

precession, constant of The ratio of the luni-solar precession to the cosine of the obliquity of the ecliptic. Its value is 54·94 seconds of arc per year.

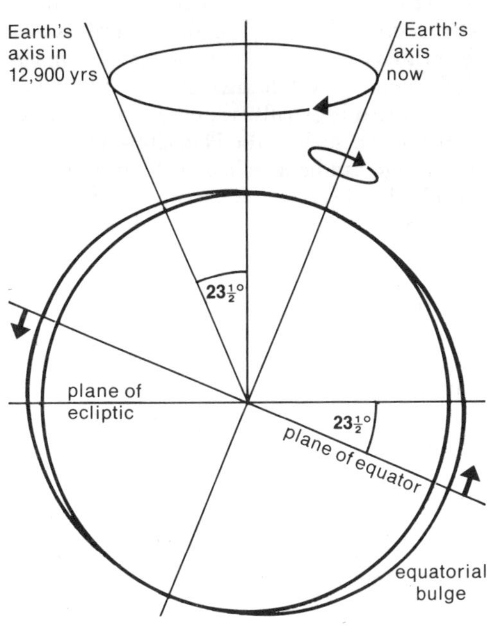

Precession of the Earth's axis.

predictor telescope Essentially a 7 × 50 or a 5 × 25 monocular designed for military use. It is a refracting telescope having an aperture of 48 mm and a focal length of 150 mm. Its distinctive feature is the adjustable eye lens at right angles to the tube, located above a prism with the field lens cemented to the upper surface. It forms an erect image.

pre-nova A star which will become a **nova**. A rapid increase in luminosity is the prelude to the final outburst. Observations of the pre-nova stage are therefore retrospective, and can only be made with reference to photographic material already in existence.

pressure The force exerted on unit area. The derived unit of pressure in the **SI** is the **pascal**.

pressure broadening see **spectrum**

primary The name given to a planet which revolves round the Sun or to the more massive component of a binary star. The bodies which revolve round the primaries are sometimes referred to as secondaries.

prime focus The position at the open end of a reflecting telescope, on the central axis of the primary mirror, where the primary image is formed. In small instruments the **Newtonian focus** is usually adopted. In very large instruments, such as the Hale 200-inch reflector, a cage can be installed at the prime focus to hold the observer with additional necessary equipment, without significant loss of aperture.

prime meridian That meridian which is defined as the zero of longitude, and from which all others are reckoned. In the case of Earth, it is the meridian passing through the centre of the Airy Transit Circle at the Royal Greenwich Observatory, Greenwich.

primeval fireball The original condensate of matter which, according to the **big bang theory**, exploded to produce the present expanding universe. The density of the universe was 10^{93} g cm^{-3} and the temperature had fallen to 10^{12} K at 10^{-43} of a second after the explosion.

primeval nebula The nebula of gas and dust from which our solar system (Sun and planets) are thought to have been formed.

prime vertical The vertical circle which is perpendicular to the meridian and which passes through the east and west points on the horizon.

Principia Short form of reference to *Philosophiae Naturalis Principia Mathematica* by Isaac Newton.

prism A small block of glass or other transparent material, of triangular section, used in optical apparatus to deviate and/or disperse a beam of light. Prisms are used to produce spectra in a **spectrograph**. The **objective prism** has permitted great advances to be made in the study of stellar spectra.

prismatic astrolabe One of the most accurate of astrometrical instruments. It is used for making fundamental determinations of star positions by observing the transit of stars at a fixed altitude of 60°. As designed by A. Claude and L. Driencourt, the instrument consists of a pool of mercury to provide an artificial horizon, an equilateral prism with its edges horizontal, and a telescope mounted horizontally. Parallel rays of light from the star fall on the artificial horizon and prism, and are then directed to the telescope, in the field of which two images are seen simultaneously. One moves downwards and the other moves upwards. The two images coincide at the moment of transit at an altitude of 60°. The instrument has been improved by A. Danjon by the use of a double Wollaston prism to form the two images and of automatic timing devices.

probable error The figure appended to the arithmetical mean (adopted as the measured result) of a series of determinations which indicates the error that is not exceeded by 50% of the cases. For example, the radial velocity of a star in km s^{-1} reported as $+4{\cdot}7 \pm 0{\cdot}2$ means that the velocity lies between $+4{\cdot}9$ and $+4{\cdot}5$ km s^{-1}.

prolate spheroid see **spheroid**

Functioning of a prismatic astrolabe.

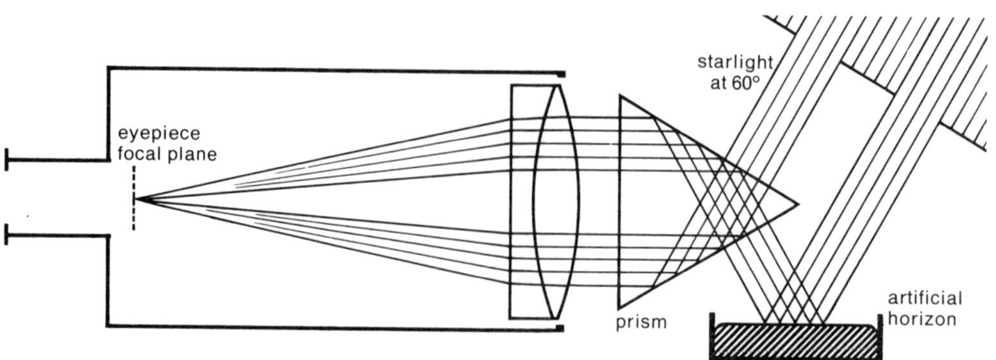

prominence see **solar prominence**

proper motion The continuous apparent motion of a star across the celestial sphere, measured by the true motion of the star relative to the motion of the Sun. It is determined by comparing the star's position on photographic plates exposed on widely separated occasions, sometimes 50 or more years apart. For most stars the proper motion is negligible; in most cases it amounts only to about 0·1 seconds of arc per year. **Barnard's star** has the largest known proper motion.

proton A positively charged elementary particle which forms part of the nucleus of all atoms. Its mass is 1·007,28 amu, i.e. $1{\cdot}6726 \times 10^{-24}$ g, and is 1836 times greater than that of the **electron**.

proton-proton reaction One of the nuclear reactions thought to be the source of solar and stellar energy: the other is the **carbon-nitrogen cycle**. The proton-proton reaction involves three stages in which hydrogen is converted into helium with the liberation of an enormous amount of energy. In the first stage two protons (hydrogen nuclei) combine to form a deuterium nucleus with release of a positron, a neutrino, and radiation. In the second stage the deuterium nucleus combines with a proton to form an isotope of helium (He^3) with release of radiation. The third stage involves the reaction between two He^3 nuclei which combine to form the normal helium nucleus (He^4) with release of two protons and radiation. Four hydrogen nuclei are therefore needed to form one helium nucleus. The reaction requires a temperature of about 10^7 K and a density of about 100 g cm^{-3}. The conversion of hydrogen into helium in this reaction results in a loss of mass of about 0·7%, which is equivalent to approximately 6×10^{11} joules per gram of converted hydrogen radiated as energy.

protostar An embryonic star. A very dense condensation of interstellar matter in a nebula which is believed to develop into a new star by continued contraction.

Proxima Centauri The star which is nearest to the Sun. It is the faint component of the triple star Alpha Centauri. It has an apparent magnitude of 11 and its distance is 4·3 light-years.

p-spot see **sunspot**

Ptolemaic system The geocentric world system devised by Ptolemy and based on the work of **Hipparchos of Nicaea**. The Earth is fixed at the centre about which Mercury, Venus, Sun, Mars, Jupiter, and Saturn revolve in complicated orbits. Beyond these planets lies the sphere of fixed stars. Each of the bodies revolves on a small circle called the epicycle, which in turn revolves around the Earth on a larger circle called the **deferent**. This complicated system of eccentric circles and epicycles adequately describes the daily planetary motions, but has no use as a theory. The Ptolemaic system remained unchallenged until the revival of the heliocentric theory by Copernicus (see **Copernicus, Nikolaus**) in the 16th century.

Ptolemy (Claudios Ptolemaeos) (*Fl.* in Alexandria A.D. 125–150) An Alexandrian

A 17th-century engraving of Ptolemy, the last great astronomer of the Ancient World.

astronomer, geographer and mathematician. His main work is the *Great Astronomical System* whose Greek title has become the *Almagest* in the garbled version of the Arabic translation. His principal contributions to astronomy are an elaborate geocentric theory of planetary motions involving the idea of the **deferent** and **equant**, and the discovery of an inequality (now called **evection**) in the Moon's motion, which he accounted for by the use of eccentrics and epicycles.

pulsar An object first noticed in August 1967 by Jocelyn Bell, working under A. Hewish (see **Hewish, Antony**) at Cambridge. A pulsar is characterized by the emission of electromagnetic radiation at radio frequency in pulses of great regularity. The first pulsar to be discovered was emitting pulses every 1·337 301 13 seconds. Pulsars are believed to be rapidly rotating **neutron stars**, one of the final stages in the life of massive stars produced by supernovae. In a manner identical to that by which a lighthouse beam is seen to flash, a beam of radio waves emitted from some point on or near the rotating pulsar sweeps past the Earth and is received in the form of pulses. A considerable number of pulsars have now been discovered. In 1969 the first optical identification of a pulsar was made in respect of a star in the **Crab Nebula**.

pulsating star A type of variable star whose fluctuating brightness arises from a regular pulsation of the star resulting from its expansion and contraction, i.e. from a periodical conversion of radiant energy into gravitational energy and back again. Cepheid variables and RR Lyrae stars are examples of pulsating stars.

quadrant An astronomical instrument used for measuring the altitude of celestial bodies. It consists of a quarter circle with a sighting arm pivoted at the centre so that it can move over the arc, which is graduated in degrees. For use the instrument is mounted vertically, when it is known as a MURAL QUADRANT. The sighting bar was at a later stage replaced by a refracting telescope. Notable mural quadrants include those used by Tycho Brahe and those erected at the Royal Observatory, Greenwich.

Quadrantids A meteor shower which appears during the first week of January. Its radiant is in the constellation Boötes.

quadrature see **aspect**

quantum The smallest discrete unit of energy (or some other physical property) of a system. The quantum of electromagnetic radiation is the **photon**.

quantum mechanics The system of mechanics which has its origin in the **quantum theory** and has displaced Newtonian mechanics for the interpretation of physical phenomena occurring on a very small scale, e.g. on the atomic scale.

quantum theory The theory developed by M. Planck from his concept of the discontinuity of the emission of radiation. Radiation is emitted, and absorbed, in discrete units called quanta whose energy is proportional to the frequency.

quarks Hypothetical elementary particles, together with their corresponding antiparticles, postulated by M. Gell-Mann (1929–) as the basis of all matter. At present there are thought to be at least four quarks and four antiquarks, which may be combined in a variety of ways to form and give the properties of other known particles (with the exception of electrons, muons, and neutrinos). They have not as yet been discovered. Gell-Mann came across the name in the novel *Finnegan's Wake* by James Joyce, who coined the word in order to signify the least bit of matter.

quartz-crystal clock A clock, of greater accuracy than a pendulum clock, which is controlled by the oscillations of a quartz crystal in an electrical circuit. It has an extremely low daily clock rate. As quartz-crystal clocks show ageing after some years, the more stable atomic clock is used when even greater precision is needed over a long period of time.

quasar Acronym formed from *quas*i-stell*ar* object (*abbrev.:* QSO). A compact object which is an extremely strong source of energy, radiated over a wide range of wavelengths from X-ray through optical to radio wavelengths. Only a small proportion are radio sources. The spectra of quasars show a very large red shift, which, if it results from expansion of the universe, indicates that they are the oldest and most distant objects in the universe. The exact nature of these objects has not been resolved. They were first recognized by T. A. Matthews and A. R. Sandage in 1962.

quiet Sun The Sun when it is not perturbed by activity. This occurs when the 11-year cycle is at its minimum.

radar Acronym from *ra*dio *d*etection *a*nd *r*anging. A method of measuring the distance of an object by observing the time for a radio signal to travel to the object and back again. Short powerful pulses of electromagnetic radiation are sent out on a wavelength between 1 mm and 10 m travelling at the speed of light. The time taken for the radio signal to be received back is noted. Radar was developed for military purposes during World War II, but since then has been adopted for navigation, tracking satellites, guiding moving ships, missiles, etc. The system has found application in astronomy for tracking meteors, determining the heights of various ionized layers of the upper atmosphere, and obtaining radar echoes from the planets for the measurement of distance and period of rotation and for determining surface profiles.

radial velocity The component of a body's velocity in the direction of the observer's line of sight. The radial velocity is determined from the **Doppler effect**. If the displacement in wavelength of spectral lines is positive, the object is moving away from the observer; if it is negative, the object is approaching the observer. The velocity is measured in km s^{-1}. The radial velocities of 15,107 stars (all those known up to 1950) is contained in R. E. Wilson's *General Catalogue of Stellar Radial Velocities*, Washington, D.C., 1953.

radian (*abbrev.*: rad) A supplementary unit in the **SI**. It is the plane angle subtended at the centre of a circle by an arc of the circle having a length equal to the radius of the circle.

$$2\pi \text{ radians} = 360°; 1 \text{ radian} = 57°{\cdot}296$$

radiant That point on the celestial sphere from which a shower of meteors appears to radiate. A meteor shower is identified by the constellation in which the radiant is situated.

radiation pressure The mechanical force exerted by electromagnetic radiation by transfer of its momentum to the surface on which it is incident. Small particles in interplanetary space are subjected to considerable pressure by radiation from the Sun. For example, the tail of a comet always points away from the Sun.

radiation temperature see **celestial bodies, temperature of**

radiation zones or **radiation belts** Areas of charged particles surrounding the Earth, Jupiter and Saturn. See **Van Allen belts**; **magnetosphere**.

radiative braking A slowing down of the speed of a star's rotation on account of radiation.

radiative equilibrium The state in which a given volume of stellar gas emits the same quantity of energy per second as it receives.

radio astronomy That branch of astronomy which studies radiation of radio frequency emitted or reflected by celestial bodies and the universe. Its beginning dates from the time when K. G. Jansky (see **Jansky, Karl**) discovered the emission of radio noise from the Milky Way (his paper on the subject was published in 1932). The first radio telescope purposely built to study such emission was made by Grote Reber (his paper was published in 1940), who prepared a radio map of the heavens. In 1942 J. S. Hey found that the Sun was an emitter of radio waves. No great advance in the subject was possible until after World War II. Since then special observatories devoted to the study of this subject have been established throughout the world.

Several thousand radio sources have been discovered. These include the Sun and Jupiter, radio galaxies, quasars, spiral galaxies, pulsars, supernova remnants, interstellar clouds of hydrogen (producing the 21-cm emission) and molecules such as ammonia and formaldehyde. Some radio sources have been identified with optical ones, as with 19N4A or Cygnus A, one of the strongest radio sources. A nomenclature has been adopted by the I.A.U. for radio sources. It consists of the hour of Right Ascension, the tens digit of degrees of Declination N or S, and a serial letter. For example, 05N2A signifies the object (the Crab Nebula) in the following position, R.A. $05^{h}31^{m}$, Dec. 21° 59′N. A general way of referring to such sources is by the constellation followed by a capital letter, thus, Taurus A for the above example.

radio galaxy A galaxy that is an abnormally high emitter of radiation at radio frequencies and of synchrotron radiation. Such a galaxy is usually a giant elliptical, e.g. M87 in the constellation Virgo.

The equatorially mounted 42·7-m radio telescope at the National Radio Astronomy Observatory, Green Bank, West Virginia.

radio interferometer A type of radio telescope which operates on the same principle as an optical interferometer. It consists of two or more antennae which are spaced along a baseline and receive radiation of radio frequency from the same radio source. The output signals from pairs of antennae are fed to a common radio receiver. Several different lengths of baseline are possible, by varying the distance between the pairs of antennae. The receiver signals measured at each baseline value are analysed, and information is thus obtained on the intensity, position, and possible structure of radio sources.

radio source A celestial object that emits radiation of radio frequency of higher intensity than the background radiation. It does not necessarily correspond with any visible object.

radio star The name formerly used for a **radio source**.

radio telescope Apparatus constructed to detect, receive, and record radiation of radio frequency from celestial bodies and space. The apparatus consists of antennae, receivers, and recorders. One common type of radio telescope is the RADIO DISH. The antenna consists of a dipole having a length equal to half the wavelength of the radiation it is desired to receive. It is placed at the focus of a parabolic reflector, which in the preferred construction is steerable so that it can be directed to any particular area of the heavens. The signal received by the antenna is analysed, converted to a lower frequency and amplified before being recorded for intensity. Other types of radio telescopes employ an array of antennae arranged for maximum resolution of the incoming signal and suitability for the work in hand. The antennae can be arranged in parallel rows, as a cross, or any other appropriate form. See also **radio interferometer**.

radio window The gap in the Earth's atmospheric absorption which allows electromagnetic radiation of radio frequency to penetrate to the Earth's surface. The range of radio wavelengths to which the Earth's atmosphere is transparent is from a few millimetres to about 20 metres. See also **optical window**.

radius vector An imaginary straight line joining the centre of a celestial body to the focus of the orbit in which it revolves, e.g. the line connecting a planet and the Sun.

Ramsden disc Another name for **exit pupil**.

Ranger The name given to a series of lunar space probes launched by the United States. The first two failed; the third missed the Moon and went into solar orbit as a planetoid; the fourth crashed on the Moon and failed to transmit any data; the fifth became another planetoid; the sixth crashed on the Moon and failed to transmit data. The last three, Rangers 7, 8, and 9, made successful hard landings and transmitted thousands of photographs of the lunar surface, thus greatly adding to our previous knowledge. Ranger 7 was launched on July 28, 1964, Ranger 8 on February 17, 1965, and Ranger 9 on March 21, 1965.

raster The area of the screen of an oscilloscope on which the image is formed.

R.A.S. thread A screw thread found on components of some telescopes. It is of Whitworth formation, with 16 threads per inch on a diameter of 1¼ inches.

Rayleigh limit The minimum resolvable angle between the wavelengths of two lines in a spectrum. It is equal to $1{\cdot}22\lambda/D$ radians, where λ is the wavelength and D is the aperture of the objective in centimetres.

Rayleigh scattering The preferential scattering of the short wavelengths of light by very small particles. The blue of the sky results from the scattering of the shorter blue wavelengths by suspended particles in the Earth's atmosphere, or by molecules of the air. The scattering is inversely proportional to the fourth power of the wavelength of the light.

reading microscopes Microscopes, four or six in number as required, placed at equal distances around the circle of a transit circle for accurately determining the inclination of the tube. It is now usual for cameras to be fitted in order to obviate time-consuming measurements by the observer.

rectangular co-ordinates Any system of co-ordinates based upon axes perpendicular to each other.

recurrent nova see **nova**

reddening A phenomenon exhibited by starlight as it is selectively absorbed and scattered on passing through clouds of interstellar dust and the Earth's atmosphere. The longer the path and the greater the density of the clouds of dust, the greater the amount of reddening exhibited by the light. A correction factor for interstellar reddening must be introduced when spectra are used for estimating the distances of remote stars, otherwise they will seem to be more distant than they really are.

red dwarf A star of small diameter, relatively high mean density, and low absolute magnitude. Red dwarfs lie at the lower end of the main sequence of the **Hertzsprung-Russell diagram**.

red giant A giant star in a late stage of evolution, having exhausted the hydrogen fuel of its core. It is so called because the maximum intensity of its radiation lies in the red part of the spectrum, as the temperature is low (about 4000 K). The diameter is about 100 times that of the Sun, the luminosity 100–10,000 times that of the Sun, and the density is extremely low (10^{-4} to 10^{-7} times that of the Sun). Red giants lie in the upper right hand part of the **Hertzsprung-Russell diagram**. Betelgeuse in the constellation Orion is an example.

red shift A displacement of the absorption and emission lines in a spectrum towards longer wavelengths, i.e. towards the red end of the spectrum for optical lines. If this phenomenon is accepted as being the result of a **Doppler effect**, then the red shift means that the object in question is receding. See **Hubble constant; gravitational red shift.**

Red Spot or **Great Red Spot** An oval area about 40,000 × 15,000 km on the southern hemisphere of **Jupiter**. It is now known to be a vortex of cold red-coloured clouds rotating in an anticlockwise direction and lying about 8 km above the surrounding clouds, i.e. it is an atmospheric phenomenon.

reduced proper motion The corrected observed proper motion (see **restricted proper motion**) of a star in seconds of arc per year converted to absolute proper motion as a linear velocity in kilometres per second.

reduction of observations The mathematical analysis of observed astronomical data in order to derive the required astronomical information. Much work of this kind is now performed immediately by electronic computers connected directly to the telescope.

reduction of star places The conversion of the apparent position of a star, as obtained by observation, to the **mean position** for a given epoch. It involves applying numerous corrections to the observed values for Right Ascension and Declination.

reference stars Certain stars whose positions and proper motions have been accurately determined, and so can be assumed in connection with observations for the differential positions of other stars.

reflecting telescope or **reflector** A telescope which produces an image by the reflection of light from a primary concave mirror, usually parabolic, to a secondary mirror; this in turn reflects the light to an eyepiece placed in a position that depends on the design of the instrument (Gregorian, Newtonian, Cassegrain, and their modifications). Reflecting telescopes can be constructed of larger sizes than refracting telescopes because the mirror can be supported at the back. In contrast a lens has to be supported around the rim, and is consequently liable to distort under its own weight. As reflectors can be made much larger than refractors, their light-gathering power is much greater. Reflectors also have the advantage of being free from chromatic aberration.

reflection One of the properties of rays of light by which they are returned or thrown back from the surface on which they fall. If a ray of light is incident upon a plane surface, then the angle of reflection measured from the normal to the surface is equal to the angle of incidence. If a ray of light falls upon a convex reflecting surface, such as a mirror, then it will diverge and form a virtual image behind the mirror. If the surface is concave, then the ray will converge and form a real image in front of the mirror. Reflection plays a great part in optical systems. Reflecting telescopes use concave mirrors to gather light.

reflection nebula see **nebula**

reflector A telescope which utilizes a curved mirror for gathering light. See **reflecting telescope.**

refracting telescope or **refractor** A telescope which produces an image by the use of lenses. The earliest type of this instrument, as used by Galileo, was limited in scope and suffered from considerable chromatic and spherical aberrations. Improvements in this type of instrument were suggested by J. Kepler. They were achieved in practice by a succession of opticians, including J. Dolland and J. v. Fraunhofer, who devised means of reducing aberrations in the objective, and others, such as C. Huygens and J. Ramsden (1735–1800), who devised better eyepieces. A refractor produces an inverted image. The era of refractors of large aperture is past, for it is more economical to build reflectors when increased light-gathering power is required. The largest refractor, that at the Yerkes Observatory, has an aperture of 40 inches and was built in 1897; the largest two reflectors have apertures of 236 inches and 200 inches: they are at Zelenchukskaya Stanitsa (USSR) and the Palomar Observatory (United States) respectively.

refraction One of the properties of light whereby a ray is bent or deflected from its

The 40-inch refracting telescope at Yerkes Observatory, Wisconsin.

direction of propagation on entering another medium of different refractive index. Refraction plays an important part in optical apparatus, which may contain prisms of various shapes and convex or concave lenses, made of suitable transparent material such as glass, quartz, or plastic. A ray of light in passing from one medium into a denser one is bent, or refracted, towards the normal. The REFRACTIVE INDEX of a medium is a constant and is given by the ratio of the sine of the angle of incidence to the sine of the angle of refraction, the light passing from vacuum into the medium. This constant is equal to the inverse ratio of the velocity of propagation of light in the two media. The index of refraction varies with the wavelength of light, and so is the cause of various chromatic phenomena.

refractor A telescope which utilizes a lens for gathering light. See **refracting telescope**.

Regiomontanus (properly Johann Müller; Königsberg, June 6, 1436 – Rome, July 6, 1476) A German astronomer and mathematician. He observed the comet of 1472 (now known as Halley's Comet), invented the method of lunar distances for determination of longitude at sea, and assisted in the reform of the Julian calendar.

regolith The surface dust and debris covering the surface of the Moon, Mars, and possibly other bodies in the solar system. It is produced by the disintegration of rocks through continual heating and cooling, as well as by the impact of meteorites on the surface.

regression of the Moon's nodes see **node**

regular variable A **variable star** whose luminosity fluctuates in a regular cyclic manner.

relative orbit The apparent orbit of the fainter component of a binary system, relative to the brighter component.

relativistic Denoting or involving the motion of particles, approaching the velocity of light.

relativistic red shift see **gravitational red shift**

relativity Two theories produced by A. Einstein. The SPECIAL THEORY OF RELATIVITY was published in 1905. It introduces the concept of space-time, by which is to be understood that the concept of space cannot be separated from that of time. Time is local and varies from one system to another; it is not therefore possible to verify the simultaneity of two events taking place in two systems in motion relative to each other. One of the basic principles of the Special Theory is that all systems, or frames of reference, in uniform motion relative to each other are equivalent in respect of the physical laws governing phenomena.

Another basic principle is that the velocity of light *in vacuo* is constant: it is the same for all observers, no matter in what direction they move relative to each other. This velocity is the maximum for the transport of energy, and cannot be attained by any material body. From this it can be shown that the mass of a body increases with its velocity according to the formula

$$m = m_o \ / \ \sqrt{(1 - v^2/c^2)}$$

Furthermore mass and energy are related by the formula

$$E = mc^2$$

m is the mass of the body, m_o is the mass of the body at rest, v is the velocity of the body, c is the velocity of light, and E is the energy.

The Special Theory of Relativity has been verified in a number of ways. The **Lorentz-Fitzgerald contraction** is a natural consequence of the theory, which also explained the failure of the famous experiment conducted by A. A. Michelson (see **Michelson, Albert**) to measure the velocity of the Earth relative to the aether, and so caused the concept of the aether to be abandoned.

The GENERAL THEORY OF RELATIVITY was published in 1916. It extends the physical laws to phenomena seen by observers not in uniform motion relative to each other; acceleration is now considered relative, and gravitation is introduced. Gravitation is considered not as a force but as a consequence of the curvature of space-time. This curvature is brought about by the presence of matter. The theory, which is concerned with the generalization of Newton's Law of Gravitation, has contributed much to the development of present astronomical theories. Verification of the theory has been afforded by: an explanation of the phenomenon of the advance of the perihelion of the orbit of Mercury; by the experimental measurement of the deviation of light by a large mass (as by the Sun during an eclipse); by the gravitational red shift of spectral lines in light emitted by atoms in a very strong gravitational field; by the cosmological abundances of helium and deuterium which are consistent with those predicted by the **big bang theory**; by the time-delay of electromagnetic radiation on passing the limb of the Sun; by laser ranging of the Moon; and by the behaviour of the binary pulsar PSR 1913 + 16. Within experimental error all these astronomical tests confirm the General Theory of Relativity.

residuals The differences between predicted and observed values, frequently found on analysis of astronomical observations. These differences are expressed as 'observed minus computed' (O – C). Such residuals, relating to a given phenomenon, if collected over a sufficiently long period of time and then analysed statistically, may reveal some long-term trend.

resolving power The ability of an optical system, such as a telescope or the human eye, to distinguish close but separate objects as separate observable images. Such objects include the components of a close binary or the surface features on the Moon. The resolving power of a telescope in seconds of arc is given by the quotient of 11 divided by the aperture of the objective in centimetres, or by 4·56 divided by the aperture in inches (DAWES' LIMIT). The resolving power of the human eye is about one minute of arc. See also **Rayleigh limit**.

restricted proper motion The observed angular displacement of a star, relative to other stars, corrected for various factors, such as precession and nutation. See **reduced proper motion**.

retrograde motion Motion in the opposite direction to the prevailing one. The motion of most bodies in the solar system is from west to east. Some satellites, such as the outermost ones of Jupiter, have a retrograde motion. See also **apparent retrogression**.

revolution The term used in astronomy to denote the movement of a planet or other celestial object around its orbit, as distinct from rotation of the object on its axis.

Rhea Satellite V of **Saturn**. See also **satellite**.

Rheticus (properly Georg Joachim; Feldkirchen, 1514 – Kaschau, 1576) An Austrian astronomer and mathematician, remembered as the assistant to Copernicus. He published the *Narratio Prima* (Gedani, 1540), which is an outline of the great work written by Copernicus, namely *De Revolutionibus . . .*, with the editing and publishing of which he (Rheticus) was entrusted by Copernicus before he died.

Rigel The star β Orionis. It is the brightest star in the constellation Orion and marks the left foot in the quadrilateral figure of Orion. It is a blue supergiant and the primary component of a multiple star, and has a visual magnitude of 0·14.

Right Ascension (*abbrev.:* R. A., symbol: α) One of the two equatorial co-ordinates (the other is **Declination**) used to define the position of a celestial object. It is the angular distance measured along the celestial equator from the First Point of Aries (see **Aries, First Point of**) eastwards to the point where the declination circle containing the object intersects the equator. R.A. is usually expressed in hours, minutes, and seconds, although it may be expressed in angular measure.

rille A broad cleft resembling a trench observed on the surface of the Moon. Photographs taken by lunar probes have revealed some to be long chains of small craters. See **rima**.

rima (*Latin:* 'rift, cleft') The name adopted by the I.A.U. in 1961 for **rille**.

ring galaxy A galaxy having the form of an elliptical ring, often with luminous matter present in its interior. They are thought to form as the result of a collision between a normal spiral galaxy and intergalactic gas clouds.

ring micrometer see **micrometer**

ring mountain A characteristic lunar formation consisting of a crater with a high circular wall which slopes steeply inwards to the floor of the crater, and slopes gently outwards to the surface of the Moon.

ring nebula A nebula that has the appearance of a ring when viewed through a telescope. See **planetary nebula**.

Ring Nebula The ring-shaped planetary nebula (M57, NGC 6720) in the constellation Lyra.

ring plain The largest lunar formation, consisting of a walled plain or crater of diameter from 50 to 300 km.

Ritchey-Chrétien telescope A modified form of the **Schwarzschild telescope**, produced by G. W. Ritchey (1864–1945) and H. Chrétien (1879–1956), in which the primary mirror is hyperbolic and the secondary is ellipsoidal. A Cassegrain arrangement is adopted. The system is free from chromatic aberration and practically free from spherical aberration and coma, though there is some astigmatism and the field is curved.

Roche limit The minimum distance at which a satellite in orbit about its primary can be stable. A satellite, held together by its own gravitation and revolving in a circular orbit about a larger central mass of the same mean density, will disrupt if it comes within the critical distance of 2·44 times the radius of the central mass. The Roche limit (R) is given by the expression

$$R = 2{\cdot}44\, r \sqrt[3]{d_1/d_2}$$

where d_1 and d_2 are the densities of the central mass and the satellite respectively and r is the radius of the primary. The expression was discovered by E. A. Roche (1820–83), Professor of Astronomy at Montpellier.

rocket see **surge**

rockoon A small rocket which is carried aloft by a balloon before being fired at a great height. The technique avoids the drag experienced by a rocket fired from the ground. Consequently, it can fly as high as a larger more expensive device. The technique was initiated by J. A. Van Allen (see **Van Allen, James**).

Rømer, Ole Christensen (Aarhus, September 25, 1644 – Copenhagen, September 19, 1710) A Danish astronomer who invented the meridian circle. In order to explain the apparent inequality in the times of the eclipses of Jupiter's satellites, he assumed that light was propagated with finite velocity: he found that light takes about 11 minutes to travel from the Sun to the Earth.

Rosse, third Earl of see **Parsons, William**

Rosse, fourth Earl of see **Parsons, Laurence**

rotation The turning of a celestial body on its imaginary axis, as opposed to its revolution in orbit. Data for the periods of rotation of the planets, Sun, and Moon are given under the relevant entry. Rotation of the stars can be determined by spectroscopic observations. Rotation of the **Galaxy** was discovered by J. H. Oort (see **Oort, Jan**) in 1926. Rotation of galaxies beyond the Milky Way has been detected, for example, in the Andromeda Galaxy.

Royal Astronomer for Ireland The holders of this office, who were also Directors of the Dunsink Observatory, near Dublin, are listed in the following table.

Name	*Held office*
Rev. John Brinkley (1763–1835)	1791–1827
Sir William Rowan Hamilton (1805–1865)	1827–1865
Franz Friedrich Ernst Brünnow (1821–1891)	1865–1874
Sir Robert Stawell Ball (1840–1913)	1874–1892
Arthur Alcock Rambaut (1859–1923)	1892–1897
Charles Jasper Joly (1864–1906)	1897–1906
Sir Edmund Taylor Whittaker (1873–1956)	1906–1912
Henry Crozier Plummer (1875–1946)	1912–1921

Appointments to this office have not been made since the establishment of the Republic of Ireland in 1919.

RR Lyrae The first cluster variable star to be discovered. It is the prototype of stars exhibiting a regular variation in brightness and whose period of light change is less than 1 day. These RR LYRAE STARS are very old giant stars and are found in globular clusters in the halo of our Galaxy and also in its central region. They are **pulsating stars**.

Rudolphine Tables Planetary tables calculated by J. Kepler on the basis of observations made by Tycho Brahe but in accordance with his own laws of planetary motion. They were dedicated to his patron, Emperor Rudolf II, and were published at Ulm in 1627. Their accuracy was such that they continued in use till about the mid-18th century.

runaway star A star of spectral type O or B which has an unusually high velocity of motion in space.

Russell, Henry Norris (Oyster Bay, New York, October 25, 1877 – Princeton, New Jersey, February 18, 1957) An American astronomer who became Director of Princeton Observatory. He is distinguished for his contributions to astrophysics and stellar evolution, and for his determination of orbits of binary stars. With E. Hertzsprung he originated the **Hertzsprung-Russell diagram**.

Russell mixture A mixture of chemical elements in the same proportions as found in the Sun and similar stars: it contains about 75% hydrogen. It was determined by H. N. Russell and gives a measure of the relative abundance of the elements in the universe.

Ryle, Martin (Brighton, September 27, 1918 –) A British astronomer who is Professor of Radio Astronomy in the University of Cambridge and the present Astronomer Royal (1972–). He is distinguished for his pioneering work in radio astronomy.

Sagan, Carl Edward (New York, November 9, 1934 –) An American astronomer at the Smithsonian Astrophysical Observatory. He is known for his work on planetary atmospheres and surfaces, particularly of Venus and Mars, on the origin of terrestrial life, and on exobiology.

Sagittarius A One of the 'brightest' radio sources known, situated at R.A. $17^h42'$, Dec. $-28°55'$ (Epoch 1950·0), which is within 0°·1 of the galactic centre.

Saha, Meghnad N. (Dacca, October 6, 1893 – Calcutta, February 16, 1956) An Indian physicist and astrophysicist, known for his theoretical work on spectra, the solar corona, radiation, and ionization. He derived an equation, known as the SAHA EQUATION, which at a given temperature enables the thermodynamic equilibrium between neutral atoms, ions, and free electrons to be calculated. It is particularly important for calculating the state of the gas in a stellar atmosphere: it provides an explanation for the differences in the spectra of stars of different spectral types, and permits their temperatures to be deduced.

Saiph The star κ Orionis. It marks the right foot in the quadrilateral figure of Orion. It is a blue-white supergiant of apparent magnitude 2·2.

Salpeter process or **triple-alpha process** A nuclear reaction described by E. E. Salpeter (1924–) in which three alpha particles (helium nuclei) are transformed into carbon with the release of energy. It takes place at a temperature of about 2×10^8 K and a density of 10^5 g cm^{-3} after all the hydrogen of the core has been exhausted. Heavier elements are formed by the capture of alpha particles by the carbon nuclei.

Salyut The name of a series of Soviet space laboratories placed in orbit around the Earth. They are manned by crews from Soyuz space craft which dock with the Salyut stations. The first Salyut was launched on April 19, 1971. The first docking (without transfer of crew) of the Soyuz craft took place on April 24. On June 7, 1971 a crew of three docked with the Salyut and spent 23 days in the space laboratory carrying out astronomical, biological, and geophysical observations.

Sampson, Ralph Allen (Skull, Ireland, June 25, 1866 – Bath, November 7, 1939) A British mathematician and astronomer who was appointed Astronomer Royal for Scotland. He is known for his *Tables of the Four Great Satellites of Jupiter* (1910), which are still used, and his work on stellar spectroscopy.

Sandage, Allan Rex (Iowa City, June 18, 1926 –) An American astronomer distinguished for his contributions to the theory of stellar evolution. He is the co-discoverer of **quasars**.

Saros The period of 223 lunations (6585·32 days), which is almost equal to 19 eclipse years (6585·78 days). Hence, after such a period the Sun, the Moon, and the Moon's nodes return to very nearly the same position relative to each other. This period was discovered by observation by the Chaldean astronomers, who called it the Saros. Knowledge of this period enabled ancient astronomers to predict eclipses for, if an eclipse of the Sun or the Moon is observed, then it is almost certain that after a complete Saros there will be a similar eclipse. This occurs because the Sun, Moon, and nodes will have returned to almost the same relative positions, and the same sequence of solar and lunar eclipses will recommence. The difference of 0·46 of a day between the number of lunations and eclipse years in one Saros means that each eclipse will occur about 140° west of the previous one. In predicting a date, allowance must be made for leap days.

satellite A celestial body which revolves in orbit about a planet. Data relating to the satellites associated with the planets of the solar system are given in the table on pages 150–151. Values for the orbits of satellites are subject to considerable variations, and so too are the diameters and densities of the satellites. The orbital inclinations are related to the plane of the primary's equator; the motion is retrograde in those cases where the inclination is greater than 90°. Satellites are numbered consecutively for each planet in the order of their date of discovery. Most have received names, officially adopted by the I.A.U.; others, including the recently discovered satellite of Pluto, have one or more proposed names. In listing satellites it is usual to put them in order according to their distance from the primary.
Mars Both satellites are irregular in shape. In 1969 Mariner 7 photographed the elliptical shadow of Phobos cast on Mars. In 1971–72 Mariner 9 detected craters on both satellites and measured their sizes. The objects are thought by some to be captured minor planets, rather than natural satellites. See photograph on page 152.
Jupiter Satellites VI, VII, X, and XIII form a group in orbits about 11·5 million km distant from the primary; their periods are about 260 days and the orbital motion is direct. Satellites VIII, IX, XI, XII, and XIV form another group in orbits about 22·5 million km distant from the primary; their periods are about 700

Planet and satellite	*Name of satellite*	*Discovered by*	*Year of discovery*	*Mean distance from primary* miles × 10^3	km × 10^3	*Diameter* miles	km	*Mean sidereal period* d
Mercury	No satellite							
Venus	No satellite							
Earth								
I	Moon	—	—	238·853	384·4	2172	3476	27·321661
Mars								
I	Phobos	Hall	1877	5·81	9·3	17	27	0·318910
II	Deimos	Hall	1877	14·59	23·5	9	15	1·262441
Jupiter								
V	Amalthea	Barnard	1892	113·0	181·0	100	160	0·498179
I	Io	Galileo	1610	263·5	421·8	2290	3660	1·769138
II	Europa	Galileo	1610	417·0	670·0	1800	2880	3·551181
III	Ganymede	Galileo	1610	665·1	1065·0	3130	5000	7·154553
IV	Callisto	Galileo	1610	1170·0	1885	2900	4640	16·689018
XIII	Leda	Kowal	1974	6900	11110	5	10	240
VI	Himalia	Perrine	1904	7125	11450	60	100	250·5662
VII	Elara	Perrine	1905	7300	11740	20	30	259·6528
X	Lysithea	Nicholson	1938	7310	11750	10	20	259·2188
XII	Ananke	Nicholson	1951	13100	21000	10	20	631
XI	Carme	Nicholson	1938	14050	22500	15	25	692
VIII	Pasiphaë	Melotte	1908	14600	23500	20	32	744
IX	Sinope	Nicholson	1914	14700	23950	15	25	758
XIV		Kowal	1975					
Saturn								
X	Janus	Dollfus	1966	98	159	190	300	0·749
I	Mimas	Herschel	1789	115	185	310	500	0·942422
II	Enceladus	Herschel	1789	148	238	380	600	1·370218
III	Tethys	Cassini	1684	184	294	620	1000	1·887803
IV	Dione	Cassini	1684	235	377	800	1300	2·736916
V	Rhea	Cassini	1672	328	527	950	1500	4·517503
VI	Titan	Huygens	1655	762	1221	3600	5800	15·945448
VII	Hyperion	Bond	1848	926	1482	300	480	21·276657
VIII	Iapetus	Cassini	1671	2225	3558	740	1200	79·33085
IX	Phoebe	Pickering	1898	8920	12946	160	250	550·337
Uranus								
V	Miranda	Kuiper	1948	83	120	250	400	1·41349
I	Ariel	Lassell	1851	120	192	870	1400	2·520384
II	Umbriel	Lassell	1851	167	267	620	1000	4·144183
III	Titania	Herschel	1787	273	438	680	1100	8·705876
IV	Oberon	Herschel	1787	366	586	1000	1600	13·463262
Neptune								
I	Triton	Lassell	1846	222	355	2300	3700	5·876844
II	Nereid	Kuiper	1949	3750	6000	190	200	359·881
Pluto								
I	Charon*	Christy	1978	10?	17?	750?	1200?	6·39?

*Suggested name

Mean synodic period d h m s	Inclination of orbit °	Eccentricity	Density (water = 1)	Apparent magnitude at opposition	Mass g × 10^{24}	Name of satellite
						Mercury
						Venus
						Earth
29 12 44 02·9	23·4	0·0549	3·34	−12·7	73·5	Moon
						Mars
7 39 26·6	1·1	0·0210		11·6		Phobos
1 06 21 15·7	1·8	0·0028		12·8		Deimos
						Jupiter
11 57 27·6	0·4	0·003		13·0		Amalthea
1 18 28 35·9	0·0	0·000	3·8	4·8	73	Io
3 13 17 53·7	0·5	0·0001	3·4	5·2	47·5	Europa
7 03 59 35·9	0·2	0·0014	2·4	4·5	154	Ganymede
16 18 05 06·9	0·2	0·0074	2·3	5·5	95	Callisto
254	26·7	0·147		20		Leda
265 22 43	28	0·158		13·7		Himalia
276 04 56	28	0·2072		16·0		Elara
275 17 09	29	0·1074		18·6		Lysithea
551	147	0·169		18·8		Ananke
597	163	0·207		18·1		Carme
635	148	0·410		18·8		Pasiphaë
645	157	0·275		18·3		Sinope
						Saturn
00 17 58 30	0·0	0·000		14·0		Janus
00 22 37 12·4	1·5	0·0202	1	12·1	0·04	Mimas
1 08 53 21·9	0·0	0·0045	1	11·8	0·07	Enceladus
1 21 18 54·8	1·1	0·000	1·1	10·3	0·65	Tethys
2 17 42 09·7	0·0	0·0022	3·6	10·4	1·0	Dione
4 12 27 56·2	0·3	0·0010	1·1	9·8	2·3	Rhea
15 23 15 31·5	0·3	0·0292	2·3	8·4	137	Titan
21 07 39 05·7	0·6	0·1042	3	14·2	0·31	Hyperion
79 22 04 59	14·7	0·0283	3	11·0	1	Iapetus
523 13	150	0·1633		16·5		Phoebe
						Uranus
1 09 55 31	3·4	0·017	5	16·5	0·1	Miranda
2 12 29 39·0	0·0	0·0028	5	14·4	1·2	Ariel
4 03 28 25·8	0·0	0·0035	4	15·3	0·5	Umbriel
8 17 00 01·2	0·0	0·0024	6	14·0	4	Titania
13 11 15 36·5	0·0	0·0007	5	14·2	2·6	Oberon
						Neptune
5 21 01 49·7	159·9	0·000	5·1	13·5	140	Triton
362 01	27·7	0·7493		18·7	0·03	Nereid
						Pluto
?	?	?	?	15–16	?	Charon*

Phobos, the inner satellite of Mars, photographed by Viking Orbiter 2.

days and the orbital motion is retrograde. G. P. Kuiper (see **Kuiper, Gerard**) suggests that these two groups have resulted from the disintegration of two bodies captured by Jupiter in the past. It has also been suggested that they are captured minor planets.

The American spacecraft Voyagers I and II have discovered several very active volcanoes on Io, whose activity can be explained as being caused by the gravitational pull of Jupiter and neighbouring satellites. The same spacecraft have photographed the surfaces of the other Galilean satellites, Europa, Ganymede, and Callisto. Europa exhibits a smooth surface criss-crossed by thin cracks; Ganymede and Callisto are both greatly cratered.

Saturn. Satellites I to VII and X are all within 1·5 million km of the primary, and the inclination of their orbits to the planet's equator is about 1° or less; in the case of satellites VIII and IX they are more than 3 million km from the primary, and the orbital inclinations are much greater. It has been suggested that these two satellites may not have been part of the original Saturnian system, but may have been subsequently captured by the planet.

Neptune. The diameter of Triton is uncertain though it seems to be the largest satellite in the solar system. It has retrograde motion. Nereid has the most eccentric orbit of any natural satellite in the solar system; its distance from the primary varies from 140,000 km to 9,500,000 km.

satellite galaxy A small galaxy physically associated with a larger one. For example, the Andromeda Galaxy has two, M32 and NGC 205.

Saturn One of the giant planets of the solar system, orbiting the Sun between Jupiter and Uranus. Viewed through the telescope it appears as a flattened golden yellow disc with white rings around it. The disc has surface markings and belts similar to those on Jupiter. White spots of short life occasionally appear. The density of the planet, $0{\cdot}7\,g\,cm^{-3}$, is much lower than that of water. It is therefore assumed to be composed largely of hydrogen, some helium, and is thought to have a rocky core. Hydrogen, methane, and ethane have been detected in the upper atmosphere. At the low temperature prevailing at the visible surface of the atmosphere, water and ammonia will have been frozen out.

Saturn is remarkable for the magnificent system of rings by which it is encircled. They were first observed by Galileo in 1610. Their nature was described by C. Huygens (see **Huygens, Christiaan**) in his *Systema Saturnium* (1659). The system lies in the equatorial plane of Saturn and consists of three bright rings and some fainter rings only recently discovered. The bright rings, starting from the outside, are Ring A, which is greyish-white and noticeably darker than Ring B, which is quite white, and Ring C which is faint and blue-grey in colour. The faint rings are Ring D, which is the innermost of Saturn's rings and was discovered in 1969, and Rings F and G, which lie outside Ring A and were discovered in 1979 by the Pioneer 11 spacecraft. The rings are relatively thin and become invisible when viewed edge-on. The transparent rings consist of innumerable particles of icy dust and fragments of rock, all circulating round the planet like satellites. The spectrum of the rings reveals that ice particles at about −190°C are present. The main data concerning Saturn and its rings are given in the table. Between Rings A and B there is a gap about 2800 km wide, which was discovered by G. D. Cassini (see **Cassini**) in 1675, and is known as the CASSINI DIVISION. J. F. Encke (see **Encke, Johann**) noted a dark circular line on Ring A in 1837. It is known as ENCKE'S DIVISION. Although it has been observed

Globe	
Diameter (equatorial)	119,300 km
Diameter (polar)	107,700 km
Density (water = 1)	0·71 g cm^{-3}
Mass	$5{\cdot}81 \times 10^{29}$ g
Volume	$8{\cdot}2 \times 10^{17}$ km^3
Sidereal period of axial rotation	10^h 14^m
Escape velocity	36·26 km s^{-1}
Albedo	0·76
Inclination of equator to orbit	26° 44′
Surface temperature	127 K (maximum)
Surface gravity (Earth = 1)	1·159

Orbit	
Semi-axis major	9·539 A.U. = 1427×10^6 km
Eccentricity	0·0556
Inclination to ecliptic	2° 29′ 21″
Sidereal period of revolution	10,759·20 d
Mean orbital velocity	9·65 km s^{-1}
Mean synodic period	378·1 d

Rings		
Ring A	Outer diameter	272,300 km
	Inner diameter	239,600 km
Ring B	Outer diameter	234,000 km
	Inner diameter	181,000 km
Ring C	Inner diameter	149,300 km

Saturn and its rings.

since then there is no certainty that it is a true gap like the Cassini division. Ring C, also known as the CRÊPE RING, was discovered by W. C. Bond and G. P. Bond in 1850. It is difficult to observe because it is much less dense than the outer two bright rings. The thickness of the system of rings is estimated to be between 10 and 25 km. The rings lie within the **Roche limit**. Consequently the matter of which they are composed will not condense to form a satellite.

Satellites. Saturn has ten known satellites, although several more have been reported following the Pioneer 11 fly-by. TITAN, the largest, is the only satellite in the solar system to have a significant atmosphere, namely, methane. This discovery was made in 1944 by G. P. Kuiper (see **Kuiper, Gerard**). Pioneer 11, which flew past Saturn at 220,000 miles, found Titan to be covered by a reddish atmosphere. PHOEBE, the outermost satellite of Saturn, was the first to be discovered from the southern hemisphere as well as the first to be found photographically. Its motion is retrograde.

In 1900 W. H. Pickering (see **Pickering, William**) reported the discovery of a new satellite of Saturn, for which the name Themis was proposed. It was about 1·5 million km from the primary, had a sidereal period of $20^{d}\ 20^{h}\ 25^{m}$, and a magnitude of 18. As this object has not been seen since that time, it has been suggested that what was observed was in fact a minor planet that happened to be in the field at the same time as Saturn.

scattering of light The random deflection of light rays which occurs as a result of collision with particles in a medium. See **Rayleigh scattering**.

Scheiner, Christoph (Wald, July 25, 1575 – Neisse, July 18, 1650) A German astronomer who was one of the earliest observers of sunspots, which initially he thought to be small planets revolving around the Sun. Like Galileo he supported the geocentric theory of the universe. He made improvements to the telescope, helioscope, and pantograph and wrote works on optics and the Sun.

Scheiner, Julius (Cologne, November 25, 1858 – Potsdam, December 20, 1913) A German astrophysicist known for his spectroscopic studies of the stars. He devised a sensitometer to measure the sensitivity of photographic emulsions. The Scheiner sensitometer scale was extensively used in Germany and Austria for measuring the speed of photographic emulsions.

Schiaparelli, Giovanni Virginio (Savigliano, March 14, 1835 – Milan, July 4, 1910) An Italian astronomer who was an assiduous observer of planetary surfaces. He prepared maps of the surface features on Mars and Mercury. His nomenclature for the Martian features has been subsequently modified. A specially prepared map with designations of features has now been adopted by the International Astronomical Union. His use of the term **canali** for one of them was the cause of much subsequent controversy. He also made observations of Venus, discovered that meteor streams may be the debris of disintegrated comets, and contributed to the history of Babylonian astronomy.

Schmidt, Bernhard Voldemar (Insel Nargen, Estonia, March 30, 1879 – Hamburg, December 1, 1935) An Estonian instrument-maker and astronomer, remembered for his development about 1931 of the camera which bears his name. The achievements of modern astronomical photography have been possible because of the availability of this camera. Smaller instruments, of particular use for the spectrography of very faint objects, have been developed from the original Schmidt camera.

Schmidt camera A special design of telescope developed in about 1931 by Bernard Schmidt, having a small focal ratio and a very wide field with sharp definition. It consists of a spherical (rather than a paraboloidal) primary mirror, with a correcting plate placed at the centre of curvature of the mirror. The correcting plate is figured so as to minimize optical aberrations in the final images, although there is some chromatic aberration and a little spherical aberration and coma. The final images are formed on the curved focal surface situated in front of the mirror. The shape of the focal surface necessitates the use of a special plate-holder to constrain the

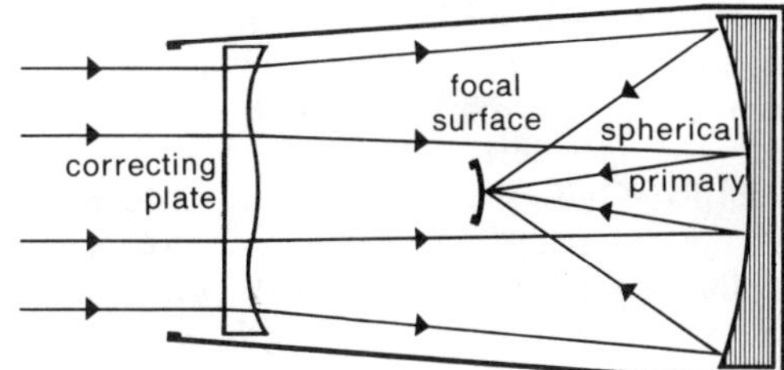

Light path in a Schmidt camera.

photographic plate/film to its curved contour. The description '48 in/72 in/121 in Schmidt', means that the diameter of the correcting plate is 48 inches, that of the spherical mirror is 72 inches, and the focal length is 121 inches.

The Schmidt camera and its numerous variations have been responsible for significant advances in photographic astronomy. The world's largest four Schmidt cameras are at the observatories at Tautenberg, East Germany (54 in/80 in/160 in), Palomar, United States (48 in/72 in/121 in), Siding Spring, Australia (48 in/72 in/180 in), and Kvistaberg, Sweden (40 in/54 in/119 in). Of the many modifications to the Schmidt camera, the most important are the **Maksutov telescope** and the **Super-Schmidt telescope**.

Schmidt-Cassegrain telescope One of the variations of the Schmidt camera in which the correcting plate is combined with two mirrors in a Cassegrain arrangement.

Schmidt, Johann Friedrich Julius (Eutin, Germany, October 26, 1825 – Athens, February 7, 1884) A German astronomer who became Director of the National Observatory, Athens. He was an authority on selenography: in 1878 he published his map of the Moon, the most accurate of its time; he was the first to report significant changes in the crater Linné and other lunar features. He observed also comets, variable stars, and the zodiacal light.

Schmidt lens The aspherical correcting plate of a Schmidt camera.

Schmidt, Maarten (Groningen, December 28, 1929–) A Dutch astronomer now settled in the United States. He is Professor of Astronomy at California Institute of Technology and Director of the Hale Observatories. He is known for his work on **quasars**, in particular for his discovery (1963) of the immense red-shift of their spectral lines. He also ascertained the spiral structure of the Galaxy and the distribution of mass within it.

Schmidt-Maksutov telescope see **meniscus-Schmidt telescope**

Schmidt telescope see **Schmidt camera**

Schwabe, Samuel Heinrich (Dessau, October 25, 1789 – Dessau, April 11, 1875) A German apothecary who began a systematic daily survey of the Sun for sunspot activity in 1826 and continued it for over 40 years. He concluded that there was a periodicity of about ten years in the number of sunspots. His published results did not attract attention until they were reproduced by Alexander von Humboldt (1769–1859) in *Kosmos* (Stuttgart, 1850). The citation on his being awarded the Gold Medal of the Royal Astronomical Society in 1857 says: 'Twelve years he spent to satisfy himself, six more to satisfy, and still thirteen more to convince mankind' of his discovery; 'the energy of one man has revealed a phenomenon that has eluded the suspicion of astronomers for 200 years'.

Schwarzschild black hole A non-rotating, spherically symmetrical **black hole**. Karl Schwarzschild in 1916 derived an exact solution to Einstein's vacuum field equations, and demonstrated that very massive bodies can become black holes.

Schwarzschild, Karl (Frankfurt am Main, October 9, 1873 – Potsdam, May 11, 1916) A German astronomer who became Director of the Astrophysical Observatory, Potsdam. He worked on photographic photometry and did fundamental studies on stellar atmospheres, on stellar motions, and in theoretical physics. He derived an exact solution for certain of A. Einstein's field equations. His work on preferential stellar motions led to his ellipsoidal hypothesis of star streaming. He designed various pieces of equipment.

Schwarzschild, Martin (Potsdam, May 31, 1912–) A naturalized American astrophysicist, son of Karl Schwarzschild and Professor of Astronomy at Princeton. He is distinguished for his work on the constitution of stars and on stellar evolution.

Schwarzschild radius The critical radius, in accordance with the General Theory of Relativity, at which a very massive body under the influence of its own gravitation becomes a black hole. It is the radius of the event horizon of a black hole, from which nothing can escape, not even light. The radius is given by the expression $R = 2GM/c^2$, where G is the gravitational constant, M is the mass of the body, and c is the velocity of light. It was deduced by Karl Schwarzschild.

Schwarzschild's law The law, formulated by Karl Schwarzschild, which states that the density produced in a photographic emulsion by exposure to light is proportional to It^p, where I is the intensity of the light, t is the time of exposure, and p is a constant. It has been found that p is constant only over small ranges of intensity.

Schwarzschild telescope A reflector designed by Karl Schwarzschild in 1905. It has two concave, almost spheroidal mirrors which give a photographic image free from spherical aberration and coma, but with some astigmatism. The instrument is cumbersome and difficult to handle. Apparently, only two have come into use.

scintillation A twinkling or tremulous motion of the light coming from the stars. When observed by the naked eye, stars appear to change in brightness and colour, particularly if they are low down in the sky; through the telescope, the positions of stars are seen to undergo short rapid changes. The effect is caused by the non-uniform density of the Earth's atmosphere, which produces uneven refraction of the light. Planets do not exhibit this phenomenon (though Mercury and Mars when in certain phases may do so slightly when close to the horizon) because the scintillations from different points on the surface are not in phase, and the fluctuations are lost in the general illumination. For astronomical purposes the inconvenience of the effects of scintillation is reduced by siting observatories at high altitudes, or almost entirely avoided by using telescopes carried by balloons, rockets, or artificial satellites.

It has also been found that radio sources scintillate. In this case it is the non-uniformity of the refractive indices of the interstellar medium, the interplanetary medium, and the ionosphere that cause the strength of the radio waves to fluctuate.

seasons In the northern hemisphere there are four seasons which are recognized for agricultural purposes, and have become so through the peculiarities of the climate. In other parts of the world, other seasonal distributions are made. Astronomically, the seasons are defined by the different positions of the Sun with respect to the ecliptic. In the northern hemisphere spring is reckoned from the vernal equinox (March 21) to the summer solstice (June 22) and lasts approximately 92^d 19^h; summer is reckoned from the summer solstice to the autumnal equinox (September 22) and lasts approximately 93^d 15^h; autumn is reckoned from the autumnal equinox to the winter solstice (December 22) and lasts approximately 89^d 19^h; winter is reckoned from the winter solstice to the vernal equinox, and lasts approximately 89^d 1^h. The seasons are also influenced by the inclination of the Earth's axis of rotation to the orbital plane. The northern hemisphere is tilted towards the Sun at the summer solstice, and hence experiencing summer when the Earth is near aphelion. On the other hand, when it is summer in the southern hemisphere, the Earth is at the winter solstice and near perihelion; the southern hemisphere is thus tilted towards the Sun and acquires a higher average temperature. Spring in the northern hemisphere corresponds to autumn in the southern hemisphere, summer to winter, autumn to spring, and winter to summer.

Secchi, Angelo (Reggio, June 29, 1818 – Rome, February 26, 1878) An Italian astronomer who became Director of the Roman College Observatory. He was a specialist in spectroscopy, as well as in solar and stellar physics, and is remembered as the pioneer of classifying stars by their spectral types. He was one of the first to use photography in astronomy.

Secchi classification The first practical system of stellar spectral classification, devised by Angelo Secchi between 1863 and 1867. His classification comprised four groups, and is still useful for a first approximate assessment. His pioneering work paved the way for the more detailed system now in use, and which was developed at Harvard by A. J. Cannon (see **Cannon, Annie**).

second (*symbol:* s) The unit of time in the **SI**. It is defined as the duration of exactly 9 192 631 770 periods of the radiation corresponding to the transition between the two hyperfine levels of the ground state of the caesium-133 atom.

For astronomical purposes the second has been defined by the I.A.U. as the fraction 1/31 556 925·9747 of the tropical year for 1900, January 0, 12 hours ephemeris time (i.e. 1899, December 31, 12 hours at noon). When the symbol s is encountered as a superior letter, ^{s}, it means that the time is given according to the astronomical fashion.

second of arc or **arc second** (*symbol:* ″) One-sixtieth part of a minute of arc or 1/3600 of a degree.

secular acceleration The long-term acceleration of the Moon in its orbital path round the Earth. Examination of data relating to lunar eclipses has revealed that the Moon's period is less now than what it was in former times; in fact, its motion is accelerating by about 11 seconds of arc per century in geocentric longitude. The phenomenon results from a number of factors, such as the perturbative effects of other planets and an increase in the period of the Earth's axial rotation.

secular parallax The apparent displacement of a celestial body in the sky as a result of the Sun's motion in space.

secular variation A small perturbation in the motion of the Moon, caused by variations in the gravitational attraction of the Sun on the Earth-Moon system during the synodic month.

seeing The quality of the image produced by a telescope at a given time in respect of clarity and steadiness as assessed by the observer. Seeing is good on a clear calm night when **scintillation** is absent. Evaluation of seeing is largely subjective, though numbers assigned according to the **Antoniadi scale** can be of use to ensure a certain uniformity in reporting conditions under which observations were made.

seismic waves Waves propagated through the Earth during and after an earthquake, explosion, or large impact. The two main types of wave studied are the PRIMARY WAVE (called the p-wave) and the SECONDARY WAVE (called the s-wave). The p-wave is a longitudinal compression wave which travels faster than the s-wave and can penetrate the Earth's core. The s-wave is a transverse shear wave which is unable to penetrate the Earth's core: it is totally reflected at the discontinuity which occurs at a depth of 2900 km. By mathematical analysis of seismological data, the behaviour of the p- and s-waves has provided considerable information about the nature of the Earth's interior.

selective absorption A phenomenon observed when light coming from distant stars passes through interestellar gas and dust clouds. The light is reddened as a result of the short wavelengths (blue light) being scattered more than the long wavelengths (red light).

selenographic co-ordinates A system of co-ordinates designed for fixing the position of features on the surface of the Moon. They may be Cartesian rectangular co-ordinates, or may be referred to selenographic latitude and longitude.

selenography The observational study of the Moon in respect of topography and surface phenomena.

selenology The study of the Moon's crustal rocks and the nature and origin of its surface features. The term is the analogical counterpart of geology, which relates to the Earth.

semi-regular variables Variable stars whose changes in brightness and form of light curve are roughly periodic but erratic. The cause of these phenomena is thought to be irregular pulsations produced by changes in the radius of the star.

An 18th-century sextant.

Seven Sisters see **Pleiades**

sextant An instrument designed for measuring the angular distance of a celestial body above the horizon, primarily intended for navigational purposes at sea, and so-called because its shape is one-sixth of a full circle.

sextile see **aspect**

Seyfert, Carl Keenan (Cleveland, February 11, 1911 – July 13, 1960) An American astrophysicist who was Director of the Barnard Observatory, Nashville, Tennessee. He is known for his work on the spectral and luminosity function of stars and his studies of galactic structure. **Seyfert galaxies** are named after him.

Seyfert galaxy A class of galaxies which have extremely bright nuclei whose spectra show strong emission lines. Most are spiral galaxies. The gas clouds in these galaxies move at very high velocities, being accelerated by unexplained but very powerful sources of energy in the nuclei. Many are radio sources and some emit X-rays. About 1% of all galaxies belong to this class. M77 (NGC 1068) is one of the brightest, besides being a radio source.

shadow bands An atmospheric phenomenon which occurs during total eclipses of the Sun. Ripples, or rapid alternations of light and shade, are observed on the ground, or on any light-coloured surface. They are caused by variations in the density of the atmosphere, being similar to **scintillation**.

Shapley, Harlow (Nashville, November 2, 1885 – October 20, 1972) An American astronomer who was Director of the Harvard College Observatory. He is distinguished for his work on photometry and spectroscopy, and for his investigations on the Galaxy in respect of size, shape, and the position in it of the solar system. These investigations followed his calibration (1916–17) of the **period-luminosity law** using Cepheid variables in globular clusters. He also made notable contributions to our knowledge of galaxies and cosmogony.

shell star or **envelope star** A hot B-type star which has an extensive envelope of gas surrounding a relatively inert core. It has been suggested that they are protostars. Energy is being released in the gaseous shell, the spectrum of which shows bright emission lines. Such stars are believed to represent a stage in the development of stars in the main sequence.

shooting star see **meteor**

short-period comet A **comet** whose apparitions are sufficiently frequent to permit correlation of orbital data.

Shortt clock A **pendulum clock** of very high accuracy, made by the Synchronome Company. A master pendulum oscillates *in vacuo* under strictly controlled conditions of temperature and humidity, and receives an impulse every half minute so as to maintain the oscillation; otherwise it is entirely free from mechanical contacts, for it does no work. Its oscillations control electrically a slave clock which performs all the necessary work of indicating time on dials, etc. When properly controlled the Shortt clock is capable of a rate of ±0·003 seconds per day, and for this reason was installed and used in many observatories until the advent of even more accurate means of time-keeping.

SI Abbreviation of Le Système International d'Unités. The units of the system were established by the Conférence Générale des Poids et Mesures in 1960. For each physical quantity there is one sole SI unit. There are seven base units: *metre* (m) for length, *kilogram* (kg) for mass, *second* (s) for time, *ampere* (A) for electric current, *kelvin* (K) for thermodynamic temperature, *mole* (mol) for amount of substance, and *candela* (cd) for luminous intensity. There are two supplementary units: *radian* (rad) for plane angle and *steradian* (sr) for solid angle. Other units are derived from combinations of the base units. There are special names for some derived SI units, including *newton* (N) for force, *pascal* (Pa) for pressure, and *joule* (J) for energy.

sidereal Of or pertaining to the stars. Measured or determined by the apparent motions of the stars.

sidereal clock A clock regulated to measure **Sidereal Time**.

sidereal day The time interval between two successive passages of a given star across the observer's meridian. It is equal to 23^{h} 56^{m} $04^{s}{\cdot}091$ of mean solar time.

sidereal month The time required for the Moon to complete one revolution around the **barycentre** with respect to the stars. It is equal to 27·32166 mean solar days.

sidereal period The time required for a planet or other celestial body to make one complete revolution with respect to a background star.

Sidereal Time Local time reckoned according to the rotation of the Earth with respect to the stars. The time is zero hours when the First Point of Aries (see **Aries, First Point of**) crosses the observer's meridian. The 24-hour system is used. The sidereal day is a trifle shorter than the mean solar day. Observatory clocks usually show Sidereal Time, as it is equal to the **Right Ascension** of an object which lies on the observer's meridian at that time.

sidereal year The time required for the Earth to complete one revolution through a point in its orbit fixed with respect to the stars. It is equal to 365·25636 mean solar days.

siderite A **meteorite** consisting mainly of iron (90–95%) and some nickel with traces of other metals.

siderolite A **meteorite** consisting of approximately equal proportions of stony material and metal, mostly iron and nickel.

siderostat An optical instrument comprising a mirror mounted equatorially and clock-driven to counteract the Earth's axial rotation. The mounting is so devised as to keep the image of one star stationary at the centre of the field, so that it may be seen continuously by a telescope, whilst the remainder of the sky rotates around the selected star. See also **coelostat**.

singularity In space-time, a region where the known laws of physics break down and the equations cease to have meaning. Theory predicts that a **black hole** contains a singularity and that at this point the mass of the black hole is compressed into an infinitely small volume with an infinitely large density.

sink A region where energy is received, as opposed to a source where energy is released.

Sinope Satellite IX of **Jupiter**. See also **satellite**.

Sirian stars Former name for stars of spectral type *A*, of which Sirius is the type star.

Sirius The star Alpha Canis Majoris. It is the brightest star in the sky, its visual magnitude being −1·47. It forms a binary system with a companion, SIRIUS B, whose existence was deduced by F. W. Bessel in 1844 from the perturbation it exerted on Sirius. The companion was optically discovered by A. G. Clark on January 31, 1862. It is a **white dwarf** with a mass equal to that of the Sun, but with a diameter only 0·022 that of the Sun. Its period of revolution is 50 years. Sirius B was the first star to be recognized as a white dwarf.

61 Cygni The first star to be measured (1838) for trigonometrical parallax (by F. W.

Bessel). It is a binary system containing a K5 star and a K7 star with a period of 720 years. There is a third non-luminous component whose mass is given as 8–16 times that of Jupiter.

Skylab The first American experimental space station or orbital workshop. It was launched under the control of NASA from the Kennedy Space Center on May 14, 1973. After some initial mishaps, it remained functional until February 9, 1974, when it was shut down. During these nine months it was manned by three separate teams of three men each. Telescopes and a coronograph were used to observe the Sun and the comet Kohoutek. Tests of Earth resources were carried out using remote sensing equipment and techniques to gather information on the Earth's geography, geology, ecology, forestry, etc. Medical and biological experiments were performed to find the effects on human beings of a long stay in space. Various technological experiments were made, including the manufacture of crystals of germanium selenide. The teams brought back about 70 km of magnetic tape recordings, 200,000 photographs of the Sun, and more than 8 km of vector cardiogram data. Although Skylab had been thought to have a lifetime of ten years, its orbit became unstable and it crash-landed on July 11, 1979 in the Indian Ocean, with fragments landing in Western Australia.

Slipher, Earl C (Mulberry, Indiana, March 25, 1883 – August 7, 1964) An American astronomer at the Lowell Observatory, brother of V. M. Slipher. He is known for the excellence of his photographic work relating to the planets, especially of Mars, which he photographed at four close oppositions: he was the first to notice the **blue clearing** on Mars. He took over 300,000 photographs of the planets.

Slipher, Vesto Melvin (Clinton County, Indiana, November 11, 1875 – November 8, 1969) An American astronomer who became Director of the Lowell Observatory. He was a specialist in astronomical spectroscopy, especially of galactic systems. He discovered the rotation and space velocities of galaxies (or extragalactic nebulae as they were then called) and measured the radial velocities of many of them. He showed that light from certain nebulae was reflected from stars embedded in them, thus discovering the first reflection nebulae.

Small Magellanic Cloud see **Magellanic Clouds**

solar activity The variable phenomena observable on the face of the Sun. They include faculae, solar flares, solar prominences, filaments, sunspots, and surges. Apart from being observable visually, solar activity manifests itself by the radiation of radio frequency which the Sun emits, and by magnetic storms on Earth which lead to radio interference and to increased auroral activity.

solar constant The amount of solar energy falling vertically on one square centimetre of the Earth's surface per second, assuming that the Earth has no atmosphere and is at its mean distance from the Sun. The mean value of the solar constant is 1·94 calories per square centimetre per minute, or 136·0 mW/cm^2. The value varies from day to day, and also with the **solar cycle**. Additionally there are some short-term fluctuations.

solar cycle The periodic fluctuations in the number of sunspots, first remarked upon by S. H. Schwabe (see **Schwabe, Samuel**) in 1843. The length of the cycle between successive minima of solar activity is on average 11·1 years. As there is a reversal of the Sun's magnetic polarity in the same period, it would seem that the basic period is more probably 22 years. The periodicity is very neatly illustrated by the **butterfly diagram**. The distribution of sunspots in latitude was observed by R. C. Carrington and studied in detail by G. Spörer (see **Spörer, Gustav**). R. Wolf devised a system for recording the variation in sunspot activity which is still used. Data are collected from observatories throughout the world, and after collation and analysis are presented as **Wolf relative sunspot numbers**.

solar eclipse An example of the obscuration of a luminous body. It occurs when the Moon

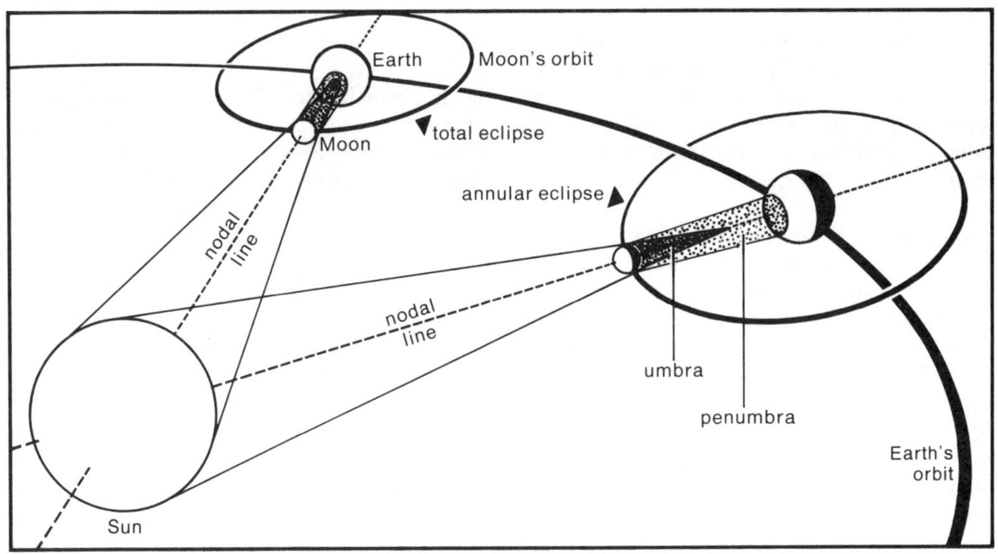

Total and annular solar eclipse.

lies between the Sun and Earth. As the Moon's orbital plane is inclined to the ecliptic, an eclipse can occur only when the Moon is at conjunction (i.e. at New Moon) and at the same time at or near one of its nodes; in addition the Sun's angular distance from one of the Moon's nodes at conjunction will determine if an eclipse can or cannot occur. It so happens that the apparent diameters of the Sun and the Moon as seen in the sky are almost the same, so the distance between

A total solar eclipse.

Earth and the Moon at the time of an eclipse will determine what kind of eclipse is produced. A TOTAL SOLAR ECLIPSE occurs at places where the **umbra** of the Moon's shadow-cone falls on and moves over the Earth's surface; at the same time the eclipse will appear PARTIAL to observers on either side of the central track of totality. Shortly before and after totality the phenomenon of **Baily's beads** is observable. The duration of totality cannot last more than 7·5 minutes. This short time is fully utilized for photography of the corona and for other studies. An ANNULAR ECLIPSE occurs when the vertex of the Moon's shadow-cone does not reach to

the Earth. A rim of light is then seen around the darkened face of the Sun. The maximum number of solar eclipses that can occur in a year is five. The total number of solar and lunar eclipses in a year is seven. If only two eclipses occur in a year, then they will both be solar.

solar flare A spectacular chromospheric eruption first noted by R. C. Carrington in the vicinity of a large newly formed sunspot. Flares appear suddenly, reach a maximum in some few minutes, and usually disappear within an hour or two. Short-lived prominences, known as surges, spring from them and are often observable near the solar limb. Flares represent an immense outburst of intense radiation and energetic particles that can cause radio interference, magnetic storms, and aurorae on Earth.

solar motion The motion of the Sun and other members of the solar system relative to that of neighbouring stars. Observation of the so-called fixed stars over long periods of time revealed that they have proper motions, and that their directions of motion have a well-defined trend towards the same point in the sky. The Sun's neighbouring stars being taken as a local standard of rest, it is found that the Sun, and hence the entire solar system, is moving: this causes the observed trend in the directions of proper motions to a specific part of the sky called the **apex**. The solar motion with respect to the local standard of rest is $19{\cdot}4\,\mathrm{km\,s^{-1}}$ towards R.A. 18^h, Dec. +30° in the constellation Hercules.

solar parallax The Sun's mean **equatorial horizontal parallax**. It is calculated by first determining the distance of a body close to the Earth. For this purpose close approaches of the minor planet Eros have been used. The solar parallax is one of the most important astronomical constants because it determines the mean distance of the Earth from the Sun: this distance is the astronomical unit.

solar plage A bright cloud of hot gas in the Sun's chromosphere, seen on a spectroheliogram taken in calcium light. Plages, also known as bright flocculi, are intimately connected with the development of sunspots, which they surround.

solar prominence A flame-like eruption of gaseous matter projected for an enormous distance from the Sun's chromosphere. They were first recorded at the total solar eclipse of 1842 and observable only on such occasions until the **coronograph** had been designed. Prominences consist of high-density gas at a temperature of 10^4 K suspended in the low-density corona at a temperature of 10^6 K. They are classified as ACTIVE or QUIESCENT. The former last for no more than an hour or so, but the latter can endure for weeks or months before erupting. All prominences give spectra showing lines of neutral hydrogen, helium, and ionized calcium. The forms and behaviour of prominences are very varied, and their direction of flow is controlled by the Sun's magnetic field.

solar stars The name sometimes applied to stars of spectral class G, the Sun being of this type.

solar system The Sun and all the celestial bodies which revolve round it, and are maintained as a physical unit by gravitational attraction. The orbiting bodies comprise the nine planets, their satellites, and thousands of minor planets, comets, and meteoroids. Data relating to the planets are given under the individual entries. For data on the satellites see table at **satellite**.

The planetary system is no more than the nucleus of the space in which the periodic motions of all known solar-system bodies take place. This space could be a sphere of radius between 40,000 to 100,000 A.U., or more. Our planetary system is isolated in space. All the planets revolve around the Sun in the same direction as the Sun itself rotates; the orbital motion of most of the satellites is direct too. The Sun is by far the most massive component of the solar system, its mass being 330,000 times that of the Earth. The nine planets have a total mass equal to 448 times that of the Earth, of which Jupiter accounts for 70%. The interplanetary space contains cosmic dust and ionized gas (solar-wind particles) in an extremely tenuous state. The solar system came into being in the Galaxy nearly 5000 million years ago, probably as the end-product of a contracting cloud of interstellar gas and dust. The exact mechanisms by which the Sun and its retinue were formed remains a major problem for cosmogonists.

solar-terrestrial relationships A term which covers all the effects observed in the atmosphere and on the surface of the Earth that are attributed to the Sun's activity, and result from ultraviolet radiation and corpuscular emission. Ultraviolet radiation ionizes molecules of gas in the atmosphere, and this ionization produces the ozonosphere and affects the Earth's magnetic field strength. It also alters the properties of the ionosphere and so affects radio transmission. The corpuscular emission is responsible for the **aurora**. Certain meteorological and biological phenomena may be related to solar activity.

solar tower A tower with a **coelostat** at the top and other equipment at the base, specially designed for research in solar physics. The advantage of this arrangement is that only the coelostat needs to be moved; all the other heavy apparatus remains fixed in position. G. E. Hale (see **Hale, George**) designed the first solar tower, which was erected at the Mount Wilson Observatory in 1907. Since then an installation of this type has become normal for a solar observatory.

solar wind The steady flow of charged particles (protons and electrons) from the solar corona into interplanetary space. It produces the magnetic field in the region between the Sun and planets. Some particles get trapped in the Earth's magnetic field and the outer **Van Allen radiation belt**. Others reach the upper atmosphere in the region of the magnetic poles and cause aurorae. The solar wind carries away about 10^{-13} of the Sun's mass per year. Its intensity increases during periods of solar activity.

Solid Schmidt telescope A constructional variation of the **Schmidt camera** in which the optical system is made from a solid block of glass, the surfaces of the corrector plate and primary being figured on its faces.

solstice The time when the Sun in its apparent annual motion along the ecliptic reaches its greatest declination of 23½°N or 23½°S. The times are approximately June 22 for the SUMMER SOLSTICE and December 22 for the WINTER SOLSTICE.

solstitial colure The declination circle passing through the north and south celestial poles and the poles of the **ecliptic**. The solstitial and equinoctial colures (both great circles) cut each other at right angles at the celestial poles.

solstitial points The two points on the ecliptic half way between the **equinoxes** where the Sun lies at the time of the summer and winter solstices.

Sothic cycle A period of 1460 years in the Egyptian calendar, named after Sothis (Sirius, the Dog Star), at whose heliacal rising the year was supposed to commence. During this period the New Year retrogressed through a complete cycle of seasons, thus accumulating an error of one whole year; this occurred because the year was taken as 365 days, which amounts to about one quarter of a day shorter than the tropical year.

Southern Cross The constellation Crux in the southern hemisphere.

south point That point on the celestial sphere, due south of the observer, where the meridian intersects the horizon.

south polar distance (*abbrev.:* S.P.D.) The angular distance, measured in the meridian, between an object and the south celestial pole. It is equal to 90° minus the **Declination** of the object.

South Tropical Disturbance A feature in the same latitude as the Red Spot on Jupiter. It was first noted by P. B. Molesworth (1867–1908) in 1901 and continued to be observed until 1940, when it disappeared. It appeared as a dark marking, rapidly increased in size, and exhibited considerable changes in size, shape, and motion before disappearing.

Soyuz A series of Soviet manned spacecraft consisting of three main sections: the orbital module, the re-entry module to carry the crew during flight into and descent from space, and the propulsion/instrument section. The first Soyuz was launched on April 23, 1967. Soyuz 11 performed the first docking and transfer of crew to the orbiting space laboratory **Salyut** 1.

space probe An unmanned craft containing scientific apparatus, launched by rocket beyond the Earth's atmosphere into space in order to make observations and experiments. It carries equipment for telemetering data and/or photographs back to Earth. It can be controlled and manoeuvred from one or more ground-based stations.

space reddening see **reddening**

space, temperature of The temperature of interstellar space measured as black-body radiation is about 3 K. See **microwave background radiation**.

space-time Development of the Special and General Theory of Relativity involves the concept of a four-dimensional space-time continuum. That is to say, three-dimensional space (length, breadth and height) and time are not independent but are linked together.

space velocity The true velocity of a star in space with respect to the Sun. It is the hypotenuse of the right-angled triangle formed by its **tangential velocity** (T) obtained from observation of its proper motion, and its **radial velocity** (R) obtained by measurement of the Doppler shift in its spectrum. The space velocity is then $\sqrt{(T^2+R^2)}$.

spectral-luminosity classification A system of classifying stellar spectra devised by W. W. Morgan (1906–), P. C. Keenan (1908–), and E. Kellman of Yerkes Observatory for their *Atlas of Stellar Spectra*. It permits a more accurate determination of the distance of remote stars. It relates the spectral characteristics of the star to its luminosity, thus using two parameters to describe the three variables of temperature, luminosity, and abundance. The different LUMINOSITY CLASSES are designated by a roman numeral, there being two classes for supergiants (Ia and Ib), two for giants (II and III), and one each for subgiants (IV), main-sequence stars (V), and subdwarfs (VI). The classification is also referred to as the MKK SYSTEM, the MK SYSTEM or the YERKES SYSTEM.

spectral type The group in which a star is classified according to its spectrum. See **stars, spectral classification of**.

spectrogram A photographic record of the spectrum of an object, obtained by means of a spectrograph.

spectrograph An instrument used to obtain a permanent photographic record of the spectrum of an object. It is a spectroscope whose telescope is replaced by a camera. For astronomical purposes the spectrograph is generally designed for use with a specific telescope. The instrument is capable of various modifications in design, depending on the use to which it is to be put. One example is the **spectroheliograph**.

A SLIT SPECTROGRAPH incorporates a narrow slit through which light from the object under observation first passes. It prevents overlapping of the images of different wavelengths, and ensures a sharply defined spectrum. Where spectra of whole fields of stars are required, a SLITLESS SPECTROGRAPH is used. For example, an **objective prism** can be mounted in front of the objective of the telescope. However, large prisms of sufficient accuracy are not available for use with very large instruments. In these circumstances, a negative collimating lens is placed just within the prime focus of the telescope to produce a parallel beam of light which is directed into the spectrograph proper, where it is dispersed either by a prism or a grating before being photographed.

spectroheliocinematography The continuous recording of solar phenomena, such as the development and changes in appearance of prominences, by cinematography of the image produced in the usual spectrograph.

spectroheliogram A photograph of the Sun taken by scanning the disc in a selected wavelength with a spectroheliograph. Usually, either the red line of hydrogen (Hα) or the violet line of calcium (K) is used for this purpose.

spectroheliograph An instrument which is a modification of the **spectrograph** and is used to photograph the Sun in the light of one particular wavelength only. The development of the instrument owes much to G. E. Hale (see **Hale, George**) and H. A. Deslandres (1853–1948).

spectrohelioscope A spectroheliograph adapted for visual use. The disadvantage of the spectroheliograph is that one does not know the state of the solar surface until the photographic plate has been developed, by which time conditions may have altered considerably.

spectrometer A spectroscope fitted with some facility for measuring the deviation of light passing through it and for precise measurement of the positions of spectral lines. It is used mainly in the laboratory.

spectrophotometer Equipment incorporating a photoelectric device for scanning the spectrum produced by a dispersing element, such as that at a telescope. It records immediately the intensity-wavelength graph for the object being observed.

spectrophotometry The study of the distribution of radiation from a body by measuring the intensity of light in the spectrum as recorded by a spectrophotometer. The graph produced by the instrument consists of LINE PROFILES, which are plots of intensity against wavelength.

spectroscope An instrument for producing spectra for the analysis of light. Three types of instrument may be distinguished: the straightforward instrument including the direct-vision or pocket spectroscope, in which the spectra are viewed by the eye; the **spectrometer**; and the **spectrograph**. For use in the ultraviolet part of the spectrum a quartz prism and lenses are substituted for glass components. For greater resolution of the spectrum the prism is replaced by a diffraction grating.

spectroscopic binary A **binary** system that cannot be resolved visually, but whose existence can be deduced from the periodic Doppler shift of the spectral lines.

spectroscopic parallax The distance of a star determined from its absolute magnitude (derived from its luminosity class) and its apparent magnitude (see **distance modulus**). It is the usual method for determining stellar distances.

spectrum (*plural:* spectra) An orderly arrangement displaying how the intensity of a beam of light or other electromagnetic radiation is distributed over a spread of wavelengths. When a beam of light is passed through a prism, or other dispersing agent such as a diffraction grating, the radiation is spread out fanwise in an orderly manner, and in the case of white light produces a spectrum in the form of a coloured band. The existence in white light of distinct coloured rays of differing refrangibility was discovered by Newton, who started his optical researches in 1666 and published them in book form as *Opticks* (London, 1704).

Spectra are of various types. That of white light, as received from a star such as the Sun, is continuous and is known as a CONTINUUM. A LINE SPECTRUM contains lines corresponding only to certain wavelengths or frequencies, which are characteristic of the atomic nature of a given material. An EMISSION SPECTRUM contains discrete lines produced by substances in an incandescent state at low pressure. An ABSORPTION SPECTRUM contains discrete dark lines superimposed on a continuous spectrum. It is produced when radiation from a hot source emitting a continuous spectrum passes through a layer of cooler gaseous material, as in the case of the Sun. BAND SPECTRA result from the emission or absorption of radiation by molecules. STELLAR SPECTRA are mostly continuous and exhibit numerous absorption lines. Some show emission lines.

The intensity of an emission line in a spectrum indicates the amount of energy present in the radiation. The width of the line indicates the wavelength band over which it is distributed. The intensity of the emission is greatest at the centre of the line and decreases towards the edges; the reverse is the case for an absorption line. These variations in the appearance of a line are known as the LINE CONTOURS. The pressure of the gas which is producing the spectral line can affect its width; this effect is known as PRESSURE BROADENING. Temperature conditions in the gas cause a thermal Doppler effect which produces DOPPLER BROADENING of the line. See also **Stark effect; Zeeman effect**.

speculum General term for the concave mirror of a reflecting telescope. Such mirrors were formerly made of SPECULUM METAL, an alloy composed of 67–69% copper and 31–33% tin. Mirrors are now made of glass or Pyrex, the concave surface being silvered or aluminized to give a reflecting surface.

spherical aberration see **aberration**

spherical astronomy An alternative name for **astrometry**.

spherical triangle A portion of the surface of a sphere bounded by the arcs of three great circles. The properties of the spherical triangle are important astronomically because the mathematical problems that arise in positional astronomy are based on the concept of the **celestial sphere**.

spherical trigonometry That branch of trigonometry which deals with the relations existing between the sides and angles of spherical triangles. The formulae are basic for the mathematics of celestial mechanics.

spheroid The solid generated by the revolution of an ellipse about one of its axes. If generated about the minor axis the result is an OBLATE SPHEROID. If generated about the major axis the result is a PROLATE SPHEROID.

spicules Narrow jets or tiny prominences seen protruding from the chromospheric layer of the Sun when the chromosphere is viewed edge on. They are short-lived (about five minutes each) and congregate at the edges of **supergranulation cells**. They are possibly caused by chromospheric gases shooting into the lower corona. They are visible only by means of a **Lyot filter**.

spiral galaxy A galaxy or stellar system consisting of a nucleus of stars from which spiral arms emerge, winding around the nucleus and forming a disc-shaped region. The arms contain gas, dust, and young stars, and exhibit various forms which have been classified by E. P. Hubble (see **Hubble classification**).

Spörer, Gustav Friedrich Wilhelm (1822–95) A German astronomer who determined the position of the Sun's equator and the period of rotation for various zones of the Sun. He formulated the law which bears his name.

Spörer's Law The relationship between the position of sunspots and their 11-year cycle: the average position drifts from mid-latitudes (about ±30°) towards the equator as the cycle goes from one minimum to the next. The phenomenon was first noticed by R. C. Carrington (see **Carrington, Richard**), but was studied in detail by G. F. W. Spörer. See also **butterfly diagram**.

spot-group A complex of sunspots.

spring tide see **tides**

sputnik A Russian word meaning in a general sense a 'travelling companion' and in astronomy a 'satellite'. It was the name given to the first series of artificial Earth satellites, which were launched from the USSR. Sputnik 1 was launched on October 4, 1957; Sputnik 2, launched on November 3, 1957, carried the first terrestrial living creature, the dog 'Laika', into space; Sputnik 26 (the last) was launched on April 2, 1963.

SS Cygni star An alternative name for a **U Geminorum star**.

S star see **stars, spectral classification of**

standard atmosphere The unit of atmospheric pressure defined as 101,325 pascals (1,013,250 dyn/cm^2) exactly. It corresponds to the presure exerted by a column of mercury 760 mm high, within one part in seven million.

standard co-ordinates In order to determine the angular separation between two stars from the linear separation of their images on a photographic plate, it is necessary to know the positions of the latter with reference to rectangular co-ordinates drawn on the plate. The scale of the plate must be known, as well as the Right Ascension and Declination of the point to which the optical axis of the telescope was directed during exposure of the plate. The centre of the plate serves as the origin for the rectangular co-ordinates, the axes of which are drawn perpendicular and parallel to the meridian; the former is the ξ-axis, the latter the η-axis. If the optical axis of the telescope does not coincide with the plate centre, a correction must be made to allow for the condition. The co-ordinates (ξ, η) of each image are the standard co-ordinates, and correspond to the differences in Right Ascension and Declination between the position of each star referred to the position of the plate centre. See **plate centre**; **plate constants**.

standard epoch Certain astronomical quantities, such as the co-ordinates of stars, are found to have altered when determined on different occasions. The change is caused by proper motions of the bodies concerned, or by other factors. In order that observational data may be compared, it is established practice to reduce all observations made over a long period of time to one date, which is known as the standard epoch. The point of time used as fixed reference is at present 1950·0.

star A celestial body that shines through being self-luminous, as opposed to planets which shine by reflected light. Stars are gaseous bodies whose radiant energy is produced by nuclear reactions, mainly through the conversion of hydrogen nuclei into helium nuclei.

star atlas A chart or map of various parts of the heavens, drawn on a small scale and assembled in book form or as a portfolio.

star catalogue A systematic listing, for a given epoch, of stellar co-ordinates, magnitude, spectral type, proper motion, and other data. The oldest surviving catalogue is the **Almagest**. Modern catalogues may be divided into GENERAL SURVEY CATALOGUES, such as the *Bonner Durchmusterung*, FUNDAMENTAL CATALOGUES giving precise data on a restricted selection of stars, and SPECIALIST CATALOGUES devoted to particular subjects, such as double stars or stellar spectra. Numerous catalogues of stars have been published since the 17th century. Some of them are still cited in the literature by an established abbreviation. These abbreviations are given in the list of abbreviations on pages 204–207.

star charts Large scale charts of the sky, now produced photographically, for professional astronomical work. The following are important examples: **Carte du Ciel**, **Franklin-Adams charts**, **Palomar Sky Atlas**, and E. Delporte, *Atlas Céleste* (Cambridge, 1930). The last shows the boundaries of the constellations adopted by the I.A.U., and includes a catalogue of stars down to magnitude 4·5, as well as other items.

star clouds Regions in various parts of the Galaxy where there is a noticeably higher density of stars than in the surroundings, as for example in the constellation Sagittarius.

star cluster A spatial concentration of stars of approximately the same distance from us and sharing the same direction of motion in space. See **cluster**. See also **stellar association**.

Stark effect The splitting or broadening of the lines in a line spectrum into a number of components when the atoms which are emitting or absorbing light are subjected to the influence of a sufficiently powerful electric field, which alters the energy level of the atoms. The broadening is proportional to the density of ions in an ionized gas, and gives an indication of pressure in a stellar atmosphere as well as of stellar luminosity. The Stark effect was discovered by J. Stark (1874–1957). It has some analogy with the **Zeeman effect**.

star names In former times people gave proper names to the brightest stars, e.g. Dog Star or Sirius to Alpha Canis Majoris, Betelgeuse to Alpha Orionis; such names are still used. An exhaustive account of the names of stars is to be found in R. H. Allen, *Star Names and their Meanings* (New York, 1899; Dover Reprint, New York, 1963).

Star of Bethlehem The nature of the Messianic star seen by the Magi has prompted enquiry for centuries on account of its relevance to the date of Christ's birth. The Biblical account is now thought to be legendary; the phenomenon has been ascribed to a miracle, to fireballs or shooting stars, to the comets of 5 B.C. or 4 B.C., to a bright nova of 4 B.C., to the planet Venus which could be fifteen times brighter than Sirius at Bethlehem, and to the conjunction of two or three planets. Taking into account historical facts, astronomical data, and astrological significances, it would seem that the triple conjunction of Jupiter and Saturn in the constellation Pisces in 7 B.C. provides the most probable answer to the problem.

stars, brightest Some of the brightest stars are listed, according to their **apparent magnitude**, in the table. The apparent magnitude indicates the brightness that a star appears to have as observed from Earth; its value depends on the intrinsic luminosity of the star, its distance from the Earth, and on the amount of light lost through absorption.

Star	*Name*	*Apparent magnitude*	*Spectral type*	*Distance (light-years)*
α Canis Majoris	Sirius	−1·47	A1	8·7
α Carinae	Canopus	−0·7	F0	100
α Centauri	Rigil Kent	−0·28	G2	4·3
α Boötis	Arcturus	−0·06	K2	36
α Lyrae	Vega	0·0	A0	26
α Aurigae	Capella	+0·06	G8	35
β Orionis	Rigel	0·14	B8	700
α Canis Minoris	Procyon	+0·4	F5	11
α Eridani	Achernar	+0·48	B5	100
β Centauri	Hadar	+0·61	B1	500
α Orionis	Betelgeuse	+0·8 (av.)	M2	500

stars, chemical constitution of Stars are globes of incandescent material. Spectral analysis of the light radiated from them, though relevant to the stellar atmosphere, is considered to give valid information as to the constitution of stars, as they are regarded as being homogeneous. Most of the nearby stars are found to have very nearly the same chemical constitution as the Sun. The main constituent is hydrogen (over 70% by mass); helium is next (about 25% by mass); the balance is made up of heavier elements.

stars, colour of Some of the brightest stars can be seen by the naked eye to have a recognizable colour. All stars, being radiant bodies at very high temperatures, will have a colour depending on the wavelengths of the light which is emitted. The higher the temperature of a star, the more will the maximum of intensity be shifted towards the blue end of the spectrum. See **colour index**.

stars, diameter of If the angular diameter and parallax of a star are known, then its diameter can be calculated from the formula

$$D = \frac{214d}{\pi}$$

where D is the diameter of the star in terms of the Sun's diameter, d is the angular diameter of the star as measured, and π is its trigonometrical parallax. The range of stellar diameters is enormous. Betelgeuse has a diameter one thousand times that of the Sun, whereas some highly evolved stars have diameters less than that of Earth.

stars, distances of The distances from Earth of even the nearest stars are extremely great. For those not more than several hundred light-years away, the distance can be determined by measuring the star's **trigonometrical parallax**. For more distant stars the Doppler shift of spectral lines (see **Doppler effect**) can be used to determine the radial velocity of individual stars in a moving cluster, and this in conjunction with the decrease in angular diameter of the cluster with time enables the distance of the star to be found. The distance can also be determined from the apparent and absolute magnitudes of a star (see **apparent magnitude**). The distances of some of the nearest stars are listed in the table on page 169.

stars, energy production in The main sources for the continuous production of the enormous amounts of energy that are emitted by stars for millions of years are considered to

Star	Parallax (″)	Distance (light-years)	Magnitude apparent	Magnitude absolute
Proxima Centauri	0·762	4·3	10·7	15·1
α Centauri A	0·751	4·3	0·0	4·4
α Centauri B	0·751	4·3	1·4	5·8
Barnard's star	0·545	6·0	9·5	13·2
Wolf 359	0·402	8·1	13·5	16·5
Lalande 21185	0·398	8·2	7·5	10·5
Sirius A	0·375	8·7	−1·5	1·4
Sirius B	0·375	8·7	8·5	11·4

lie in the transformation of hydrogen into helium according to the **carbon-nitrogen cycle** or the **proton-proton reaction**. Reactions between helium nuclei are also possible, forming heavier nuclei (of carbon, oxygen, etc.) with release of further energy. In addition, contraction of stars during certain stages of their evolution can result in the production of energy.

stars, galactic concentration of The concentration of stars in the Galaxy towards the galactic plane. The stars are also symmetrically distributed about the galactic plane. The galactic concentration is therefore a measure of the progressively decreasing stellar density with increasing galactic latitude.

stars, luminosity of The absolute brightness of a star, being the total radiant energy emitted per second. It is expressed in joules per second, i.e. watts, in ergs per second, or in magnitudes. The **Draper classification** for stellar luminosity, developed at Harvard College Observatory, was modified by workers at the Mount Wilson Observatory who found that the variation in sharpness of spectral lines exhibited by stars of the same spectral type could be used to determine the luminosity of a star and to fix its position on the **Hertzsprung-Russell diagram**. The prefix *d*, *g*, or *c* before a spectral type indicated that the star was a dwarf, giant, or supergiant respectively. The MKK SYSTEM, in which the spectral type receives a Roman figure as suffix to show its **luminosity class**, is now in general use. See also **spectral-luminosity classification**.

stars, masses of The mass of a star can be deduced only when the effect of its gravitational attraction upon the motion of some other body can be observed. This can be done in the case of binary stars. It is possible also in the case of single stars whose spectra exhibit a detectable gravitational **red shift**. The mean masses of stars in accordance with the **mass-luminosity relationship** are as follows:

Spectral type	*Mean mass* (Sun = 1)
O	30
B	10
A	2·2
F	2
G	1·1
K	1·1

The theoretical upper limit for stellar mass is about 60 solar masses, but very few stars exceed 30 solar masses. The star UV Ceti B has the smallest known mass, namely 0·035 that of the Sun.

stars, nomenclature of Stars are grouped together into constellations. Within each constellation the stars are, for convenience in general use, identified by a letter or number followed by the name of the constellation in the genitive form of its Latin name. The brightest stars are indicated by Greek letters, as in α Lyrae or ε Eridani. These letters are known as BAYER LETTERS because they were assigned by J. Bayer in his *Uranometria* (Augsburg, 1603). Stars that are not so bright are prefixed by a number, as in 61 Cygni. These numbers are known as FLAMSTEED NUMBERS because they relate to the definitive edition of J. Flamsteed's *Historia Coelestis Britannica*, (London, 1725). Yet fainter stars are referred to by their designation in the catalogue appropriate to their type, e.g. double-star, variable-star catalogues, etc. Variable stars are indicated by a prefix of one or more capital Roman letters, from 'R' onwards, as in T Canis Majoris or TY Orionis. 'An Index to the Constellations' by F. Schlesinger and A. H. Pond is in *Popular Astronomy* (1925) **53**, 82–98.

stars, places of The positions as given by co-ordinates on the celestial sphere. Positions are usually given in **Right Ascension** and **Declination**, which are referred to the celestial equator and to the equinox at a specified epoch, e.g. 1900·0, 1950·0, 2000·0. The APPARENT PLACE of a star is that directly observed from the Earth, after correction for refraction, and is its position on the geocentric celestial sphere, referred to the true equator and the equinox of date. The TRUE PLACE of a star is its position on the heliocentric celestial sphere, referred to the true equator and equinox of date, as it would be if observed from the Sun. The MEAN PLACE of a star is its position on the heliocentric celestial sphere, referred to the mean equator and equinox for the beginning of the year of observation.

stars, rotation of The only star whose speed of rotation can be determined by observation of markings (sunspots) on its surface is the Sun. Its mean velocity of rotation is about 1·0 km s^{-1} at the equator. In other cases recourse must be made to the **Doppler effect**, which produces a broadening of absorption lines in the stellar spectrum. The speed of rotation can be determined from the amount of broadening. The velocity of equatorial rotation varies with the spectral type of the star; values between 25 and 200 km s^{-1} are usual, but exceptionally can be up to 500 km s^{-1}.

stars, spectral classification of The physical conditions of stars determine the nature of their radiation, which can be examined spectroscopically. From the study of the characteristics of stellar spectra since the middle of the 19th century, a comprehensive range of SPECTRAL TYPES has been evolved. P. A. Secchi (1818–78), in the 1860s, made the first spectral classification of stars by dividing them into four classes: (I) white, showing lines of hydrogen; (II) yellow, showing lines of metals; (III) and (IV) red, showing strong absorption bands. Group V was added by M. F. J. C. Wolf (1863–1932) and G. A. P. Rayet (1839–1906) to include white stars whose spectra had enhanced emission lines.

As a result of work at the Harvard College Observatory, the *Henry Draper Catalogue of Stellar Spectra* was published (1918–24). It uses a spectral classification which was evolved from the original *Draper Catalogue* of 1890. The sequence of spectral types is arranged according to the prominence or absence of certain lines in the spectra. The original sequence had six types bearing the letters *B*, *A*, *F*, *G*, *K*, and *M*. The present sequence, which is arranged in order of decreasing stellar surface temperature, is as follows:

W—*O*—*B*—*A*—*F*—*G*—*K*< (*M*—*S* / *C*)

or

W—*O*—*B*—*A*—*F*—*G*—*K*—*M* (with *C* branching from *G*—*K* and *S* branching from *K*)

Nearly all stars can be assigned a position in this classification. The surface temperature and the main features of each spectral type are given in the first table on page 171

O, *B*, and *A* stars are conventionally called 'early-type' stars; *K* and *M* stars are called 'late-type' stars. This description relates only to their position in the sequence and has no relevance to the stage of their evolution. In addition to the classes listed in the first table on page 171, the Harvard system includes the special classes *Q* for novae (spectrum variable) and *P* for planetary nebulae (line emission, no continuum). They are placed before spectral type *W*. Classes *W* to *M* are divided into 10 numerical subclasses. A Roman figure is now added to the spectral classification to indicate the type of star (see **luminosity class**). Further refinements include the use of additional letters as suffixes (or prefixes in the case of *d* and *wd*) to the spectral type, giving more information about the star. These letters are listed in the second table on page 171.

stars, spectral distribution of The *Henry Draper Catalogue* gives the spectral type of 225,300 stars. The *apparent* distribution of the stars among the spectral classes is *B* 3%, *A* 27%, *F* 11%, *G* 17%, *K* 36%, *M* 8%. The true distribution, i.e. on the basis that all the stars are situated in the same region of space, must be found by the statistical treatment of the data presented.

Type	*Surface temperature*	*Main features*
W	50,000–100,000 K	Wolf-Rayet stars, formerly in Class O. Intense continuum. Broad emission bands of highly ionized helium, carbon, oxygen, and nitrogen.
O	30,000–50,000 K	Blue-white stars. Intense continuum. Absorption lines of ionized helium; emission lines of ionized hydrogen, helium, oxygen, and nitrogen.
B	12,000–25,000 K	Blue-white stars. Intense continuum. No emission lines. Lines of ionized silicon and magnesium. Absorption lines of hydrogen.
A	8000–10,000 K	White stars. Helium absent. Absorption lines of hydrogen. Lines of ionized calcium, magnesium, and other metals.
F	6000–7500 K	Yellow-white stars. Strong lines of calcium and other metals. Hydrogen weak.
G	4000–6000 K	Yellow stars. Spectrum similar to that of Sun. Strong lines of calcium. Lines of iron and molecular bands of CH and CN.
K	3000–5000 K .	Orange stars. Lines of hydrogen weak. Lines of metals stronger. Molecular bands stronger. Lines of titanium oxide (TiO) first appear.
M	3000–3500 K	Orange-red stars. Lines of metals. Bands of TiO. Molecular bands of CH and CO.
S	about 2500 K	Red stars. Spectra similar to Class *M*. Emission lines of hydrogen. Strong bands of zirconium oxide (ZrO).
C	about 2500 K	Orange-red and deep red stars. Prominent band spectra. Lines of iron, sodium, and calcium. Peculiar bands of carbon compounds. Molecular bands of CN and CO.

Affix	*Indication*
d	a dwarf star
e	emission lines present
f	emission lines of certain O-type stars
k	interstellar lines present
m	metallic lines present
n	lines are nebulous or diffuse
p	a peculiar spectrum
s	sharp lines
w	weak lines
wd	a white dwarf

star streaming A phenomenon discovered by J. C. Kapteyn in 1904 whereby the proper motions of stars (corrected for the Sun's motion) are not random, but have two preferred directions of motion in exactly opposite directions: one vertex is in the constellation Orion, the other in Scutum. Participation in streaming is most noticeable with stars of spectral type A.

star test A test of the optical quality and adjustment of a telescope by examination of the in-focus and extra-focus images of a suitable star. Abnormalities in the diffraction pattern indicate defects in the optical system and/or its adjustment.

stationary point The point in the apparent path of a planet when it appears motionless, because the orbital motion of the planet lies in the line of sight. At that point the planet's motion changes from direct to retrograde (or vice versa).

steady-state theory A cosmological theory put forward by H. Bondi (see **Bondi, Hermann**) and T. Gold (see **Gold, Thomas**) in 1948, and further developed by F. Hoyle (see **Hoyle, Fred**) and others. According to this theory the universe has always existed; it had no beginning and will continue for ever. Although the universe is expanding, it will maintain its average density through the continuous creation of new matter, in the form of hydrogen, at the rate of $2{\cdot}8\times10^{-46}$ g cm^{-3} s^{-1}. Most cosmologists now reject the theory because it has certain weaknesses, and particularly because of the discovery of **microwave background radiation**.

Stefan-Boltzmann constant The constant σ in the formula expressing the **Stefan-Boltzmann Law**. Its value in the **SI** is $5{\cdot}67032\times10^{-8}$ W m^{-2} K^{-4}.

Stefan-Boltzmann Law A law deduced experimentally by J. Stefan (1835–93) and L. E. Boltzmann (1844–1906) which states that the total amount of radiation emitted by a black body is directly proportional to the fourth power of its absolute temperature. Thus $E = A\sigma T^4$, where E is the total energy emitted, A is the surface area of the radiator, σ is the **Stefan-Boltzmann constant**, and T is the absolute temperature.

stellar association A sparsely populated cluster of stars having masses in the range of 100 to 1000 solar masses, which are slowly dispersing. They are divided into two groups: OB-ASSOCIATIONS, so called by V. A. Ambartsumyan (1908–) who pointed out in 1949 that many *O* and *B* stars collected together in groups; and T-ASSOCIATIONS, consisting, as was noted by A. H. Joy (1882–1973) in the 1950s, very largely of T Tauri stars. The approved way of designating an association is by the name of the constellation followed by an arabic numeral, e.g. Cas OB5 in Cassiopeia.

stellar evolution The development or change in physical characteristics of a star over long periods of time. The changes are so slow that they cannot be observed in the case of an individual star. J. C. F. Zöllner (1834–82) in 1865 suggested that the colour of a star was indicative of its temperature. A. Secchi recognized that his spectral classification (1867) corresponded to a scale of temperature and believed that it corresponded also to stages in stellar evolution. H. K. Vogel (1841–1907) introduced subgroups into Secchi's classification and saw it as a scale of stellar evolution. In his view white or blue stars were born from gaseous nebulae at high temperature. The work of J. N. Lockyer led him to put forward (1902) an evolutionary interpretation of stellar spectra. The fundamental work of A. J. Cannon and E. C. Pickering, which culminated in *The Henry Draper Catalogue of Stellar Spectra*, and the derivation in 1912 of the **period-luminosity law** by H. S. Leavitt led to the **Hertzsprung-Russell diagram** in 1913. The H-R diagram has prompted several theories of stellar evolution.

It is now believed that stars are formed from dense interstellar clouds of gas (mainly hydrogen) and dust. The clouds are triggered into a state of contraction by some as yet uncertain mechanism. Under the action of gravitation, protostars form and continue to contract. The pressure finally becomes so great that stars are formed: thermonuclear reactions (**carbon-nitrogen cycle, proton-proton reaction**) can take place in the core, where hydrogen is converted into helium with release of energy. At this stage a star enters the main sequence of the H-R diagram. It is estimated that stars are being formed at the rate of about ten a year.

Stars will remain on the main sequence until the cores are about 12% of the total mass (**Chandrasekhar-Schönberg limit**). The cores then start to contract. More energy is released to cause the outer layers of the stars to expand and the stars become **red giants**. The internal temperature increases until another nuclear reaction (**Salpeter process**) sets in during which helium is converted into carbon. This reaction gives out less energy than the fusion of hydrogen to helium so that at this stage stars begin to contract again and luminosity falls. Pulsation may then occur with the production of pulsating **variable stars**.

The final stage in the evolution of stars depends on their mass. Stars with a mass less than or not much more than that of the Sun will become **white dwarfs**. In more massive stars further thermonuclear reactions can take place in the core, with the production of elements heavier than carbon. In a few million years these stars become unstable: it is thought that they undergo a catastrophic collapse which leads to a **supernova** explosion. Any stellar core surviving the explosion will then most likely be a **neutron star** or a **black hole**.

stellar interferometer An instrument using the principle of **interference**, dividing and recombining beams of light so as to produce interference patterns. It is employed for measuring small angles, such as the angular diameter of a stellar disc, or the angular separation of two stars. The instrument, which exists in various forms, makes it possible to discover and measure double stars whose separation is insufficient to be revealed visually by telescopes. Measurements of stellar diameters by the interferometric method have confirmed the correctness of the values obtained by calculation theoretically from magnitude and spectral type.

stellar light That part of the background illumination of the sky contributed by stars which are too faint to be observed by the naked eye.

stellar magnitude, determination of Photoelectric and photographic photometry have displaced visual methods for the accurate determination of stellar magnitudes. It is usual to make measurements in three separate wavebands, and to compare the results with the accurately known values for stars in the North Polar Sequence.

The following systems are in use. The PHOTOVISUAL SYSTEM (PV SYSTEM) determines magnitudes from photographs taken on special plates through a filter passing wavelengths between 5000 and 6000 Å. The INTERNATIONAL PHOTOVISUAL SYSTEM (IPV SYSTEM) is based on the work of F. H. Seares (1873–1964), covering the wavelengths 4900 to 5900 Å. The PHOTOELECTRIC VISUAL SYSTEM (V SYSTEM) is the photoelectric equivalent of the PV system. It was developed at the Yerkes Observatory by W. W. Morgan (1906–). It involves the use of a photoelectric photometer with a filter passing wavelengths between 4800 and 6500 Å. The PHOTOGRAPHIC SYSTEM (PG SYSTEM) determines magnitudes from photographs taken on special plates sensitive to blue light with a filter to cut off the ultraviolet radiation. The INTERNATIONAL PHOTOGRAPHIC SYSTEM (IPG SYSTEM) is based on the work of F. H. Seares covering the wavelengths 3420 to 5170 Å. The B SYSTEM is the photoelectric equivalent of the PG system. It involves the use of a photoelectric photometer in conjunction with a special blue filter passing wavelengths between 3800–5400 Å. Magnitudes in the U SYSTEM are measured in the ultraviolet part of the spectrum. They can be obtained photographically using plates sensitive to blue and ultraviolet light in conjunction with a filter passing ultraviolet light, or photoelectrically using a photoelectric cell in conjunction with a special ultraviolet filter. It uses wavelengths between 3400 and 4000 Å. The R SYSTEM obtains photographic magnitudes from photographs taken on panchromatic plates used with a red filter. It uses wavelengths between 6100 and 6700 Å. The G SYSTEM obtains photographic magnitudes from photographs taken on plates with a green filter. It uses wavelengths between 4500 and 5000 Å. The I SYSTEM obtains photographic magnitudes in the infrared part of the spectrum. It uses wavelengths between 7400 and 10,000 Å.

As stellar magnitudes are affected by the colour of stars, a **colour index** has been devised to allow for this effect. The Ipg and Ipv systems were originally defined to make the colour index of A0 stars in the North Polar Sequence equal to zero; however, when allowance has been made for space-reddening, the true zero of the colour index corresponds to stars of spectral type A4.

It has become the practice to measure magnitudes in three wavebands, as it has been found desirable to add ultraviolet measurements to pg and pv magnitudes. Two such THREE-COLOUR SYSTEMS have been devised using combinations of the U, B, V, G, and R systems described above. The UBV SYSTEM (ultraviolet-blue-visual) of H. L. Johnson and W. W. Morgan measures stellar magnitudes through three filters: ultraviolet (U) at 3600 Å, blue (B) at 4200 Å, and visual (V) at 5400 Å. The system is so defined that

$$B - V = U - B = 0$$

for A0 stars: the colour index has a negative value for hot stars and a positive value for cool stars. This particular system is now widely used and is likely to prevail over others. The second three-colour system is the RGU SYSTEM (red-green-ultraviolet) devised by W. Becker. Mention may be made of the six-colour system (U, V, B, G, R, I) of J. Stebbins (1878–1966), A. E. Whitford

(1905–), and G. Kron (1913–), which is defined so that

$$B + G + R = 0.$$

stellar populations see **Population I stars; Population II stars**

Stephan's Quartet A group of peculiar galaxies discovered and observed in 1877 by Edouard Stephan (1837–1923), founder of the Observatoire de Marseille. The objects (NGC 7317, 7318, 7319, 7320) lie in the constellation Pegasus and are faint and difficult to observe. They are characterized by large red shifts. The cluster is sometimes called STEPHAN'S QUINTET if NGC 7318 is divided into its components 7318A and 7318B.

stereo comparator see **comparator**

Stjerneborg The underground observatory built by Tycho Brahe in 1584 on the island of Hven between Copenhagen and Elsinore as an extension to **Uraniborg**.

stop see **lens stop**

stratopause That part of the Earth's **atmosphere** lying between the **stratosphere** and the **mesosphere**. In this region the temperature drops to about 270 K (max) at a height of about 50 km.

stratosphere The layer of the Earth's **atmosphere** immediately above the tropopause. It is a quite stable layer in which the temperature is fairly constant in the vertical direction, ranging from about 200–230 K at the bottom to about 270 K at the top. It extends from about 15 km to about 50 km above the Earth's surface.

Stratton, Frederick John Marrian (Birmingham, October 16, 1881 – Cambridge, September 2, 1960) An English astronomer who was Professor of Astrophysics and Director of the Solar Physics Observatory, Cambridge. He is known for his study of spectra of the Sun's chromosphere and of novae.

Strömgren, Bengt Georg Daniel (Göteborg, January 21, 1908–) A Danish astronomer, Professor of Astrophysics at the University of Copenhagen. He developed the theory of luminous clouds of ionized hydrogen (known as STRÖMGREN SPHERES) surrounding very hot stars. He is known also for his work on the internal nature of stars.

Strömgren, Elis (Helsingborg, May 31, 1870 – Copenhagen, April 5, 1947) A Swedish astronomer who became Director of the observatory at Copenhagen. He is known for his work on double stars, cometary orbits, perturbations, and the three-body problem.

Struve, Emile (District of Ryasan, September 9, 1874–19??) Great-grandson of Friedrich Georg Struve. A Russian astronomer who worked at the Pulkovo Observatory.

Struve, Friedrich Georg Wilhelm von (Altona, April 15, 1793 – Pulkovo, November 23, 1864) A German astronomer who was Director of the Dorpat Observatory and later of the Pulkovo Observatory. He is known for his study of double stars and his catalogues thereof. He was a pioneer in the measurement of stellar parallax in the same year (1838) as F. W. Bessel (see **Bessel, Friedrich**). He supervised the measurement of the Russo-Scandinavian arc of meridian, the results of which were published in 1860.

Struve, Georg (Pulkovo, December 29, 1886 – Berlin, June 10, 1933) Grandson of Otto Wilhelm von Struve. A Russian astronomer who became Director of the University Observatory of Berlin-Neubabelsberg. He is known for his observations of planets and satellites.

Struve, Karl Hermann von (Pulkovo, October 3, 1854 – Herrenalb, August 12, 1920) Son of Otto Wilhelm von Struve. A Russian astronomer who became Director of the observatory at Königsberg. He founded (1913) the new University Observatory of Berlin-Neubabelsberg. He is known for his observations of planets and satellites.

Struve, Otto (Kharkov, August 12, 1897 – Berkeley, California, April 6, 1963) Son of Wilhelm Ludwig von Struve. A naturalized American astronomer who became Director of the Yerkes and McDonald Observatories, and also of the National Radio Astronomy

Observatory at Green Bank. He is distinguished for his work in stellar spectroscopy and evolutionary problems of the universe.

Struve, Otto Wilhelm von (Dorpat, May 7, 1819 – Karlsruhe, April 16, 1905) Son of Friedrich von Struve. A German astronomer who succeeded his father as Director of the Pulkovo Observatory (1862–89), and then went to Karlsruhe. He is known as an observer and discoverer of double stars.

Struve, Wilhelm Ludwig von (Pulkovo, November 1, 1858 – Simferopol, November 4, 1920) Brother of Karl Hermann von Struve. A Russian astronomer who became Director of the observatory at Kharkov. He investigated the proper motion of the solar system.

style or **gnomon** That part of a sundial which, by its shadow on the dial plate, indicates the hour of the day.

subdwarf A star of smaller radius and lower luminosity than a normal dwarf star of the same spectral type. Subdwarfs occur mainly in spectral types *F*, *G*, and *K*, and lie below the main sequence of the **Hertzsprung-Russell diagram**. They are mainly **Population II stars**.

subgiant A star of smaller radius and lower luminosity than a normal giant star of the same spectral type. They occur mainly in spectral types *G* and *K*, and lie between the giant sequence and the main sequence on the **Hertzsprung-Russell diagram**.

subpulse The weaker component of the pulse of a **pulsar**.

subsolar point That point on the surface of the Earth at which, at a given instant, the Sun is exactly in the zenith.

substellar point That point on the surface of the Earth at which, at a given instant, any specified star is exactly in the zenith.

summer solstice see **solstice**

Summer Time An Act of Parliament in 1916 decreed that legal time for general purposes in Great Britain should be one hour in advance of Greenwich Mean Time. Since then there have been other SUMMER TIME ACTS decreeing the duration of Summer Time. Particulars regarding the duration in past years will be found in various volumes of *Whitaker's Almanac.*

Sun The central body of the solar system around which all other members revolve in their orbits. The apparent motion of the Sun in the **ecliptic** results from the fact that the Earth rotates on its axis and revolves in its orbit. The Sun does however have a motion of about 20 km s^{-1} in the galactic system towards the **apex**. It also participates in general rotation about the galactic centre with a velocity of about 250 km s^{-1}, one revolution taking about 250 million years. The movement of sunspots reveals that the solar sphere rotates. Data relating to the Sun are given in the table.

Distance from the Earth	
mean (the astronomical unit)	149·6 × 10^{6} km
maximum (at aphelion)	152·1 × 10^{6} km
minimum (at perihelion)	147·1 × 10^{6} km
Diameter	1·392 × 10^{6} km
Density (mean)	1·409 g cm^{-3}
Mass	1·989 × 10^{30} kg
Volume	1·412 × 10^{18} km^{3}
Period of axial rotation	
at equator	24^{d} 6^{h}
at poles	about 35^{d}
Inclination of axis of rotation to pole of ecliptic	7° 15′
Surface gravity	27,398 cm s^{-2}
Spectral type	G2V
Luminosity	3·86 × 10^{26} watts
Magnitude (mean visual)	−26·86 (apparent), +4·71 (absolute)
Rotational velocity (mean)	1·9 km s^{-1}
Escape velocity	618 km s^{-1}
Temperature of surface	nearly 6000 K
Temperature of core	about 14 × 10^{6} K
Magnetic field strength	1–2 gauss, but much higher in active regions.
Age (minimum)	4·5 × 10^{9} years
Chemical composition (by mass)	about 73% hydrogen, 25% helium, balance, all other elements.

The Sun is a star consisting of gaseous matter held together by gravitational attraction. It is of spectral type G and appears in the main sequence of the Hertzsprung-Russell diagram. Its temperature, pressure, and density increase towards the centre, where the values are about $15–20 \times 10^6$ K, about 10^{11} atm, and 155 g cm^{-3} respectively. The energy radiated by the Sun results from nuclear processes, i.e. the **proton-proton reaction** and **carbon-nitrogen cycle** which convert hydrogen into helium. The loss of mass in this process is $4{\cdot}3 \times 10^9$ kg s^{-1}. Even at this enormous rate the loss amounts to only 0·07% of the total mass in 10^{10} years.

The only parts of the Sun that are accessible to direct observation by various means are the **photosphere**, the **chromosphere**, and the **corona**. The photosphere is the boundary of the luminous solar sphere surrounding the interior nuclear furnace. Most of the energy emitted by the Sun is radiated from this surface, whose temperature is about 6000 K; this is also virtually the Sun's mean effective temperature. Observed under suitable conditions the surface of the gaseous photosphere, whose pressure is about 10^{-2} atm, has a granulated appearance. This granulation is probably caused by the upward movement of convection currents from the unstable layer below, where the temperature progressively rises towards the core. Also apparent are **sunspots**. The continuous spectrum of the Sun, with its dark **Fraunhofer lines**, results from reactions in the photosphere.

The atmosphere above the photosphere becomes more rarefied with increasing altitude, and the temperature rises to about 10^6 K. The chromosphere, so-called because of its rosy tint when seen during a total solar eclipse, lies above the photosphere. It consists of hot gases, and extends for thousands of kilometres. It is generally in a state of turbulence, the temperature rising extremely rapidly with height. The upper chromosphere consists of atomic hydrogen and calcium. It is the realm of **solar flares**, which are associated with sunspots, and **solar prominences**.

Extending outwards for millions of kilometres into space is the extremely tenuous gaseous corona. Until the invention of the **coronograph** by B. F. Lyot it was not possible to observe the corona except during a total solar eclipse. The temperature of the corona is about 10^6 K and is responsible for the highly ionized state of calcium, iron, and nickel; their radiation produces spectral emission lines that were at one time ascribed to an unknown chemical element, **coronium**. The lower corona is the region responsible for the emission of electromagnetic radiation of long wavelength known as SOLAR NOISE. There is continuous emission from the corona of corpuscular radiation, known as the **solar wind**. The **Zeeman effect**, observed in the lines of the solar spectrum, indicates the existence of magnetic fields in the Sun.

The Sun must *never* be viewed through a telescope, even if a dark glass is interposed between the eye and the eyepiece. For observation of the Sun's disc, its image should be projected on to a white screen. Eyesight can be severely damaged if the Sun is observed directly through a telescope or through binoculars.

sundial An instrument for showing the time of day by a shadow cast by the Sun as its altitude varies through the course of a day. There are numerous possible constructions. The hour lines may be drawn on a vertical, horizontal, or inclined plane. The shadow is cast by the STYLE (or GNOMON). The sundial shows apparent solar time: in order to convert this to mean time, a correction, known as the **equation of time,** must be applied.

sunrise The moment of first appearance of the Sun's upper limb above the horizon. It is defined as the moment when the Sun's **zenith distance** is 90°50′.

sunset The moment when the Sun's upper limb disappears below the horizon. As in the case of sunrise it is defined as the moment when the Sun's zenith distance is 90°50′; this value is arrived at by making allowance for the Sun's semi-diameter (16′) and atmospheric refraction (34′). For meteorological purposes sunrise and sunset are assumed when the centre of the solar disc is on the horizon; allowance is made for refraction, so that the zenith distance is 90°34′.

sunshine recorder An instrument for recording the time during which the Sun is visible and not obscured by cloud, etc. The Campbell-Stokes recorder, devised by J. F. Campbell in 1853 and improved by G. G. Stokes (1819–1903) in 1880, is widely used for meteorological purposes. It consists of a glass globe which acts as a lens and focuses the Sun's image on a special card, accurately marked in hours and fractions of hours. The Sun's rays burn a track on the card, from which the number of hours of sunshine can be determined.

sunspot A region of temporary disturbance in the Sun's photosphere characterized by a dark patch composed of two parts: a dark central **umbra** and its surrounding lighter, greyish **penumbra**, which has a filamentary structure radiating outward from the umbra. Sunspots appear dark because they are cooler than the surrounding photosphere, the temperature being 4500–5000 K. Spots are usually formed in pairs of opposite magnetic polarity between latitudes 30–40° north or south of the equator; they move across the face of the Sun as it rotates.

Several spots may form in the same area and produce a complex group. The main spot leading (preceding) a group as it moves across the Sun is known as the P-SPOT; the main spot following is known as the f-spot. Spots vary in diameter from several thousand kilometres up to about 200,000 km for the largest, which are sometimes visible to the naked eye. The number of sunspots present at one time varies, depending on the Sun's activity during the 11-year **solar cycle**. The spectrum of a sunspot differs from that of the photosphere as a result of its lower temperature. The motion of gases in the penumbra was detected by the Doppler shift of spectral lines (**Evershed effect**). The presence of the **Zeeman effect** observed by G. E. Hale in 1908 showed the existence in some sunspots of strong magnetic fields, the polarity of which reverses for each solar cycle. Although no satisfactory theory has so far been advanced as an explanation for sunspots, it is thought that these intense magnetic fields affect (probably suppress) the convective motion of the photosphere by which hot gases are brought to the solar surface.

Sunspots in the hydrogen alpha line – each about the size of the Earth.

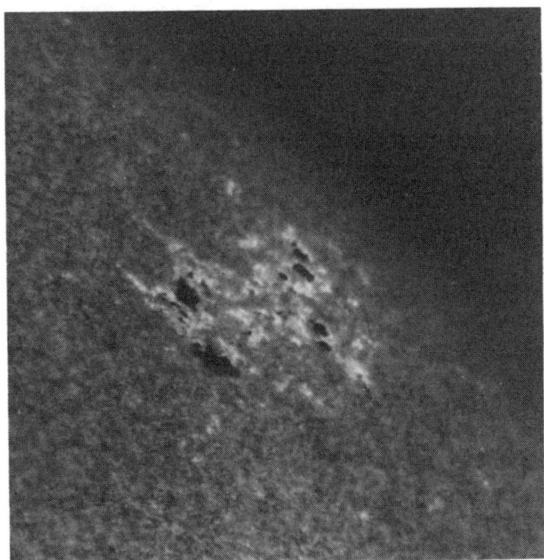

supergiant An extremely luminous star of large diameter and low density, Betelgeuse and Rigel being examples. The luminosities of supergiants are several magnitudes greater than those of giant stars, and so lie above giants on the **Hertzsprung-Russell diagram**. They are **Population I stars**, situated in the spiral arms of the Galaxy.

supergranules or **supergranulation cells** Convective cells which are distributed fairly uniformly over the solar disc, and are much larger than ordinary photospheric granules (see **granulation**). Material has been detected flowing from the centre to the edge of the cells, where most of the magnetic flux coming from the photosphere is concentrated, and it is believed that the magnetic field at the edge of the cells permits the formation of **spicules**.

superior conjunction One of the planetary **aspects**. It is the position of an inferior planet (Mercury or Venus) when the Sun is between it and the Earth. The planet is then virtually invisible.

superior culmination see **culmination**

superior planet Any of the planets which lie beyond the Earth's orbit and so are more distant from the Sun. They are Mars, Jupiter, Saturn, Uranus, Neptune, and Pluto.

supernova A star whose luminosity increases suddenly by as much as 19 magnitudes (about 7 magnitudes more than is the case with a **nova**) as the result of an extremely violent explosion; the outburst of energy is as much as 10^{43} joules. Analysis of ancient records identifies several such occurrences in our own Galaxy, before the invention of the telescope. One was observed in 1054 by Chinese astronomers. The **Crab Nebula** and its associated pulsar are the remains of this event, which was visible in broad daylight. Another was observed in 1572 when Tycho Brahe and others saw a new star in the constellation Cassiopeia. It was then as bright as Venus at maximum brightness, and was observable for 18 months. It cannot now be identified. A third, seen in 1604 in Ophiuchus, was observed for over a year by many astronomers, including J. Kepler. Like the 1054 event, it has left identifiable evidence. Several hundred supernovae have been observed in external galaxies, one of the most noteworthy being that of August 1885 in the central region of the Andromeda Galaxy. It reached an apparent magnitude of +6 and disappeared after six months.

The light curves of supernovae show that there are two types. TYPE I SUPERNOVAE at the maximum of the outburst attain a luminosity with a mean absolute magnitude of −14 to −17; the magnitude may reach −18 or −19. The spectrum is continuous, but weak in the ultraviolet region, and of the non-hydrogen kind. They are rare events, with perhaps one per galaxy in 500 years. TYPE II SUPERNOVAE are more common. They attain a lower luminosity at maximum with a mean absolute magnitude of −12 to −13·5, possibly reaching −16. The spectrum is strong in the ultraviolet region, and is of the hydrogen kind. Type I supernovae occur in spiral and elliptical galaxies and so probably belong to Population II stars. Type II supernovae occur in the spiral arms of galaxies among the young, more massive stars of Population I, and have a lower velocity of expansion (about 5000 km s^{-1}). It is estimated that supernovae of Type II are produced in the Galaxy at the rate of about one to five per century. Another estimate gives an average production of one supernova per spiral nebula in 360 years. A relation between the outburst

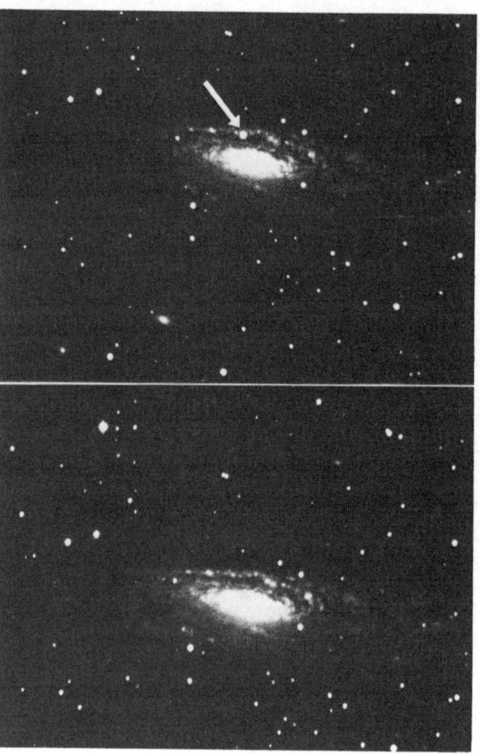

The supernova explosion which was observed in 1959 in the galaxy NGC 7331.

of a supernova and radiation of radio frequency has been definitely established.

A supernova is thought by most astronomers to be the beginning of the final stage in the evolution of a massive star. The explosive outburst results in most if not all of the star's mass being blown away. Any remains are considered to be in the form of a very dense collapsed stellar core which may be a **neutron star** or possibly a **black hole**.

supernova remnant A gaseous emission nebula, being the expanding shell of matter thrown off into space during the outburst of a supernova. These remnants are often strong radio and X-ray sources.

Super-Schmidt telescope A **Schmidt camera** designed by J. G. Baker especially for recording meteor trails. The corrector plate consists of an achromatic doublet situated between two opposing meniscus lenses whose spherical surfaces are concentric with the mirror. It has a wide field and works at f/0·67.

surge or **rocket** A type of **solar prominence** observed in many cases over or close to a **solar flare**. In contradistinction to a flare, which shows little or no vertical motion, a surge may have a velocity of 200 km s^{-1} or more and will extend 40,000 km or more above the chromosphere, before falling back along the same path. The whole event lasts only about 20 minutes.

Surveyor The name given to a series of U.S. spacecraft which made soft landings on the Moon for the purpose of taking photographs and making chemical analyses of the lunar surface material. Surveyor I was launched on May 30, 1966. It and the six others in the series obtained many successful photographs of the lunar terrain.

synchronous rotation or **captured rotation** Rotation of a celestial body whose period is equal to the orbital period. Consequently, it will always present the same face towards the body about which it revolves.

synchrotron radiation The radiation emitted by high-energy particles as a consequence of being accelerated in a strong magnetic field. The energy of the charged particles (electrons) decides the frequency at which the emission occurs. Electromagnetic radiation of radio frequency is emitted by 1 MeV electrons. This type of emission is believed to come from interstellar gas clouds in radio galaxies.

synodic month The Moon's synodic period during which its elongation increases by 360°: it is the same as one **lunation**. Its length is 29·5306 days.

synodic period The period of apparent revolution of one body about another as observed from the Earth, measured from opposition to opposition.

system noise The unwanted sound that occurs in any system. In a radio telescope it is compounded of noise from the sky and noise produced in the receiver.

syzygy The configuration occurring when two celestial bodies are at conjunction with or at opposition with a third. It is used also of the points in the orbit of a planet or the Moon when that linear relationship occurs. Syzygy occurs at Full Moon and New Moon, and is important in connection with tides, which are then greatest.

tail of a comet The appendage or streamer of varying length (up to 10^8 km) and extremely low density formed behind the head of a comet. It develops as the comet approaches perihelion and gradually disappears as the comet recedes from the Sun. Straight tails of gaseous matter are driven by the **solar wind**. Curved tails of dust are driven by solar radiation pressure. The tail always points away from the Sun.

tangential velocity or **transverse velocity** That component of a star's velocity which is at right angles to the line of sight. It is given by the equation

$$T = 4{\cdot}74\,\mu p^{-1}\ \text{km s}^{-1}$$

where μ is the annual proper motion of the star expressed in seconds of arc per year and p is its distance in parsecs. If the **radial velocity** (V) is known then the velocity of the star in space is given by the formula

$$v = \sqrt{(V^2 + T^2)}$$

T association A **stellar association** containing many T Tauri stars.

Taurids A meteor shower which lasts about a month, the maximum occurring during the first week of November. It comes from a double radiant between the Hyades and the Pleiades. The β-Taurids shower is a daylight stream which was discovered by radar. Both showers are thought to originate with **Encke's Comet**.

Taurus A constellation of the zodiac which contains many important objects, including the two open clusters, the Hyades and the Pleiades, and the **Crab Nebula**, which is the strong radio source Taurus A and the strong X-ray source Taurus X-1.

tektites Small glassy objects, about the size of gravel, found only in certain specific areas. They were first discovered by Charles Darwin in Australia, and so were originally called AUSTRALITES. They also occur in Moldavia (hence MOLDAVITES), parts of Africa, the East Indies, and North America. They consist mainly of silica, with small quantities of metallic oxides. Their chemical and physical characteristics and ages are different according to the areas in which they are found, and they have nothing in common with the geological characteristics of their surroundings. They clearly originated at a high temperature of about 1500–2000 K, and then solidified. Their true origin is still unknown, though one suggestion is that they are the by-products from the fall of very large meteorites.

telemetering The recording of any event at a distance. In space research it is the transmission of encoded observational data from space back to Earth by radio transmission, e.g. data from artificial satellites or space probes.

telephoto lens Properly, a lens whose focal length is considerably greater than its back focus, i.e. the distance between the back surface of the lens and the image of an object at infinity. It is used in cameras, the advantage being that less camera extension is required than with a normal lens of the same focal length. The front element of a telephoto lens is always converging, the back element is diverging.

telescope The principal instrument used by astronomers to collect light from celestial objects and form a concentrated image of them. It comprises an objective for gathering the light and forming an image at a focus, and an eyepiece to magnify this image. If the objective is a lens, the telescope is known as a refractor. If it is a mirror then the telescope is known as a reflector. A telescope may be used visually or photographically; in the latter case auxiliary equipment such as cameras, photoelectric devices, and spectrometers will on occasion be needed. Various types of mounting (see **telescope, mounting of**) are in use for telescopes. The image produced by an astronomical telescope is inverted, as opposed to that formed in a terrestrial telescope, which is erect.

The refractor installed in the Yerkes Observatory in 1897, which has an objective 40 inches in diameter, is still the largest of its

A view of part of the world's largest optical telescope at Zelenchukskaya Stanitsa in the Northern Caucasus, USSR.

The 200-inch Hale reflector at Palomar Mountain, California, seen from the south.

kind. It is unlikely that refractors of this size will be made in the future, because reflectors with mirrors of larger diameter are easier to make and handle. The largest reflector outside Europe is at Palomar Observatory, California, and has a mirror 200 inches in diameter. The world's largest reflector has a mirror with a diameter of 236 inches and is at Zelenchukskaya Stanitsa, in the Caucasus Mountains, USSR. With the development of spacecraft, it has become possible to send telescopes into space above the Earth's atmosphere, and so obtain observations unaffected

The McMath Solar Telescope, Kitt Peak National Observatory, Arizona: The Sun's light is passed down a sloping tunnel from a coelostat at the top of the tower.

by atmospheric absorption. **Radio telescopes**, which have been developed from radar used in World War II, have opened up a new branch of astronomy.

telescope, mounting of Rigidity, steadiness, and ability to be pointed at all parts of the sky are essential if a telescope is to make useful astronomical observations. The heavier the instrument the greater the engineering problems that are involved and have to be solved. Some instruments are normally mounted on a portable tripod.

An ALTAZIMUTH MOUNTING allows the telescope to be moved about two axes at right angles to each other; the vertical axis gives rotation in **azimuth**, the horizontal one gives rotation in altitude. The great disadvantage of this mounting is the difficulty in moving the instrument simultaneously in azimuth and altitude in order to correct the Earth's diurnal motion; this must be done if it is required to counteract the motion of the body under observation across the field of view. Since the introduction of computer-controlled drive mechanisms, the altazimuth mounting is more reliable and has been used, for example, with the 6-metre reflector in the USSR.

The EQUATORIAL MOUNTING, also called PARALLACTIC MOUNTING, is used for almost all serious observational and photographic work on celestial bodies. One axis of rotation of the instrument (the polar axis) is parallel to the Earth's axis of rotation, and so points to the celestial pole; the other axis of rotation (the **declination** axis) is perpendicular to the polar axis. The object under observation can be kept permanently in the field of view by driving the telescope around the polar axis, in the direction opposite to that of the Earth's **diurnal motion**, at the rate of one revolution per sidereal day. There are various forms of equatorial mounting. The GERMAN MOUNTING developed by J. von Fraunhofer is used for small instruments. The BENT PILLAR MOUNTING, developed by the firm of Carl Zeiss of Jena, is used for instruments of medium sizes. The ENGLISH or YOKE MOUNTING, the FORK MOUNTING and the HORSE-

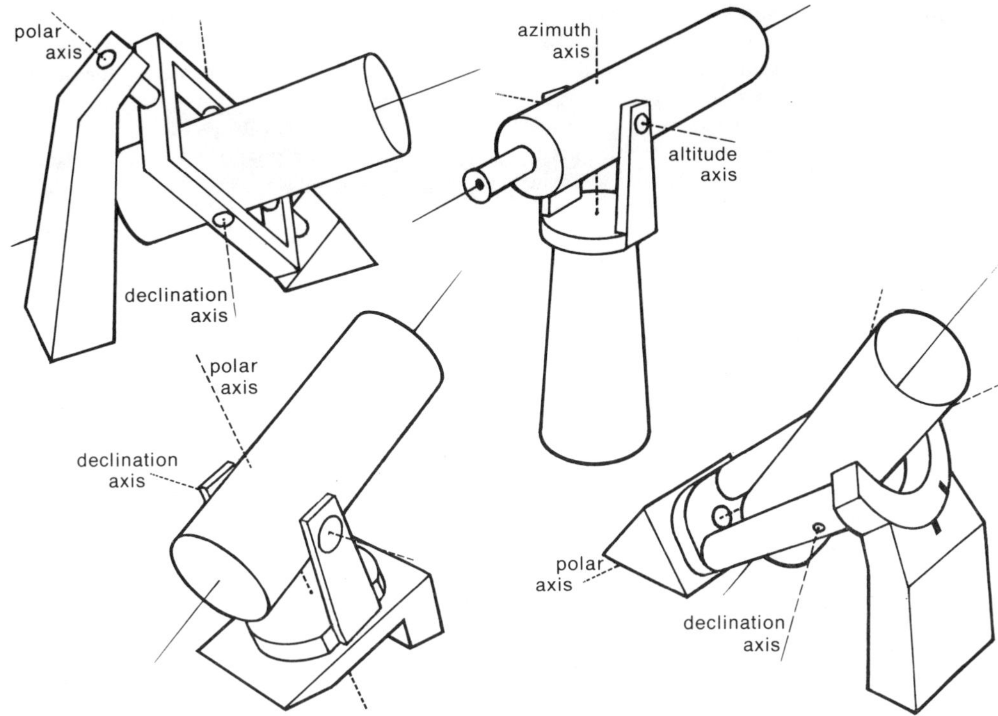

Telescope mountings. Top right: altazimuth mounting. Top left: equatorial yoke mounting. Above: equatorial fork mounting. Above right: equatorial horseshoe mounting.

SHOE MOUNTING, together with the altazimuth mounting, are shown in the illustrations. Large instruments require specially designed bearings to support the great weight involved and to reduce friction as much as possible.

temperature The condition of a body which determines its power to transfer heat to, or receive it from, another body. Temperature indicates the availability of heat (energy). See also **celestial bodies, temperature of**.

temperature scale A means of assigning a numerical value to the temperature of a body. A thermometer is an instrument which indicates variations in the quantity of heat present in a system. The indications given by a perfect thermometer would always be proportional to the quantities of heat absorbed. Ordinary thermometers do not satisfy this requirement because their temperature scales depend on the properties of the substance used to provide the indications. The FAHRENHEIT SCALE is an empirical one commonly used in Great Britain and the USA. The degree Fahrenheit (°F) is 180th part of the temperature difference between the ice point (the temperature of equilibrium between ice and water under a pressure of one normal atmosphere), fixed at 32°F, and the boiling point of water under a pressure of one normal atmosphere, fixed at 212°F. The CELSIUS SCALE is also an empirical scale. The degree Celsius (°C), which was called the degree centigrade before 1948, is 100th part of the temperature difference between the ice point (fixed at 0°C) and the boiling point of water under a pressure of one normal atmosphere, fixed at 100°C. The KELVIN SCALE, based on thermodynamic considerations, is an absolute scale. It removes the limitations of a thermometer from the concept of temperature because 'its characteristic is quite independent of the physical properties of any specific substance'. A temperature interval of one **kelvin** (K) is equal in magnitude to 1°C.

terminator The dividing line between the illuminated and the dark parts of a planet or satellite which can exhibit a phase effect. The terminator of the Moon is irregular on account of the mountainous nature of the lunar surface. The terminator is observable in the case of Mercury, Venus, Mars, and also the Earth (when viewed from space).

terrestrial meridian see **geographical meridian**

terrestrial planets The planets Mercury, Venus, Earth, and Mars, so called because they have rather similar characteristics in respect of size, density, and few or no satellites.

terrestrial telescope The essential requirement for this type of instrument is that it must produce an erect image. The inverted image produced by the optical system of a conventional refracting telescope can be made erect by introducing an erecting lens before the eyepiece. An erecting lens is not included in an astronomical telescope because it would increase the loss of light.

tesla (*symbol:* T) The derived unit in the **SI** for magnetic flux density. It equals one weber per square metre of circuit area. One weber is the magnetic flux which, linking one turn, will produce in it an electromotive force of one volt if reduced to zero at a uniform rate in one second. 1 gauss (c.g.s unit) $= 10^{-8}$ Wb.

Tethys Satellite III of **Saturn**. See also **satellite**.

Thales of Miletos (*c.* 624–548 B.C.) Called the founder of Greek science and philosophy, he provided the first general theory of the universe: everything developed from a primordial mass of water. He knew of the **Saros** and is credited with having used it to predict the eclipse of the Sun which occurred probably on May 28, 585 B.C. during a battle between the Lydians and the Medes.

Themis A reputed satellite of Saturn, reported by W. H. Pickering in 1900 as being in orbit between Titan and Hyperion. See **Saturn**.

theodolite An instrument for measuring angles in the horizontal and vertical planes, but better adapted for the accurate measurement of the former. It is used primarily for surveying, though it has been used for determining the positions of stars in azimuth and altitude and for observations of artificial satellites. For the latter purpose a **kinetheodolite** gives greater accuracy.

thermocouple An instrument for measuring temperature. It consists of wires of two dissimilar metals welded at both ends. When one end is heated, whilst the other is kept cold, an electric current flows in the wires (SEEBECK EFFECT). The electromotive force developed in the circuit depends on the difference in temperature between the two junctions. When used in conjunction with large telescopes, the image of the body under observation is focused on one joint of the thermocouple and the surface temperature of the body can thus be measured. The **thermopile** is preferable to the single element of a thermocouple.

thermodynamics The subject that deals with all physical and chemical changes in which heat effects play a part. In a narrow sense it deals with states and changes of state in solid, liquid and gaseous substances under the action of mechanical and thermal influences. The LAWS OF THERMODYNAMICS are as follows.

LAW I: the LAW OF THE CONSERVATION OF ENERGY. In a closed system the total energy (the sum of the thermal and mechanical energies) is constant. This law is concerned with the balance of energy.

LAW II: Heat cannot pass from a colder to a warmer body, unless it is assisted by some external agency. In terms of **entropy** the second law states that only those processes in which the entropy increases or remains constant can take place. This law is concerned with the direction of the transfer of energy. Laws I and II are the result of experience.

LAW III: the NERNST HEAT THEOREM. The entropy of every pure substance is zero at the absolute zero of temperature. One consequence of this law is that the absolute zero of temperature is unattainable.

All the above laws are capable of being expressed in various other ways, according to the physical property under consideration.

thermopile A sensitive instrument for detecting and measuring radiation of long wavelength, i.e. heat. It consists of a number of junctions of dissimilar metals (such as antimony and bismuth) connected alternately in series. The thermoelectric current, which is produced when the instrument is exposed to heat, is measured in the same way as for a thermocouple.

thermosphere see **atmosphere**

Thomson, William (Baron Kelvin of Largs) (Belfast, June 26, 1824 – Largs, Scotland, December 17, 1907) A mathematician and inventor noted for his important contributions to theoretical and experimental physics, geophysics, astronomical and terrestrial dynamics, etc. He introduced the absolute scale of temperature (known as the **Kelvin scale**). He estimated the age of the Earth and the Sun, and the amount of energy radiated by the Sun. He also investigated the effects of elastic tides.

three-body problem A fundamental problem in celestial mechanics which requires determination of the motions of three bodies under the influence only of their mutual gravitational attractions. The problem is soluble only in special cases. For example, J. L. Lagrange (see **Lagrange, Joseph**) showed in 1772 that if three bodies (two of large mass and one of negligible mass) are situated at the vertices of an equilateral triangle, then they will remain in the same relative positions, whilst revolving about the centre of mass of the system. This condition is exemplified by the Trojan group of minor planets. In the case of the restricted three-body problem, investigated by K. G. J. Jacobi (1804–51), two bodies of considerable mass move in circular orbits about their common centre of mass, whilst the third body of infinitely small mass moves in the same plane as the other two. The three-body problem has been studied by most mathematicians of note.

tidal friction Friction caused by the tidal ebb and flow of water over the ocean floor and resulting in the dissipation of a small amount of energy that comes from the Earth's energy of rotation. Although the amount of energy so taken from the Earth is extremely small, nevertheless the dissipation has the effect of slowing down the Earth's rotation by an amount which lengthens the day by about one millisecond per century.

tides A phenomenon caused by the differential gravitational attractions of the Moon (mainly) and the Sun (to a lesser extent) on the Earth's hydrosphere, and which is evident from the alternate rise and fall of the ocean, as seen on beaches, estuaries, etc. Because the Earth's axis of rotation is inclined to its orbital plane, successive tides at a given point are not equal. This effect is called the DIURNAL INEQUALITY. There are two high tides and two low tides each day. When the Sun and the Moon are exerting a pull in the same direction (as is the case at New Moon or Full Moon) their effects are additive and result in much higher tides (SPRING TIDES). When the pull of the Sun is at right angles to that of the Moon (as is the case at First Quarter and Last Quarter) the tides are much lower (NEAP TIDES). The attractions of the Sun and the Moon should raise tides also in the atmosphere. The effect of the Sun in this respect is largely obscured by its thermal effect, but the existence of a lunar atmospheric tide has been investigated.

time In ordinary language time is a measure of duration by reference to the regular occurrence of some natural phenomenon. From remote antiquity the basis of time has been the passage of the Earth once around its orbit, giving the unit of one year, and one rotation of the Earth on its axis, giving the unit of one day. The unsuitability of the Earth as an accurate and invariable standard of time has been demonstrated by modern astronomy. A. Einstein raised unexpected problems concerning the concept of time by proving that it cannot be considered apart from space (see **relativity**).

Three systems of measuring time are in use for astronomical purposes, namely, **Ephemeris Time**, **Sidereal Time**, and **Universal Time**. The last of these is used as the basis for civil time-keeping. (See also **Greenwich Mean Time**.) LOCAL TIME, which is time as measured at a given place, is based on the mean solar day, and is not the same at places of differing longitude. A difference of 15° in longitude produces a difference of one hour in local time. To avoid the inconvenience of a multiplicity of local times, the Earth has been divided into 24 zones, each of which has a standard **zone time**.

The passage of time is normally measured by means of a clock. National observatories maintain clocks which show local sidereal time and mean solar time, the former being regularly checked by observations of the fundamental stars (clock stars) when they cross the meridian. Pendulum clocks, of which the **Shortt clock** is the most accurate, have given place to even more accurate time-keepers, such as the **quartz-crystal clock** and the **atomic clock**; the latter is used for defining the second in the **SI**.

In view of the possibility that at some time in the future there could be manned lunar bases, a proposal has been made for the establishment of a Lunar Time, comparable with Universal Time, for the benefit of residents on the Moon, particularly for those on the far side who would be out of radio contact with Earth. One lunation, i.e. one complete light-dark cycle of the Moon (one lunar day), takes approximately one month of Earth time. The lunar day would be divided into 30 'lunes' (each approximately 24 hours in length), and each 'lune' into 24 'lunours' (each approximately equal to one hour of Earth time).

time signals Signals which are regularly transmitted by radio broadcasting stations for domestic purposes so that ordinary clocks may be brought to the correct hour; by observatories to provide a time-reference for astronomers; and by observatories with clocks carefully compared by the Bureau International de l'Heure in Paris to provide precision signals for navigators and others. From 1833 up to World War II a time-ball was released at the Royal Observatory, Greenwich, so that ships on the adjacent river and nearby docks could regulate their chronometers. The practice is continued only for sentimental reasons.

Tisserand, François Félix (Nuits-Saint Georges, January 13, 1845 – Paris, October 20, 1896) A French astronomer known for his work on the capture of periodic comets. He is remembered as the founder of the *Bulletin Astronomique* (1884), which he edited until his death.

Titan Satellite VI of **Saturn**. See also **satellite**.

Titania Satellite III of **Uranus**. See also **satellite**.

Titius, Johann Daniel (1729–96) Professor of Physics at Wittenberg who ought to be credited together with J. E. Bode with the empirical law usually called **Bode's Law**.

Tombaugh, Clyde William (Streator, Illinois, February 4, 1906–) An American astronomer remembered for his discovery of the planet **Pluto** in 1930. He discovered also star clusters, variable stars, nebulae, and minor planets. He was engaged in extensive searches for distant planets and satellites.

topocentric Pertaining to observations which are referred to the point of observation as origin, e.g. topocentric co-ordinates, which are measured from some point on the Earth's surface, as opposed to geocentric co-ordinates which are measured from the centre of the Earth.

torquetum An astronomical instrument of the 13th century, the invention of which has been ascribed to various Islamic astronomers. It was a complicated instrument which could only have been used by an expert astronomer. The horizontal base-plate, when properly orientated, enabled measurements of azimuth to be made. An equatorial plate was hinged to the northern edge of the base-plate so that it could be positioned at an angle equal to the **co-latitude** of the place where the observations were to be made. A third plate hinged to the second plate made an angle of 23°30′ with the plane of the equator, thereby placing it in the plane of the ecliptic. A graduated circle, placed perpendicular to and mounted at the centre of the third plate, was used to measure celestial latitudes. A plumb-line attached to a free-hanging semicircle, in turn attached to the graduated circle rising from the third plate, indicated the elevation of the star under observation. All the plates were fitted with alidades. It provided a solution to the problem of converting equatorial to ecliptic co-ordinates.

torr A unit of pressure equal to 133·322 Pa, i.e. 1 mm Hg. It is a non-SI unit accepted for use in conjunction with the SI.

totality The period during a **solar** or **lunar eclipse** when the eclipsed body is completely obscured by the umbra of the eclipsing body. During a total solar eclipse the bright face of the Sun is completely hidden from view only over a certain area of the Earth, which is known as the ZONE OF TOTALITY. Elsewhere, the eclipse is partial.

tower telescope see **solar tower**

transducer A device which receives waves of some kind (electromagnetic, acoustic, etc.) from one system and transmits them, not necessarily as waves of the same kind, to another system. The device is said to be passive or active according to whether the incoming waves are its only source of energy, or energy is supplied to it from a secondary source, respectively.

transient lunar phenomena (*abbrev.:* T.L.P.) Short-lived changes occurring in lunar features. The observation in 1958 by N. A. Kozyrev of activity in the crater Alphonsus, followed by other trustworthy observations, removed doubts about the reality of transient events on the Moon. Such events had been reported for many years but treated with scepticism. Since the manned visits to the Moon, when instruments were set up by the Apollo astronauts, it is now known that there is a connection between moonquakes and these transient phenomena, though their nature is not yet properly understood.

transistor A multi-electrode semiconductor device based on the semiconducting properties of certain elements such as silicon and germanium. Electric current flowing between two of the electrodes is modified (amplified, etc.) by the current or voltage applied to the other electrode(s). Transistors have considerable advantages over thermionic valves, which they have displaced for most purposes: they are smaller, less fragile, easier and hence cheaper to manufacture, and require only a small supply voltage. They have completely revolutionized modern electronics.

transit 1. The passage of a celestial body across the observer's meridian as a result of the Earth's axial rotation. This is known as MERIDIAN TRANSIT, and is observed with a **transit instrument**. **2.** The passage of a body between the Earth and the Sun. The planets Mercury and Venus do so on occasion, when they appear as a black spot before the solar disc. This is the TRANSIT OF AN INNER PLANET. **3.** The passage of a satellite of a planet in front of the planet, as may be observed in the case of the four large satellites of Jupiter. **4.** The passage of some feature across the apparent disc of a body as a result of that body's axial rotation. An example is the passage of Jupiter's Red Spot across the planet's central meridian (CENTRAL MERIDIAN TRANSIT).

transit circle or **meridian circle** The principal instrument of fundamental astronomy, used to determine time and the positions of celestial objects with the highest precision. The instrument must be very stable and rigid, so its size is restricted in order to avoid distortion, and its foundations must be firm. The instrument consists of a refracting telescope, rarely exceeding 23 cm in aperture, mounted with its mechanical axis east-west so that it rotates in the plane of the meridian. One or two accurately divided circles are mounted on the mechanical axis, and permit the altitude of the object under observation to be determined. Right Ascension is found by recording the time of transit of the object across the meridian by a sidereal clock.

It is usual for transit circles to be constructed so that they can be removed from their bearings, turned through 180°, and replaced. In this way systematic errors can be determined for both orientations of these REVERSIBLE TRANSIT CIRCLES. Modern instruments are very complicated, being equipped, amongst other things, with facilities for the automatic recording of observations on punched cards or on magnetic tapes.

transit circle, mirror An instrument suggested by H. H. Turner (see **Turner, Herbert**) in which a plane mirror mounted with a graduated circle on an east-west axis replaces the usual cumbersome transit telescope. This arrangement avoids errors introduced by distortion of the telescope tube, etc. The mirror directs the light received from the object into a collimating telescope, through which observations are made.

transit instrument A telescope mounted on an east-west axis and rotatable only in the plane of the meridian. It does not have accurately graduated circles like the **transit circle**, for its sole purpose is to ascertain the time of transit of bodies across the meridian. Transit instruments are not often used now, as they have been almost completely superseded for precision time-keeping.

transit of inner planet see **transit**

Trans-Neptunian planet A planet presumed to exist in orbit beyond Neptune. The search for this body was instituted in 1905 by P. Lowell (see **Lowell, Percival**) but was subject to various delays. The planet was eventually discovered by C. W. Tombaugh (see **Tombaugh, Clyde**), the discovery being announced on March 13, 1930. See **Pluto**.

Trans-Plutonian planet A planet presumed to exist in orbit beyond Pluto. It has not been discovered, though searches have been made.

transuranium elements Chemical elements which have an atomic number greater than 92 and have been created by synthetic nuclear techniques. These artificial elements now extend to one with an atomic number of 104.

transverse velocity see **tangential velocity**

triangulation The geodetic method of surveying which involves establishing a baseline of accurately known length, and then determining the elements of the triangles into which the area to be surveyed can be divided. The method has been adapted to astronomy for the determination of the parallax of astronomical bodies.

trigonal see **aspect**

trigonometrical parallax The parallax of a star determined by direct means using the principle of triangulation and a known baseline. Near stars show a small change in position with respect to more distant stars as a result of the Earth's orbital motion. This alteration in position can be measured from photographs taken six months apart. The angular displacement gives the amount of parallax, from which, knowing the distance from the Sun to the Earth, the distance of the star can be determined by trigonometry. See **parallax**.

triple-alpha process see **Salpeter process**

triplet A compound lens consisting of an assembly of three component lenses, which may be air-spaced or cemented.

triquetrum A parallactic scale used in antiquity for measuring the elevation of a celestial body. It consisted of an upright post at the top of which was a rotatable rod having two sight-vanes, along which the object could be observed. Its zenith distance was read on the graduated scale of another rod which formed a triangle with the post and the inclined sighting arm.

Triton Satellite I of **Neptune**. See also **satellite**.

Trojans A group of minor planets which move in the neighbourhood of the two stable **Lagrangian points** which are equidistant from the Sun and Jupiter, i.e. at a distance of 778 million km from both the Sun and Jupiter. The two points are situated in the planet's orbit, one being 60° ahead of, and the other 60° behind, Jupiter. This disposition is in accordance with a solution of the **three-body problem** found by J. L. Lagrange. The first Trojan was discovered by M. F. J. C. Wolf in 1906 at the Heidelberg Observatory, and so were many of the others. These minor planets have undoubtedly been captured by the gravitational attraction of Jupiter. Most are named after heroes of the legendary Trojan War. The group ahead of the planet contains

Achilles, Hector, Nestor, Agamemnon, Odysseus, Ajax, Diomedes, Antilochus and Menelaus. The group behind the planet contains Patroclus, Priamus, Aeneas, Anchises, Troilus, and the unnamed Object Arend (1950 SA). Another 34 Trojans have been discovered by T. Gehrels and others on photographic plates taken in 1973 with the 48-inch Schmidt telescope at Palomar. No doubt there are many other undiscovered members of these groups.

tropical month see **month**

tropical year The time between two successive passages of the Sun through the mean vernal equinox in its apparent path around the celestial sphere. Because of precession of the vernal equinox along the ecliptic in a direction opposite to the Sun's motion, the tropical year is slightly less than the sidereal year by about 20 minutes. The length of the tropical year is $365^{d}{\cdot}24219$.

Tropic of Cancer A small circle on the celestial sphere and the Earth at latitude +23½° from the equator. It marks the most northerly declination reached by the Sun at the summer solstice, on or about June 21.

Tropic of Capricorn A small circle on the celestial sphere and the Earth at latitude −23½° from the equator. It marks the most southerly declination reached by the Sun at the winter solstice, on or about December 21.

tropics The region between the Tropic of Cancer and the Tropic of Capricorn.

tropopause The region between the **troposphere** and the **stratosphere** of the Earth's atmosphere, where the temperature drops to about 200–230 K.

troposphere The bottom layer of the Earth's **atmosphere**, rising to a height of about 10 km at the poles and 15 km at the equator. The temperature ranges from about 290 K at the Earth's surface to about 200–230 K at the top. This is the region where meteorological phenomena occur.

true anomaly see **anomaly**

true place or **true position** The position of a star on the celestial sphere (i.e. its co-ordinates for the moment of observation) as it would appear from the centre of the Sun, with respect to the true equator and the true equinox of date.

Trümpler classification The classification of open star clusters devised by R. J. Trümpler and made in accordance with three characteristics: the number of stars in the clusters; the concentration of stars in the cluster; the distribution by apparent magnitude within each cluster.

Trümpler, Robert Julius (Zürich, October 2, 1886 – Oakland, California, September 10, 1956) A Swiss astronomer who became Professor of Astronomy at Berkeley. He is known for his studies of star clusters and his classification of them. He also carried out observational tests of general relativity.

Trümpler stars Stars of very high luminosity found by R. J. Trümpler to have masses several hundred times greater than that of the Sun. As such stars are unstable on theoretical grounds, it is now believed that his observational data have been wrongly interpreted.

T-Tauri stars A class of variable stars whose variation in brightness is connected with the surrounding interstellar matter. Such stars occur only in nebulae or young clusters, and are believed to be young stars still undergoing gravitational contraction before they enter the main sequence of the **Hertzsprung-Russell diagram**. The type star is T Tau in the nebula NGC 1555.

Tunguska Event On June 30, 1908, a most spectacular ball of fire was observed some 800 km north of Lake Baikal in Siberia. The explosion that followed was heard within a radius of more than 800 km and recorded world-wide seismographically. It devastated a forest of over 50,000 trees. The disaster was first thought to have been caused by the fall of a giant meteor, but no crater or fragments of the meteor have been found. Then it was suggested that the area had been hit by a comet, as it has been proved that the explosion took place about 5 km above the surface of the Earth. Nuclear physicists have suggested that a small amount of antimatter penetrated the Earth's atmosphere, as this would explain the intensity of the energy released in the explosion. Another theory ascribes the event to the fall of a huge snowball, weighing at least one million tonnes, that was once the nucleus of a comet; the shock-wave produced by this mass hurtling through the atmosphere would have been responsible for the devastation. Finally, and not surprisingly, the event has been attributed to the blowing up of a nuclear-powered alien spacecraft. It seems most likely that a cometary explanation must be given of this event, one of the great explosions of historic times.

turbulence A state of agitation in gases or liquids producing a swirling movement, as opposed to streamline flow. The passage of a ray of light through the atmosphere disturbed in this way affects the image of a celestial body seen through a telescope. See also **scintillation.**

turbulence theory A theory of the origin of the solar system put forward by C. F. von Weizsäcker. According to this theory, which is a development of Kant's, turbulent motion prevailed in the rotating primeval nebula and produced eddies. The primitive planets were formed as a result of compression at the points of contact of individual eddies. The size and distance of the planets from the Sun were determined by the size of the zones in which the eddies developed.

Turner, Herbert Hall (Leeds, August 13, 1861 – Stockholm, August 20, 1930) An English astronomer who is known for his work in connection with the completion of the Oxford Zone of the Astrographic Catalogue (see **Carte du Ciel**), his studies of seismological records, and his observations of variable stars.

twenty-one centimetre line The emission line of neutral hydrogen in interstellar clouds. It lies in the radio spectrum at a frequency of 1420 MHz; the wavelength is about 21 cm. Its existence was predicted by H. C. van de Hulst in 1944 and discovered by him and others in 1951.

twilight The intermediate period during which the illumination of the sky gradually increases before sunrise, and decreases after sunset. The phenomenon is caused by the scattering of sunlight by molecules of air and particles of dust in the Earth's upper atmosphere. The duration of twilight depends on the steepness of the Sun's apparent path with respect to the horizon, so that twilight lasts longer at higher latitudes.

The following forms of twilight are distinguished. CIVIL TWILIGHT ends, or begins, when the Sun attains a zenith distance of 96°, i.e. when the centre of its disc is 6° below the theoretical horizon. Normal daytime activities are regarded as not being possible during this period. NAUTICAL TWILIGHT ends, or begins, when the Sun attains a zenith distance of 102°, i.e. when the centre of its disc is 12° below the horizon, and the marine horizon is no longer visible. ASTRONOMICAL TWILIGHT begins in the morning and ends in the evening when the Sun's zenith distance is 108°, i.e. when the centre of its disc is 18° below the horizon, and is the time when the faintest stars can be seen with the naked eye. The times of commencement of the various forms of twilight are given in *The Astronomical Ephemeris, The Nautical Almanac,* and *Whitaker's Almanac.*

two-body problem or **Kepler's problem** A fundamental problem in celestial mechanics which deals with the calculation of the motions of two bodies in space under the influence only of their own gravitational attractions in accordance with Newton's Law of Gravitation. It is the only case of the **many-body problem** for which there is a complete solution.

Tycho Brahe see **Brahe, Tycho**

Tychonian planetary system A planetary theory put forward by Tycho Brahe.

Tycho's star The star observed and described by Tycho Brahe in 1572. It was a supernova which was brighter than Venus at maximum magnitude and was visible during daytime. Its remnant has been detected as a radio source and as an X-ray source. Kepler's star of 1604 is the only other supernova to have appeared in our Galaxy since that of 1572.

Type I, Type II supernova see **supernova**

UBV system A system of three-colour photographic photometry devised by H. L. Johnson and W. W. Morgan of Yerkes Observatory. Stellar magnitudes are determined at three different wavelength bands through three colour filters: ultraviolet (*U*) peaking at 3600 Å, blue (*B*) peaking at 4200 Å, and visual (*V*), yellow, peaking at 5400 Å. See also **stellar magnitude, determination of**.

UFO An acronym from *U*nidentified *F*lying *O*bject; popularly called 'flying saucer'. From time to time there are reports of objects moving in the sky. Most of these are explainable as mis-sightings of known objects (such as artificial satellites) or known phenomena (such as clouds), defects in photographic films and their development, hoaxes, and 'visions' of cultists or psychologically disturbed individuals. There still remain some observations, however, that cannot be disposed of in any satisfactory way. The possibility that they are spacecraft from alien worlds making reconnaissance of our planet is not generally accepted.

U Geminorum stars A class of **dwarf nova** which show sudden outbursts of brightness, giving an increase of about four magnitudes, after long periods at a steady magnitude. The outbursts can last from several days to weeks. The best known examples of this class are U Geminorum and SS Cygni.

Uhuru A small US artificial satellite launched off the coast of Kenya on December 12, 1970, fitted with equipment for the specific purpose of studying cosmic X-ray sources.

U line A line in the spectrum of sodium at 3302 Å.

ultraviolet radiation Electromagnetic radiation with wavelengths lying between those of the visible radiation of the spectrum and X-rays. Ultraviolet wavelengths range from about 4000 Å down to about 100 Å. Ultraviolet radiation from celestial sources, such as hot O- and B-type stars, can be detected and analysed by instruments carried in artificial satellites and also in balloons and rockets.

Ulugh Beg (Soltaniyeh, Persia, 1394 – Samarkand, October 27, 1449) The greatest of oriental astronomers. He built an observatory at Samarkand. He compiled the first star catalogue since Ptolemy's to be based on observations made with great precision.

umbra **1.** The dark central region of a shadow which receives no illumination from the source of light. An example is the portion of the shadow-cone cast by the Earth during an eclipse. **2.** The dark central area of a sunspot.

Umbriel Satellite II of **Uranus**. See also **satellite**.

Universal Time (*abbrev*.: U.T.) Standard time used universally for all civil purposes. It is Greenwich Mean Time (which term is used in navigational publications) beginning at midnight. Refinements of this time (designated U.T.0) for improved accuracy have produced U.T.1 and U.T.2 to take into account the effect of polar motion on the longitude of the place of observation and seasonal variations in the Earth's rotational period. CO-ORDINATED UNIVERSAL TIME (U.T.C.), based on the caesium atomic clock, is used for radio time signals. As the Earth loses about three thousandths of a second per day through the decrease in its rate of rotation, there is a gradual divergence of U.T.C. from G.M.T. In order to maintain time signals based on the atomic time scale with G.M.T., they are retarded when necessary by one leap second at midnight June 30/July 1 and December 31/January 1.

universe All space and everything contained in it. From remote antiquity various concepts of our planetary system and of the universe as a whole have been held, and various theories have been put forward to account for their structure, development, and changes. The total mass content of the universe is reckoned to be about 10^{23} solar masses; its radius to be about 2×10^{23} km; and its age to be about

2×10^{10} years. See also **big bang theory**; **steady-state theory**; **cosmology**; **mean density of matter**.

upper culmination see **culmination**

Urania One of the Muses, the daughter of Jupiter and Mnemosyne, who presides over astronomy. Her attributes are the globe and compasses, and since the 17th century she has been crowned with a circle of stars.

Uraniborg The residence and observatory established by Tycho Brahe on the Island of Hven, north of Copenhagen. The foundation stone was laid in 1576. He carried out an exceedingly accurate series of observations there over a number of years. See also **Stjerneborg**.

Uranometria A star atlas compiled by J. Bayer (see **Bayer, Johann**) and published at Augsburg in 1603. Its full title is *Uranometria, omnium asterismorum continens schemata nova methodo delineata, aereis laminis expressa.*

Uranus A giant planet orbiting the Sun between Saturn and Neptune. It was the first planet to be discovered since antiquity. It was first observed on March 13, 1781 by W. Herschel (see **Herschel, William**) whilst examining the star-field on the borders of the constellations Taurus and Gemini. He took it to be a comet, but the accumulated observations of Herschel and others showed that the object was a planet and not a comet. Herschel wished to call the new planet *Georgium Sidus* in gratitude for the pension which King George III had granted him. J. E. Bode suggested the name Uranus, the most ancient of all the gods and father of Saturn, thus maintaining the names of mythological characters for the planets. Though the suggestion to call the new planet by the name Herschel was not adopted, his initial is appropriately incorporated in the symbol used to designate the planet (♅).

The planet is just visible to the naked eye under ideal observing conditions. It closely resembles Neptune. It appears as a small blue-green disc when viewed in the telescope. It is believed to have a substantial core of rock and an extensive atmosphere of hydrogen, helium, and methane. Because of the low surface temperature, any water and ammonia will have been frozen. There is no pattern of zones on the disc. Uranus has five known satellites – Miranda, Ariel, Umbriel, Titania, and Oberon – all of which are in orbit in the planet's equatorial plane.

On March 10, 1977 in the course of making observations in NASA's Kuiper Airborne Observatory (KAO) of the occultation of star SAO 158687 for the purpose of measuring the diameter of Uranus, it was discovered that the planet has a system of not less than six narrow rings. They lie in the equatorial plane and are tenuous and unlikely to be detected visually from Earth. As in the case of Saturn the rings lie within the **Roche limit**. The main data relating to Uranus are given in the table.

Globe	
Diameter (equatorial)	55,800 km
Density (water = 1)	$1{\cdot}2\ \mathrm{g\ cm^{-3}}$
Mass	$8{\cdot}78 \times 10^{25}$ kg
Volume	$7{\cdot}32 \times 10^{13}\ \mathrm{km^3}$
Sidereal period of axial rotation	$10^{h}\,49^{m}$
Escape velocity	$25{\cdot}2\ \mathrm{km\ s^{-1}}$
Albedo	0·93
Inclination of equator to orbit	97° 53′
Surface temperature	90 K (maximum)
Surface gravity (Earth = 1)	1·17
Orbit	
Semi-axis major	19·182 A.U. = $2869{\cdot}6 \times 10^{6}$ km
Eccentricity	0·04726
Inclination to ecliptic	0° 46′ 23″
Sidereal period of revolution	306,84·9 d
Mean orbital velocity	$6{\cdot}81\ \mathrm{km\ s^{-1}}$

Urey, Harold Clayton (Walkerton, Illinois, April 29, 1893–) An American chemist who was awarded the Nobel Prize in Chemistry for his discovery of heavy hydrogen in 1934. He is particularly well known for his studies of the properties of chemical elements in relation to celestial bodies and the origin of the solar system.

Ursa Major The Great Bear. The large northern constellation containing possibly the best known of all groups of stars from remote times to the present. This group of the seven brightest stars constitute the PLOUGH (called the BIG DIPPER in the USA). The stars α and β are known as the POINTERS, because they are used to find the direction of the north celestial pole.

Ursa Major cluster A moving cluster of stars containing about 100 members, including the brightest five stars of Ursa Major as well as Sirius. Their velocity in space is 29·5 km s^{-1}, and the movement is towards a point in the constellation Aquila.

Ursa Minor The Lesser Bear. The northern constellation which contains the north celestial pole. The brightest star, Polaris, lies within one degree of the pole.

Ursids A meteor shower discovered by A. Bečvář (1901–65) in 1945. It is associated with Comet 1926 IV P/Tuttle and occurs about December 22 with a radiant in Ursa Minor.

UV Ceti star Another name for **flare star**.

Van Allen belts Zones of charged particles surrounding the Earth. They were discovered by J. A. Van Allen and his collaborators during investigations using **Geiger counters** in the first American artificial satellites, Explorers I, III, and IV, launched in 1958. The counters were for the purpose of measuring the intensity of cosmic rays. They functioned normally up to an altitude of about 1000 km, when they went 'wild' and choked. On returning to the lower altitude they resumed normal operation. Further information was obtained from the American Pioneer III spacecraft launched on December 6, 1958 and the Russian Lunik I launched on January 2, 1959.

The Van Allen belts consist of two concentric belts above the Earth's equator. They contain electrons and protons, at very high levels of energy, that have been trapped in the Earth's magnetic field; their number and energy vary with the solar activity. The maximum intensity of the inner belt of high energy is located at an altitude of about 3000 km. The other belt of lower energy is located at an altitude of about 16,000 km.

The Earth's Van Allen belts.

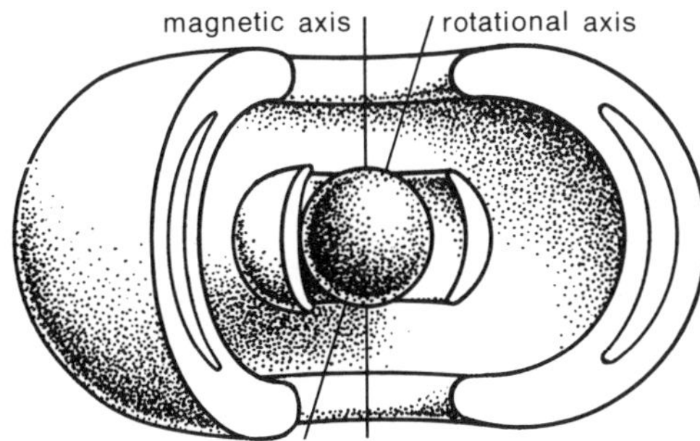

Van Allen, James Alfred (Mt Pleasant, Idaho, September 7, 1914–) An American physicist who discovered the radiation belts around the Earth, which are named after him. He is a pioneer in high-altitude research by means of rockets, satellites, and space probes and initiated the **rockoon** technique. He has investigated primary cosmic rays.

variable-focus lens A telephoto lens in which the separation between the positive (converging) and negative (diverging) elements is variable. Older types comprised a special attachment used in conjunction with the normal lens; it consisted of a negative lens that could be screwed to the back of the tube carrying the normal lens.

variable nebula A **nebula** whose luminosity is caused by a nearby variable star and hence varies in brightness.

variable stars Stars whose apparent magnitudes vary. There are numerous classes, depending on the nature of the stars, their light-curves, their absolute magnitude, and their spectral types. There are REGULAR VARIABLES, whose variation in brightness repeats in predictable cycles, IRREGULAR VARIABLES, whose variation in brightness is non-uniform and unpredictable, and SEMI-REGULAR VARIABLES. A distinction can be made between INTRINSIC VARIABLES, whose variations in brightness result from changes in the stars' emission of radiation, and EXTRINSIC VARIABLES, whose apparent variations in brightness result from some cause external to the stars, as in the case of eclipsing **binaries**. **Pulsating stars** are intrinsic variables having a regular cycle of variation in brightness, which H. Shapley ascribed (1914) to regular alternate expansion and contraction of the stars; examples include **Cepheids**, **RR Lyrae** stars, and RV Tauri stars. LONG-PERIOD VARIABLES are those whose periods exceed 100 days: they exhibit a wide variation in brightness; an example is the star Mira (o Cet), the first variable star to be discovered, having been recorded by J. Fabricius in 1596. EXPLOSIVE VARIABLES comprise novae, supernovae, and flare stars. The number of variable stars known is close on 15,000: they are listed in the *Catalogue of Variable Stars* (in Russian) by B. V. Kukarkin *et al.*, published in Moscow.

The method of designating variable stars is as follows. The first variable to be discovered in a constellation has the letter R preceding the name of the constellation. Further discoveries are denoted by S . . . Z; then by RR, RS . . . RZ, SS . . . SZ, . . . ZZ, AA . . . AZ (omitting the letter J), BB . . . BZ, . . . , QQ . . . QZ. At this point the 334th variable has been reached. Subsequent variables are numbered V335, V336, etc. The I.A.U. is responsible for the designation of variable stars.

variation A perturbation in the Moon's motion caused by the Sun's gravitational effect upon the Moon's orbital motion.

Vega (α Lyr) The brightest star in the constellation Lyra. It is the standard star of spectral type A0 and luminosity class V in the **UBV system**.

velocity curve The curve obtained by plotting the radial velocity of a component of an eclipsing binary system against time.

velocity-distance relationship see **Hubble Law**

velocity of light see **light**

Venera The name of a series of space probes launched by the USSR to the planet Venus. The first was launched on February 12, 1961 but contact with it was lost. Venera 4, launched on June 12, 1967, was the first to transmit data. Venera 5 to 12, launched between 1969 and 1978, have transmitted further data concerning the planet's atmosphere, temperature, and surface.

Venus The second planet from the Sun, with an orbit between that of Mercury and Earth. As the morning or evening star it is the most conspicuous celestial object, after the Sun and Moon; it is even visible to the naked eye during daytime at times of maximum brightness, when the Sun is high in a clear sky. It exhibits phases, as do both Mercury and the Moon. American (Mariner 10 and Pioneer Venus) and Soviet (Venera) space probes have provided considerable information about the planet's atmosphere and surface conditions. The atmosphere consists mainly of carbon dioxide (about 95%) and nitrogen (about 3%), with traces only of helium, argon, neon, oxygen, water vapour and other chemical compounds. The yellowish-white clouds enveloping the planet have been

Globe	
Diameter (equatorial, solid sphere)	12,112 km
Diameter (cloud surface)	12,200 km
Density (water = 1)	5·16 g cm^{-3}
Mass	4·88 × 10^{24} kg
Volume	9·28 × 10^{12} km^{3}
Sidereal period of axial rotation	243^{d}·0 retrograde
Escape velocity	10·36 km s^{-1}
Albedo	0·76
Inclination of equator to orbit	178°
Surface temperature	743 K
Surface gravity (Earth = 1)	0·9
Orbit	
Semi-axis major	0·7233 A.U. = 108·21 × 10^{6} km
Eccentricity	0·00678
Inclination to ecliptic	3°·4
Sidereal period of revolution	224^{d}·7
Mean orbital velocity	35 km s^{-1}
Mean synodic period	583^{d}·9

shown to comprise droplets of sulphuric acid and, in the lower layers, solid and liquid particles of sulphur. Mariner 10 established that the tops of the clouds rotate every four hours retrograde.

At the surface the temperature is about 750 K, and the pressure 100 atm or more. The temperature at the top of the clouds, which obscure the surface features, is about 250 K. It has been established by radar that the surface of the planet, though smoother than that of the Moon, is nonetheless mountainous and extensively cratered. The last transits of Venus occurred on December 8, 1874 and December 6, 1882, and were widely observed in connection with accurate determination of **solar parallax**. The next transits will occur on June 7, 2004 and June 5, 2012. Venus has no known satellite. The main data relating to Venus are given in the table.

vernal equinox see **Aries, First Point of; equinox**

vernier A device for measuring fractional portions of one of the equal spaces into which a scale is divided, invented by Pierre Vernier (1580–1637) and described by him in 1631. It consists of an auxiliary scale which can be moved in contact with the principal scale. Graduation of the auxiliary scale can be made in various ways. The commonest is for n divisions on the auxiliary scale to equal $n-1$ divisions on the principal scale, so that the distance between two divisions on the auxiliary scale is $1/n$ less than the corresponding distance on the principal scale. When the zero mark of the auxiliary scale lies between two divisions on the principal scale, then that mark on the auxiliary scale which coincides with a division on the principal scale gives a reading accurate to $1/n$ of the principal scale.

vertex The point to which the movement of a group of stars appears to converge, or from which a stream of meteors appears to radiate. See also **apex**.

vertical, angle of a The angle on the celestial sphere between the prime vertical – i.e. the vertical circle which passes through the zenith, the east point, the nadir, and the west point – and any other vertical circle.

vertical circle **1.** A **great circle** on the celestial sphere which passes through the zenith and the nadir. It is a secondary to the horizon, to which it is perpendicular. **2.** An instrument used for measuring altitude and zenith distance.

Vesta The fourth minor planet to be discovered, and the only one that is sometimes visible to the naked eye as it is the brightest of the minor planets. Its diameter is about 540 km, its mean distance from the Sun is 2·36 A.U., its mass is about $2{\cdot}4 \times 10^{20}$ kg, and the albedo is 0·24. Vesta was discovered in 1807 by H. W. M. Olbers (see **Olbers, Heinrich**).

Viking The name of two American space probes which were put in orbit around the planet Mars and from which two craft were detached and subsequently landed. The first was launched on August 20, 1975, the second on September 9, 1975. The first Lander set down on July 20, 1976, the second on September 3, 1976. The two Orbiters and two Landers were highly successful in obtaining information on the nature of the Martian surface, chemical, physical, and magnetic properties, and meteorological and seismological conditions. The Orbiters photographed the surface of the planet and that of its two small satellites, Deimos and Phobos. The search for evidence of life in some form has not yielded positive results.

violet layer A layer of the atmosphere of the planet Mars that is opaque to radiation of wavelengths shorter than 4500 Å, but transparent almost to those longer than 5000 Å. This layer therefore prevents the photography of surface details in blue light. It suddenly disappears for short periods. See **blue clearing**.

Virgo A A strong radio source associated with the peculiar galaxy M87, which has a very long luminous blue jet. The radio emission is on 18·3 MHz. The galaxy is also an X-ray emitter, as well as a powerful source of infrared radiation.

visible spectrum The range of electromagnetic radiation which is perceived by the human eye. It comprises the colours red, orange, yellow, green, blue, indigo, and violet. The wavelength range is about 7000 Å to 4000 Å.

visual binary A binary system that can be observed as a double star with a telescope. See **binary**.

visual magnitude The magnitude of a star ascertained in the colour region to which the human eye is most sensitive, i.e. about 5600 Å.

visual photometry A system of photometry which is used if other methods are not practicable and is based on the ability of the human eye to judge the slight difference in brightness between two stars. The comparison is made between the star whose magnitude is to be measured and a real star or an artificial star whose brightness can be varied by the instrument. It is difficult for the eye to judge this difference in brightness, for it can systematically overestimate one of the light sources. This gives rise to an error known as the EQUATION OF POSITION, which can amount to several tenths of a magnitude. It is necessary to be able to interchange the positions of the two stars under comparison. The equation of position is not only peculiar to each eye of an individual, but varies from one person to another.

Vogel, Hermann Carl (Leipzig, April 3, 1841 – Potsdam, August 13, 1907) A German astronomer who became the first Director of the Astrophysical Observatory, Potsdam, in 1882. He discovered spectroscopic binaries about 1890. He worked on the spectral analysis of stars, and introduced the photography of stellar spectra.

volcanic theory One of the theories put forward to explain the origin of lunar surface features as being the result of volcanic activity.

Voskhod The name of two manned artificial Earth satellites launched from the USSR. Voskhod I was launched on October 12, 1964, and was the first occasion on which three astronauts went together into space. The first space-walk was made by Alexei Leonov, one of the two-man crew of Voskhod II, launched March 18, 1965.

Vostok The name of a series of manned artificial Earth satellites launched from the USSR. Vostok I was launched on April 12, 1961 and carried the first man, Y. A. Gagarin (see **Gagarin, Yuri**), into space. Vostok VI was launched on June 16, 1963, and carried the first woman, Valentina Tereshkova, into space.

Voyager The name of two American space probes. Voyager I was launched on September 5, 1977 and Voyager II on August 20, 1977. They passed by Jupiter in March and July 1979 and relayed to Earth considerable information and many photographs of the atmosphere of Jupiter, its closest satellites – Amalthea and the four Galilean satellites – and the ring system of Jupiter, discovered by Voyager I. The two craft should reach Saturn in November 1980 (I) and August 1981 (II). Voyager II will then proceed, if all goes well, to Uranus and Neptune.

Vulcan A supposed planet within the orbit of Mercury, originally thought to be responsible by its gravitational effect for the unexplained residual in the advance of the perihelion of Mercury. This problem has since been explained by the General Theory of Relativity. A French physician, Lescarbault, on March 26, 1859, observed a small body in transit across the Sun's disc and made subsequent careful observations of it. Notwithstanding the fact that U. J. J. Le Verrier (see **Le Verrier, Urbain**) carefully examined Lescarbault's observations and integrity, and was sure of the planet's existence, no trace of it has been found since. It is now regarded as nonexistent.

walled plain Any of the largest craters surrounded by mountain ranges which are found on the Moon. The largest ring-formation of this kind is Bailly, which has a diameter of 295 km.

watt (*symbol:* W) The **SI** derived unit of power which gives rise to energy of one **joule** in one second.

wavelength The distance between two consecutive parts of a wave which are in the same phase, e.g. from crest to crest. The wavelength (λ) is related to the other characteristics of the wave motion as follows:

$$\lambda = v/\nu = vT$$

where v is the phase velocity, ν the frequency, T the period. In optical astronomy the unit of wavelength is usually the **ångström** or the nanometre (10^{-9} metre); in radio astronomy the unit is the metre (m) or centimetre (cm). Frequency is expressed in **hertz**. See also **electromagnetic radiation**; **light**; **visible spectrum**.

web see **wire**

weber (*abbrev.:* Wb) The derived SI unit of magnetic flux. It is the magnetic flux which, linking a circuit of one turn, will produce in it an electromotive force of one volt if it be reduced to zero at a uniform rate in one second. 1 weber = 10^8 maxwells.

weightlessness The result of being subjected to **zero gravity**. Astronauts experience this when their space-vehicle starts unpowered flight outside the Earth's atmosphere. Before becoming adapted to this condition some persons become disorientated and confused, have disturbed vision, and even lose consciousness. It seems that weightlessness in itself is not harmful, but can be critical in association with other physical disturbances.

Weizsäcker, Carl Friedrich von (Kiel, June 28, 1912–) A German physicist known for his researches in astrophysics and cosmology, and for his theory of the origin of the solar system and the evolution of stars and galactic systems. He postulated one of the nuclear chain reactions producing energy in the Sun and other stars, i.e. the Bethe-Weizsäcker cycle (see **carbon-nitrogen cycle**).

Werner lines Lines which are produced by molecular hydrogen in the ultraviolet part of the spectrum; they are in the same region as the **Lyman series**.

West Ford Project An American scheme to distribute an enormous quantity of tiny copper needles in a belt around the Earth in order to improve long-distance intercontinental radio communication. The first attempt was made on October 21, 1961, when a packet containing 165 kg of the needles was launched by Midas IV. Each copper needle was 17·7 mm long, had a diameter of 0·0286 mm, and weighed 0·1 milligram. This quantity was intended to give a belt of needles 8 km wide and 40 km deep. This attempt failed as the needles did not disperse for some reason. A second attempt in May 1963 was successful. The project was opposed by astronomers who feared interference with observations in radio astronomy.

west point That point on the celestial sphere, due west of the observer, where the meridian intersects the horizon.

Whipple, Fred Lawrence (Red Oak, Idaho, November 5, 1906–) An American astronomer known for his discovery of comets and his studies of meteors, planetary nebulae, stellar evolution, spectrophotometry, and the Earth's upper atmosphere.

white dwarf A member of a class of stars whose size is about that of the Earth, but whose mass is about equal to that of the Sun. These stars are of low luminosity and exhibit spectra with diffuse lines. They have exhausted most of their nuclear fuel, and probably represent the final stage in the evolution of low-mass stars. To evolve to a white dwarf, the star's mass should not exceed 1·4 solar masses (the **Chandrasekhar limit**). The high density of 10^5–10^8 g cm^{-3} results from the fact that the matter is completely ionized, with the electrons and nuclei packed extremely close together (the state known as degenerate matter). The pressure exerted by these electrons balances the inward-directed gravitational force and prevents the body from contracting any further. The first white dwarf to be discovered was the companion of **Sirius**.

white nebulae The name once applied to spiral galaxies, because they appeared white when observed through the telescope.

white spot A phenomenon sometimes observed on **Saturn** and on **Jupiter**.

wide-angle lens A lens that embraces an incident beam of very wide angle. Its use enables a large area to be photographed at a shorter working distance.

Widmanstätten pattern The distinctive pattern produced on the polished surface of an iron **meteorite** when it has been etched with acid. It is called after A. von Widmanstätten who observed the pattern on iron-nickel meteorites in 1808.

Wien's Law The Displacement Law derived by Wilhelm Wien (1864 – 1928) from his work on black-body radiation. It states that the wavelength at which maximum energy is radiated is inversely proportional to the absolute temperature of that body. As the temperature is increased, so the intensity of the radiation is displaced more into the violet part of the spectrum. Wien's Law is one of the limiting cases of **Planck's radiation law**.

Wildt, Rupert (Munich, June 25, 1905 –) A naturalized American astronomer known for his work in theoretical astrophysics, stellar spectroscopy, and on the constitution of the planets.

Williams, Arthur Stanley (Brighton, 1861 – Feock, Cornwall, November 21, 1938) A solicitor and outstanding English amateur astronomer, known for his observations of planets and variable stars. He was a pioneer in the study of Jupiter, particularly with respect to the behaviour of surface features and atmospheric currents, and also of Saturn in connection with its period of rotation.

Wilson effect A phenomenon discovered by the Scottish astronomer Alexander Wilson (1766–1813), the first Professor of Astronomy in the University of Glasgow. He noticed that when a sunspot is near to the Sun's limb it appears foreshortened by perspective, because the Sun is a sphere, and the **penumbra** then appears narrower in the direction towards the centre of the Sun than in the direction of the limb. Wilson took this as evidence that sunspots are saucer-shaped depressions, but the phenomenon is not shown by all spots, however. It is now considered that the effect is the result of greater transparency of the sunspot as compared with the surrounding photosphere, the umbra being far more transparent.

window Any of the ranges of wavelengths of electromagnetic radiation to which the Earth's atmosphere is transparent (see **optical window, radio window**). Cosmic X-rays are stopped by the Earth's atmosphere; ultraviolet radiation is absorbed for the most part by ozone in the upper atmosphere; infrared radiation is absorbed by water vapour in the atmosphere except for a few narrow wavebands; wavelengths longer than those passing through the radio window are reflected by the ionosphere.

winter solstice see **solstice**

wire The fiducial lines in the field of the eyepiece of a telescope; though they are sometimes made of very fine wire, spider's thread is the preferred material, in which case they are often called a WEB.

Wolf diagram A graph developed by M.F.J.C. Wolf and named after him. It plots the number of stars or galaxies per square degree (counted at successive limits of magnitude) against the apparent magnitude. The counts are made in obscured (dark) regions and compared with nearby unobscured regions. By comparing the two curves on the graph, the distance to and the absorption characteristics of dark nebulae may be found.

Wolf, Maximilian Franz Joseph Cornelius (Heidelberg, June 21, 1863 – Heidelberg, October 3, 1932) A German astronomer noted for his application of photographic methods for observation in astronomy. By this means he made the first photographic discovery of a minor planet (No. 323): in all he found 582. He detected several new nebulae, including the North America Nebula (NGC 7000) in the constellation Cygnus. He also discovered the periodic comet (1884 III) which bears his name, the nebulosity around the Pleiades, numerous dark nebulae, and the first cluster of galaxies (in Coma Berenices) and also the one in the constellation Virgo.

Wolf number see **Wolf relative sunspot number**

Wolf-Rayet stars More usual name for stars of spectral type W (see **stars, spectral classification of**). They were separated from class O by the I.A.U. in 1938. Their spectra contain broad emission lines from hydrogen and helium and, according to whether there is a predominance of emission bands from carbon or nitrogen, are further divided into subclass *WC* and subclass *WN*. The stars are very hot (surface temperature about 80,000 K) and luminous (average magnitude −4), and about twice the size of the Sun. Most of these stars are spectroscopic binaries and some are eclipsing binaries. They have similar characteristics to the stars found at the centres of planetary nebulae. The stars are named after their discoverers, C. J. E. Wolf (1827–1918) and G. A. P. Rayet (1839–1906).

Wolf relative sunspot number An index of sunspot activity, devised by Rudolf Wolf, which reduces the sunspot counts of different observers to a common, statistical basis, and gives the relative sunspot number for the day as

$$r = k(f + 10g)$$

where g is the number of groups of sunspots, irrespective of the number of spots contained in each, and f is the total number of spots counted in all the groups; k is a factor depending on the estimated efficiency of the observer and his telescope, and is taken as unity for observations made with the 10 cm refractor at the Zürich Federal Observatory with a power of ×64. The Wolf number averages about 5 for a year of minimum sunspot activity, and about 150 for a year of maximum activity.

Wolf, Rudolf (Fällanden, near Zürich, July 7, 1816 – Zürich, December 6, 1893) A Swiss astronomer who became director of the observatory at Zürich. He is noted for his observations of solar activity over many years and for his statistical analysis of sunspots which led to the **Wolf relative sunspot numbers**. He discovered the connection between sunspots and terrestrial magnetic disturbances.

Wollaston double image prism An optical element devised by W. H. Wollaston (1766–1828). It consists of two right-angle prisms cut from uniaxial crystals of quartz or calcite so that when cemented together to form a parallelopiped, their optical axes are mutually perpendicular. A double image prism can be used to measure the dimensions of an optical image; it is also used in certain types of spectrophotometer.

Wooley, Richard van der Riet (Weymouth, April 24, 1906–) An English astronomer who was Commonwealth Astronomer, Mt Stromlo Observatory, Canberra, Australia (1939–1955), Astronomer Royal (1956–1971), and is now Director of the South African Observatory. He is known for his research on the dynamics of the Galaxy.

Wright's phenomenon An effect discovered by William Hammond Wright (1871–1959) of Lick Observatory in 1925. He found that the diameter of Mars appeared considerably larger when photographed in ultraviolet light than when photographed in infrared light. This difference was attributed to scattering of ultraviolet radiation by the Martian atmosphere, and was used to calculate the depth of the atmosphere: a figure of about 100 km was obtained (see **violet layer; blue clearing**). Later investigation of the depth of the Martian atmosphere indicates that it is about half that value.

Wright telescope A modification of the **Schmidt camera** made by F. B. Wright. The spherical primary mirror of the standard Schmidt instrument is replaced by an ellipsoidal mirror, and the figure of the correcting plate modified accordingly. This system gives a flat field, halves the length of the tube, and can be used at a Newtonian focus.

Wright, Thomas (Biar's Green, 1711 – Biar's Green, 1786) Known as 'Thomas Wright of Durham'. An English instrument maker, natural philosopher, and mathematician. He is remembered particularly for his work *Original Theory or New Hypothesis of the Universe* (London, 1750), which gives a theory of the Milky Way as consisting of innumerable stars and that its appearance results from our being immersed (together with our Sun) in a layer of stars. He predicted that the rings of Saturn consist of very small satellites.

wrinkle ridge A certain type of lunar surface feature characterized by its irregular sinuous form. These ridges are peculiar to the maria (see **mare**), having sloping sides and reaching a height of about 200 m on average. Exceptionally, the Serpentine Ridge of Mare Serenitatis, discovered by J. H. Schröter (1745–1816), attains a height of 1 km. Sometimes the ridges are cracked open at the top.

W stars Another name for **Wolf-Rayet stars**. See also **stars, spectral classification of**.

W Ursae Majoris stars A class of eclipsing **binary stars**. See **binary**.

W Virginis stars see **Cepheid**

X-ray binary A binary system consisting of a normal star and a collapsed star, i.e. a **neutron star** or **black hole** or, in less intense sources, a **white dwarf**. The stars are very close together so that expansion of the normal star in the course of its evolution results in a flow of gas towards the collapsed star. This matter forms a disc of hot material (the accretion disc), which is a source of X-ray emission, round the collapsed star.

X-ray burster A source of intense flashes of X-rays. This kind of emission has been detected by various spacecraft, for example from Cygnus X-1 by Ariel V. Up to October 1978 five of these sources of bursts had been optically identified. They are of two types. A Type I source is probably a compact object of solar mass, and the burst is perhaps a thermonuclear one on the surface of a neutron star. A Type II burst is considered to result from instabilities in flow during accretion into a compact mass.

X-rays Electromagnetic radiation of short wavelength (about 0·01 Å to 100 Å) discovered by W. C. Röntgen (1845–1923) in 1895. The energy of X-rays is greater than that of ultraviolet radiation but less than that of γ-rays. The Earth's atmosphere is opaque to X-rays, so this radiation from outer space does not reach the Earth's surface. A new branch of astronomy, namely X-RAY ASTRONOMY, has therefore come into existence. It involves sending equipment in satellites and rockets out of the Earth's atmosphere (see **X-ray telescope**). These instruments have detected X-rays from many sources including the Sun, **pulsars, supernova remnants, X-ray binary** systems, and sources associated with clusters of galaxies. The first X-ray photograph of the Sun was obtained by H. Friedman (1916–), a pioneer in the application of rockets and satellites to astronomy, on April 19, 1960. Information on cosmic X-ray emission has been obtained from equipment carried by several satellites, including Uhuru and Ariel V. During periods of increased solar activity, there is a stronger emission of X-rays and γ-rays. For astronomical purposes X-rays are described according to the energy associated with the X-ray photons; this energy is expressed in **electron volts**.

X-ray telescope A telescope constructed for observing X-ray sources, the instrument being mounted on a space craft and functioning outside the Earth's atmosphere. X-rays at grazing incidence are reflected from the surface of a metal paraboloid, and brought to focus for detection by photoelectric means or photographed. The X-rays can be deflected by the paraboloid surface and then again by a hyperboloid surface to be brought to a second focus.

year A unit of time marked by the period of revolution of the Earth in its orbit about the Sun with respect to an arbitrary point of reference. The year is either astronomical or civil, the latter having an average of 365·2425 mean solar days (see **calendar**). In astronomy the following are distinguished:

anomalistic year	365^d	6^h	13^m	53^s	mean solar time
eclipse year	346^d	14^h	52^m	51^s	mean solar time
sidereal year	365^d	6^h	9^m	$9^s.5$	mean solar time
tropical year	365^d	5^h	48^m	46^s	mean solar time

The PLATONIC YEAR (the time during which the axis of the Earth makes one complete revolution) is about 25,800 years (see **precession**).

Yerkes system see **spectral-luminosity classification**

Z Camelopardalis stars **Dwarf novae** similar to **U Geminorum stars**, except that the range of variation in brightness is less, and there is frequently a delay before the decline from maximum brightness starts.

Zeeman effect A phenomenon discovered by P. Zeeman in 1896. When a very strong magnetic field is applied to the light emitted by a substance, the lines in the emission spectrum are widened and split into several components; this is a result of the influence of the magnetic field on the movement of the electrons in the atoms of the substance. The explanation of the effect was given by H. A. Lorentz (see **Lorentz, Hendrik**). The Zeeman splitting observed in the spectra of sunspots and stars demonstrates the existence of intense magnetic fields in celestial bodies.

Zeeman, Pieter (Zonnemaire, Netherlands, May 25, 1865 – Amsterdam, October 9, 1943) A Dutch physicist distinguished for his work in magneto-optics and particularly for his discovery of the **Zeeman effect**, which demonstrated the link between light and magnetism and was responsible for noteworthy advances in spectroscopy.

zenith The point on the celestial sphere vertically above the observer's head. The point directly opposite is the **nadir**.

zenithal hourly rate (*abbrev.*: Z.H.R.) The number of meteors in a meteor shower that would have been seen under good conditions with the radiant immediately overhead. The Z.H.R. is obtained by applying a correction to the observed hourly rate of meteors in the shower. A further correction is made in respect of moonlight. The observed hourly rate is always less than the Z.H.R.

zenith distance (*abbrev.*: Z.D.) The angular distance, measured in the meridian, of a celestial body from the zenith of the point of observation. It is equal to 90° minus the altitude of the body above the horizon.

zenith telescope A telescope mounted vertically and used for positional measurements of stars passing close to the zenith. The instrument, fitted with a declination micrometer, is used for the accurate measurement of latitude, as required in determination of latitude variation. The FLOATING ZENITH TUBE, designed by B. Cookson at the beginning of the century, permits the telescope to be reversed between two observations, the mean of which will refer to the true vertical. This improved instrument has now been replaced by the **photographic zenith tube**.

zenocentric co-ordinates A system of co-ordinates referred to the centre of Jupiter.

zenographic co-ordinates A system of co-ordinates referred to the surface of Jupiter.

zero gravity Weightlessness. A person will experience no weight inside an object, such as a space craft, in the state of FREE FALL, i.e. the normal state of motion of a body in space subjected to the gravitational attraction of a central body. The phenomenon has been noted by astronauts.

zero point energy The energy possessed by the atoms or molecules of a substance at 0 K according to quantum mechanics. According to classical mechanics the energy at 0 K would be zero.

Zeta Geminorum stars A subgroup of Cepheid variables characterized by a symmetrical light curve.

zodiac A circular zone on the celestial sphere, about 8° wide on each side of the ecliptic, which forms the background for the motions of the Sun, Moon, and planets, wherever they may be in orbit. The zodiac is divided into 12 sections, each of 30°, which are named after the constellations they contained at the time of **Hipparchos of Nicaea**. On account of **precession** of the equinoxes these constellations are not now in the same position. The constellations are usually listed in an anticlockwise direction, starting from the vernal equinox (then in Aries but now in Pisces), which is still known as the First Point of Aries.

zodiacal band An extension of the zodiacal light. It is a faint band of light around the ecliptic, joining the tips of the cones of the zodiacal light. It gradually fades until it is about 135° from the Sun, when it increases in intensity again and finally merges with the **Gegenschein**.

zodiacal light A cone of faint light, usually fainter than that from the Milky Way, visible at all seasons in the tropics in the absence of moonlight. It stretches along the ecliptic from the western horizon after evening twilight, or from the eastern horizon before morning twilight. The spectrum of the luminosity resembles that of sunlight, and so indicates that the phenomenon results from the scattering of light by particles in the plane of the ecliptic.

Zodiacal Stars, Catalogue of A catalogue compiled by J. Robertson and published in Washington, D.C., 1940. It gives data on 3539 stars for the equinox 1950·0.

Zond The name of a series of Soviet spaceprobes. Zond 1 was launched on April 2, 1964, Zond 2 on November 30, 1964; Zond 3 was launched on July 18, 1965 and sent back photographs of the hidden side of the Moon; Zond 4 was launched on March 2, 1968. Zonds 5, 6, 7, and 8, launched on September 15, 1968, November 10, 1968, August 8, 1969, and October 20, 1970, all returned to Earth after flights to the Moon.

zone A term used in geography to describe an imaginary belt surrounding the Earth, e.g. north temperate zone, torrid zone. It is used also in astronomy to describe planetary surface features, such as the light bands between darker belts on Jupiter and Saturn.

zone of avoidance A band of interstellar gas and dust along the equator of our Galaxy where the absorption is so great that no external galaxies can be seen. It is between 10° and 40° in width.

zone time The time in any of the 24 time zones of the world. Each country formerly adopted as standard civil time the local mean time of its national observatory. The development of railways made it desirable to have a system of differential time reckoning for the whole world so as to avoid confusion. At a conference held in Washington in 1884 the meridian of Greenwich was adopted as the zero of longitude, and zones of longitude, each 15° wide, were established; in each zone a common standard time is used. Standard time in each successive zone westwards is one hour slow with respect to the preceding zone. Large territories such as the USA and USSR use more than one zone time. Particulars of Standard Time throughout the world may be found in *Whitaker's Almanac*. The adoption of zone time results in a discrepancy at longitude 180° (i.e. the International Date Line), which is resolved by omitting one day from the calendar in the case of crossing from west to east, or repeating one day if the crossing is from east to west.

zoom lens A variable-focus lens used in cinematography for making a rapid transition from a distant to a close view without moving the camera.

Zwicky, Fritz (Varna, Bulgaria, February 14, 1898 – February 8, 1974) A Swiss astronomer who was distinguished for his discoveries of supernovae, dwarf galaxies, and clusters of galaxies, and for his theory of neutron stars.

Note: the dates in this book have all been expressed in the order month, day, year, although it is customary in works on astronomy to show dates in the order year, month, day.

Symbols and Abbreviations

Table 1 Symbols for units, constants and quantities used in astronomy

a	semi-axis major	H_o	Hubble constant	MHz	megahertz
a or *A*	altitude	Hz	hertz	mm	millimetre
Å	ångström	in	inch	Mpc	megaparsec
A.U.	astronomical unit	*i*	inclination to ecliptic	n	neutron
c	velocity of light	J	joule	nm	nanometre
cm	centimetre	Jy	jansky	N	newton
d	distance (in ″)	*k*	Boltzmann's constant	p	proton
e	electron	K	kelvin	Pa	pascal
e	electron charge	kg	kilogram	pc	parsec
e	eccentricity of orbit	km	kilometre	rad	radian
eV	electron volt	kpc	kiloparsec	s	second
g	acceleration of gravity	*L*	luminosity	sr	steradian
G	gravitational constant	ly	light-year	T	tesla
GHz	gigahertz	m	metre	W	watt
h	altitude	*m*	apparent magnitude	Wb	weber
h	Planck's constant	*M*	absolute magnitude		

Table 2 Greek symbols used in astronomy

α	right ascension	ζ	zenith distance	μm	micrometre
β	celestial latitude	λ	wavelength	ν	frequency
δ	declination	λ	celestial longitude	π	parallax (in ″)
ϵ	obliquity of ecliptic	μ	proper motion	σ	Stefan's constant

Table 3 Greek alphabet

Α, α	Alpha	Ε, ϵ	Epsilon	Ι, ι	Iota	Ν, ν	Nu	Ρ, ρ	Rho	Φ, φ	Phi
Β, β	Beta	Ζ, ζ	Zeta	Κ, κ	Kappa	Ξ, ξ	Xi	Σ, σ	Sigma	Χ, χ	Chi
Γ, γ	Gamma	Η, η	Eta	Λ, λ	Lambda	Ο, ο	Omicron	Τ, τ	Tau	Ψ, ψ	Psi
Δ, δ	Delta	Θ, θ	Theta	Μ, μ	Mu	Π, π	Pi	ϒ, υ	Upsilon	Ω, ω	Omega

Table 4 Decimal multiples and submultiples used with SI units

MULTIPLE	PREFIX	SYMBOL	SUB-MULTIPLE	PREFIX	SYMBOL
10	deca-	da	10^{-1}	deci	d
10^2	hecto-	h	10^{-2}	centi-	c
10^3	kilo-	k	10^{-3}	milli-	m
10^6	mega-	M	10^{-6}	micro-	μ
10^9	giga-	G	10^{-9}	nano-	n
10^{12}	tera-	T	10^{-12}	pico-	p

Table 5 Solar system symbols

☉	Sun	♅	Uranus
☾	Moon	♆	Neptune
○	Full Moon	♇	Pluto
●	New Moon	⑮	Minor planet (with appropriate number)
◐	Moon, First Quarter	☄	Comet
◑	Moon, Last Quarter	☍	Opposition
☿	Mercury	☌	Conjunction
♀	Venus	□	Quadrature
⊕	Earth	☊	Ascending node (also longitude of)
♂	Mars	☋	Descending node (also longitude of)
♃	Jupiter	♈	First Point of Aries
♄	Saturn	♎	First Point of Libra

Table 6 Abbreviations used in astronomy

A.A.	*Annuaire Astronomique*. Alternative name sometimes used for *Astronomicheskii Ezhegodnik*, the ephemeris published in the USSR.
ADS	R. G. Aitken. *New General Catalogue of Double Stars,* Washington, D.C., 1932.
A.E.	*The Astronomical Ephemeris,* published annually by the UK and US governments.
AG or AGK	*Astronomische Gesellschaft Katalog*, Leipzig, 1890.
AGK 2	*Zweiter Katalog der Astronomischen Gesellschaft . . . 1950.* Bonn, 1951.
AGK 3	*Dritter Katalog der Astronomischen Gesellschaft* . . . Hamburg-Bergedorf, 1975 –.
A.N.	*Almanaque Náutico*. The Spanish equivalent of *The Astronomical Ephemeris.* Produced annually by the Instituto y Observatorio de Marina, San Fernando, Spain.
A.U.C.	*anno urbis conditae,* 'in the year from the building of the city', i.e. Rome, in 753 B.C.; also *ab urbe condita*, 'from the foundation of the city'.
B	J. E. Bode. *Allgemeine Beschreibung und Nachweisung der Gestirne nebst Verzeichniss . . . von 17240 Sternen* . . . Berlin, 1801.
B.A.A.	British Astronomical Association. Founded 1890. *Office:* Burlington House, Piccadilly, London, W.1.
BAC	British Association for the Advancement of Science. *The Catalogue of Stars . . . reduced to January 1, 1850,* by the late Francis Baily, London, 1845.
BD	*Bonner Durchmusterung*. See entry on page 25.
BGC	S. W. Burnham. *General Catalogue of Double Stars within 121° of the North Pole,* Washington, D.C., 1906. Contains data on 13,665 double stars.
Br	A. J. G. F. von Auwers, *Neue Reduktion der Bradleyschen Beobachtungen aus den Jahren 1750–62,* St Petersburg, 1882–1903.
3C	Third Cambridge Catalogue of Radio Sources. D. O. Edge *et al.*, 'A Survey of Radio Sources at a Frequency of 159 Mc/s.' *Mem.,* (1963) *68*, 37. A. S. Bennett. 'The Revised 3C Catalogue of Radio Sources.' *Mem.,* (1963) *68*, 163.
4C	Fourth Cambridge Catalogue of Radio Sources. J. L. Caswell. 'The 4C Survey'. *Ph.D. Thesis,* Cambridge, 1966. Additions to the 4C survey are designated 4C(T). See *M.N.,* (1969) *145*, 181.
5C1, 5C2, 5C3,	Fifth Cambridge Catalogue of Radio Sources.

5C4, 5C5	
CD or CoD	*Córdoba Durchmusterung Catalogue,* by J. M. Thome and C. D. Perrine (Vol. 21), being *Resultados del Observatorio Nacional Argentino*, Vols. 16, 17, 18, 21. Contains data on 613,953 stars from declination −90° to −21°.
CGA	B. A. Gould. *Catálogo General Argentino,* being *Resultados del Observatorio Nacional Argentino*, Vol. 14. Córdoba, 1886. Gives mean positions of southern stars.
CP	Cambridge pulsar.
CPD	D. Gill and J.C. Kapteyn. *The Cape Photographic Durchmusterung for the equinox 1875,* being Vols. 3, 4 and 5 of *Annals of the Cape Observatory*, London, 1896–1900.
CT	*Connaissance des Temps,* Paris, annually.
CZ	B.A. Gould. *Catálogo de Zonas Estelares,* being *Resultados del Observatorio Nacional Argentino*, Vols 7 and 8, Córdoba, 1884.
dec.	declination.
dex.	acronym from *de*cimal *ex*ponent. A notation which indicates that the number following it is the power to which the base 10 is raised, e.g. dex (2.13) = $10^{2.13}$.
EMP	*Ephemeris of Minor Planets.* Since 1952 published at Leningrad as *Efemeridy malykh planet.*
ESA	European Space Agency.
ESO	European Southern Observatory.
E.T.	Ephemeris Time.
ETI	extraterrestrial intelligence.
FK 4	IAU. *Scheinbare Örter der Fundamental Sterne 1977. Enthaltend die 1535 Sterne des Vierten Fundamental-Katalogs (FK4),* Heidelberg, 1975.
G	B. A. Gould, *Uranometria Argentina,* being *Resultados del Observatorio Nacional Argentino*, Vol. 1, Buenos Aires, 1879.
GC	L. Boss, *General Catalogue of 33,342 Stars*, Washington, D.C., 1936.
GFH	*Geschichte des Fixstern-Himmels.* Abt. I. Der Nordliche Sternhimmel. Abt. II. Der Südliche Sternhimmel, Karlsruhe, 1922–36; 1937–56.
G.M.A.T.	Greenwich Mean Astronomical Time.
G.M.T.	Greenwich Mean Time.
Gr or GrB	S. Groombridge, *A Catalogue of Circumpolar Stars . . .* edited by G. B. Airy, London, 1838. Revised by F. W. Dyson and W. G. Thackeray as the *New Reduction of Groombridge's Circumpolar Catalogue for the Epoch 1810.0*, Edinburgh, 1905.
h	J. F. W. Herschel, *A General Catalogue of Nebulae and Clusters of Stars . . . Epoch 1860.* Supplement by J. L. E. Dreyer, Dublin, 1878. J. F. W. Herschel, *A Catalogue of 10,300 Multiple and Double Stars . . .* edited by R. Main and C. Pritchard, London, 1874.
H	*J. Hevelius, Catalogus stellarum fixarum . . .* Gedani, 1687. Positions of stars on most celestial globes of the early 18th century were based on measurements made by Hevelius. His catalogue was superseded by J. Flamsteed, *Historia coelestis Britannica*, London, 1725. The manuscript of Hevelius's *Catalogue*, having survived unscathed the vicissitudes of time is now preserved at the Brigham Young University, Provo, Utah, USA.
H	W. Herschel, *Catalogue of Double Stars. Philosophical Transactions of the Royal Society*, (1782) *72*, 112; ibid., (1785) *75*, 40. W. Herschel, *Catalogue of One Thousand New Nebulae and Clusters of Stars. Ibid.,* (1786) *76*, 457. W. Herschel, *Catalogue of Second Thousand New Nebulae and Clusters of Stars . . . Ibid.*, (1789) *79*, 212.

	W. Herschel, *Catalogue of 500 New Nebulae. Ibid., 92,* 477.
	See 'Wilhelm Herschel's 'Verzeichnisse von Nebelflecken und Sternhaufen' in *Astronomische Beobachtungen auf der Königlichen Universitäts Sternwarte zu Königsberg,* (1862) 34, 155 for full lists of items in the classifications H.I. to H.VIII.
HD	A. J. Cannon and E. C. Pickering, *The Henry Draper Catalogue of Stellar Spectra,* being *Annals of Harvard College Astronomical Observatory,* (1918–24) *91–99.*
HDE	A. J. Cannon, *The Henry Draper Extension,* being *Annals of Harvard College Observatory,* (1925–36) *100.* The former gives spectral classification of 225,300 stars; the latter adds a further 133,700 stars.
HP	E. C. Pickering *et al., Observations with the Meridian Photometer, 1879–82,* being *Annals of Harvard College Astronomical Observatory,* (1884) *14.* Gives the magnitude of 4260 stars.
HR or RHP	E. C. Pickering, *Revised Harvard Photometry,* being *Annals of Harvard College Observatory* (1908) *50.* Gives the magnitude of 9110 stars.
	E. C. Pickering, *Photometric Measures of 36,682 Stars,* being *Annals of Harvard College Observatory* (1908) *54,* A supplement to the previous item.
H-R diagram	Hertzsprung-Russell diagram.
I.A.F.	International Astronautical Federation.
I.A.U.	International Astronomical Union. Founded 1919. Established 1920. First Meeting 1922.
IC	J. L. E. Dreyer. *Index Catalogue of Nebulae found in the years 1888 to 1894,* London, 1895. *Second Index Catalogue of Nebulae . . . found in the years 1895 to 1907,* London, 1908. These are supplements to Dreyer's NGC.
IGY	International Geophysical Year.
IQSY	International Quiet Sun Year.
IR	Infrared.
I.U.A.A.	International Union of Amateur Astronomers. Founded 1969.
J.D.	Julian Date or Day.
KPNO	Kitt Peak National Observatory, Tucson, Arizona.
Lac	N. L. de Lacaille. *A Catalogue of 9766 Stars in the Southern Hemisphere, for the beginning of the year 1750. . .* Reduced at the expense of the British Association. London, 1847.
Lal or Ll	J. J. le F. de la Lalande. *A Catalogue of those Stars in the Histoire Céleste Française . . . Epoch 1800 . . .* Reduced at the expense of the British Association . . . London, 1847.
LMC	Large Magellanic Cloud.
M (followed by a number)	C. J. Messier. *Catalogue des Nébuleuses et des Amas d'étoiles.* Published in *Connoissance des Temps,* Paris, 1781.
Mem.	*Memoirs of the Royal Astronomical Society.*
MLR	mass-luminosity relationship.
M.N.	*Monthly Notices of the Royal Astronomical Society.*
NASA	National Aeronautics and Space Administration.
NGC	J. L. E. Dreyer. *New General Catalogue of Nebulae and Clusters of Stars.* London, 1888.
NHO	Northern Hemisphere Observatory.
Np	north preceding.
NP	National Radio Astronomy Observatory Pulsar.
NPD	North Polar Distance.
NPS	North Polar Sequence.
OAO	Orbiting Astronomical Observatory.
OGO	Orbiting Geophysical Observatory.
OMC	Orion Molecular Cloud.

OSO	Orbiting Solar Observatory
PGC	L. Boss. *Preliminary General Catalogue of 6199* . . . Washington, D.C., 1910.
Pi	G. Piazzi. *Praecipuarium stellarum inerrantium positiones ab anno 1792 ad annum 1802.* Panormi, 1803.
	G. Piazzi. *Praecipuarium stellarum in errantium positiones ab anno 1792 ad annum 1813,* Panormi, 1814.
PLR	period-luminosity relationship.
PZT	photographic zenith tube.
QSO	quasi-stellar object. It need not necessarily be a quasar.
QSSS	quasi-stationary spiral structure.
R.A.	Right Ascension.
RAS	Royal Astronomical Society, Burlington House, Piccadilly, London, W.1. Founded 1820.
RGO	Royal Greenwich Observatory.
SDS or IDS	R. T. A. Innes. *Southern Double Star Catalogue.* −19° to −90°. Johannesburg, 1927.
SEA	sudden enhancement of atmosphere.
SETI	search for extraterrestrial intelligence.
SI	Système International d'Unités.
SMC	Small Magellanic Cloud.
SMP	Southern Meridian Photometry. S. L. Bailey. *A Catalogue of 7922 Southern Stars observed with the Meridian Photometer, 1889–91,* being Volume 34 of *Annals of Harvard College Astronomical Observatory*, Cambridge, Mass., 1895.
SNR	supernova remnant.
Sp	south preceding.
SPD	South Polar Distance
TAI	International Atomic Time.
TLP	transient lunar phenomena.
UA	B. A. Gould. *Uranometria Argentina*, Buenos Aires, 1879.
UBV	ultraviolet, blue, visual. A system of describing stellar magnitudes.
UFO	unidentified flying object.
UO	Oxford University Observatory. *Astronomical Observations . . . under the direction of C. Pritchard.* No. 2. *Uranometria nova Oxoniensis*, Oxford, 1885.
U.T.	Universal Time.
U.T.C.	Co-ordinated Universal Time.
UV	ultraviolet.
wd	white dwarf.
Z.D.	zenith distance.
Z.H.R.	zenithal hourly rate.
Σ	F. G. W. Struve. *Catalogus Novus stellarum duplicium et multiplicium*, Dorpat, 1827.

Bibliography

A. F. O'D. Alexander, *The Planet Saturn* ... London, 1962, *The Planet Uranus* ... London, 1965.

C. W. Allen, *Astrophysical Quantities.* 3rd edition, London, 1973.

R. H. Allen, *Star Names. Their Lore and Meaning*, New York, 1963.

The Astronomical Ephemeris, H.M.S.O., London, annually.

Explanatory Supplement to the Astronomical Ephemeris, H.M.S.O., London, 1974.

D. Baker, *The Hamlyn Guide to Astronomy*, London, 1978.

B. V. Barlow, *The Astronomical Telescope*, London and Winchester, 1974.

B. J. Bok and P. Bok, *The Milky Way*, Harvard, 1974.

F. W. Cousins, *The Solar System*, London, 1972.

T. Gehrels, *Physical Studies of Minor Planets*, Washington, 1971.

O. Gingerich (ed.), *New Frontiers in Astronomy*, San Francisco, 1975.

The Handbook of the British Astronomical Association, London, annually.

V. Illingworth, *Macmillan Dictionary of Astronomy*, London, 1979.

K. G. Jones, *Messier's Nebulae and Star Clusters*, London, 1968.

Z. Kopal, *The Moon in the Post-Apollo Era*, Dordrecht, 1974.

W. H. McCrea, *The Royal Greenwich Observatory*, London, 1975.

J. H. Mallas et al., *The Messier Album*, Cambridge, 1979.

S. Mitton (ed), *The Cambridge Encyclopaedia of Astronomy*, Cambridge, 1977.

P. Moore, *The Atlas of the Universe*, London, 1970.

P. Moore, *New Concise Atlas of the Universe*, 1978.

P. Moore, *Guide to the Moon*, Guildford, 1976.

L. Motz and A. Duveen, *Essentials of Astronomy.* Belmont, California, 1979.

P. Murdin et al., *Catalogue of the Universe*, Cambridge, 1979.

T. A. Mutch, *Geology of the Moon*, Princeton, 1972.

H. W. Newton, *The Face of the Sun*, Harmondsworth, 1958.

I. Nicholson, *Astronomy, a Dictionary of Space and the Universe*, London, 1977.

M. M. Nieto, *The Titus-Bode Law and the Origin of the Solar System*, Oxford, 1972.

Norton's Star Atlas, 16th edition, Edinburgh, 1973.

A. Pannekoek, *A History of Astronomy*, London, 1961.

J. M. Pasachoff and M. L. Kutner, *University Astronomy*, Philadelphia, 1978.

B. M. Peek, *The Planet Jupiter*, London, 1958.

C. Ronan (ed.), *Encyclopedia of Astronomy*, London, 1979.

E. v. P. Smith and K. C. Jacobs, *Introductory Astronomy and Astrophysics*, Philadelphia, 1973.

R. H. Stoy (ed.), *Everyman's Astronomy*, London, 1974.

R. J. Tayler, *Galaxies: Structure and Evolution*, London and Winchester, 1978.

R. J. Tayler, *The Stars: Their Structure and Evolution*, London and Winchester, 1970.

H. P. Wilkins and P. Moore, *The Moon*, 2nd edition, London, 1961.

Periodical: The most useful general periodical devoted to astronomy is *Sky and Telescope*, which is published monthly by Sky Publishing Corporation, 19 Bay State Road, Cambridge, Mass., 02138, USA.

Acknowledgements

Photographs

Associated Press, London 82; Camera Press, London – NASA 29; Camera Press – Novosti 180; Camera Press – Rolf Schurch 117; Mary Evans Picture Library, London 39, 139; Green Bank, West Virginia 142; Hale Observatories-California Institute of Technology and Carnegie Institute of Washington front and back of jacket, 181; Hamlyn Group Picture Library 10, 16, 45, 64, 97, 110, 134, 145, 153; Istituto & Museo di Storia della Scienza, Firenze 65; Lick Observatory, Santa Cruz, California 123, 161, 178; Lockheed Solar Observatory, Burbank, California 177; NASA 52, 100, 107, 152; National Maritime Museum, London 25; National Portrait Gallery, London 74, 113; Royal Astronomical Society, London 35, 43, 46, 181; Royal Observatory, Edinburgh 79; Science Museum, London 112, 124, 158; Smithsonian Institution, Cambridge, Massachusetts 116.